Bi-Directionality in the Cognitive Sciences

Human Cognitive Processing (HCP)

Cognitive Foundations of Language Structure and Use

This book series is a forum for interdisciplinary research on the grammatical structure, semantic organization, and communicative function of language(s), and their anchoring in human cognitive faculties.

For an overview of all books published in this series, please see
http://benjamins.com/catalog/hcp

Volume 30

Bi-Directionality in the Cognitive Sciences. Avenues, challenges, and limitations
Edited by Marcus Callies, Wolfram R. Keller, and Astrid Lohöfer

Bi-Directionality in the Cognitive Sciences

Avenues, challenges, and limitations

Edited by

Marcus Callies
Johannes-Gutenberg-Universität, Mainz

Wolfram R. Keller
Humboldt-Universität zu Berlin

Astrid Lohöfer
Philipps-Universität, Marburg

John Benjamins Publishing Company

Amsterdam / Philadelphia

The paper used in this publication meets the minimum requirements of American National Standard for Information Sciences – Permanence of Paper for Printed Library Materials, ANSI Z39.48-1984.

Library of Congress Cataloging-in-Publication Data

Bi-directionality in the cognitive sciences : avenues, challenges, and limitations / edited by Marcus Callies, Wolfram R. Keller, Astrid Lohöfer.

p. cm. (Human Cognitive Processing, ISSN 1387-6724 ; v. 30)

Includes bibliographical references and index.

1. Psycholinguistics. 2. Cognitive science. 3. Interdisciplinary approach to knowledge. I. Callies, Marcus. II. Keller, Wolfram R. III. Lohöfer, Astrid.

P37.B53 2011

401'.9--dc22 2011015159

ISBN 978 90 272 2384 5 (Hb ; alk. paper)

ISBN 978 90 272 8514 0 (Eb)

John Benjamins Publishing Co. · P.O. Box 36224 · 1020 ME Amsterdam · The Netherlands

John Benjamins North America · P.O. Box 27519 · Philadelphia PA 19118-0519 · USA

Table of contents

Editors and contributors

Editors

Marcus Callies
Johannes-Gutenberg-Universität, Mainz
mcallies@uni-mainz.de

Wolfram R. Keller
Humboldt-Universität zu Berlin
wolfram.keller@staff.hu-berlin.de

Astrid Lohöfer
Philipps-Universität, Marburg
alohoefer@googlemail.com

Contributors

Beatrix Busse
Universität Heidelberg
beatrix.busse@as.uni-heidelberg.de

Marcus Callies
Johannes-Gutenberg-Universität, Mainz
mcallies@uni-mainz.de

Stephan Freißmann
Justus-Liebig-Universität, Giessen
stephan.freissmann@gmx.de

Juliana Goschler
Universität Hamburg
juliana.goschler@uni-hamburg.de

Wolfram R. Keller
Humboldt-Universität zu Berlin
wolfram.keller@staff.hu-berlin.de

Alexandra Kleeman
University of California, Berkeley
alexandra.kleeman@gmail.com

Astrid Lohöfer
Philipps-Universität, Marburg
alohoefer@googlemail.com

Wolfgang G. Müller
Friedrich-Schiller-Universität, Jena
womu@gmx.de

Susanne Niemeier
Universität Koblenz-Landau
niemeier@uni-koblenz.de

Christine Ott
Philipps-Universität, Marburg
ottc@Staff.Uni-Marburg.DE

Gerard Steen
Vrije Universiteit, Amsterdam
gj.steen@let.vu.nl

Anatol Stefanowitsch
Universität Hamburg
anatol.stefanowitsch@uni-hamburg.de

Liane Ströbel
Heinrich-Heine-Universität, Düsseldorf
stroebel@phil-fak.uni-duesseldorf.de

Gary Thoms
Strathclyde University, Glasgow
gary.thoms@gmail.com

Dirk Vanderbeke
Friedrich-Schiller-Universität, Jena
Vanderbeke@t-online.de

Stefano Versace
Strathclyde University, Glasgow
versastef@gmail.com

Yehong Zhang
Tsinghua University, Beijing
yehong.zhang@googlemail.com

Alexander Ziem
Heinrich-Heine-Universität, Düsseldorf
ziem@phil-fak.uni-duesseldorf.de

INTRODUCTION

Bi-directionality

Avenues, challenges, and limitations

Marcus Callies, Wolfram R. Keller, and Astrid Lohöfer
Johannes-Gutenberg-Universität, Mainz / Humboldt-Universität zu Berlin / Philipps-Universität, Marburg

1. On interdisciplinarity, transdisciplinarity, and bi-directionality

This book includes selected papers presented at the conference "Bi-directional Perspectives in the Cognitive Sciences" held from February 26–28, 2009, at the University of Marburg, Germany, organized by the volume editors. It brings together contributions by researchers working in the fields of linguistics, literary and cultural studies, as well as foreign language acquisition and teaching, who engage in the study of cognitive phenomena to explore the bi-directionality, that is, the mutual heuristic applicability of diverse epistemological goals, research topics, and methodologies used in cognitive approaches in their respective disciplines. The aim of this volume is to uncover hidden potentials and available prospects of transdisciplinary research in the cognitive sciences by (re-)discovering and defining common interests and objectives, and by sketching paths of bi-directionality in terms of fruitful transdisciplinary cross-fertilization, co-operative projects, and research transfer.

Definitions of what constitutes the cognitive sciences (e.g. Stillings et al. 1995: 1) show agreement as to at least three central aspects that are characteristic of the cognitive sciences in general: (i) unifying theoretical ideas and questions about the nature of the human mind, (ii) a diversity of outlooks and research methods, which are potentially synergistic, and (iii) its interdisciplinary nature. Thus, cognitive science is the science of mind, and cognitive scientists seek to understand perception, thinking, remembering, learning, using and understanding language, and other mental phenomena. As far as its interdisciplinary nature is concerned, there is still much disagreement about the exact relationship between cognitive science and other fields, and the interdisciplinary nature of cognitive science is largely both unrealized and circumscribed.

Interdisciplinarity has been *the* buzzword in the humanities for at least two decades. In the following, we discuss and clarify our perspective on interdisciplinarity and the seemingly synonymous concept of *transdisciplinarity*. We then address the notion of *bi-directionality* vis-à-vis inter- and transdisciplinarity, all of which suggest a two-way traffic between several disciplines.

Generally speaking, *interdisciplinarity* can be understood as a form of scientific co-operation that pertains to two or more disciplines or branches of learning, and ideally contributes to or benefits from two or more disciplines. Mittelstraß (2002) discusses inter- and transdisciplinarity against the background of the ever accelerating growth of knowledge in all scientific fields, and also with regard to the organizational and institutional forms of academic research. Observing an increasing particularization of disciplines and fields, he laments that simultaneously, the capacity to think disciplinarily (in terms of larger theoretical units) is decreasing. Mittelstraß reminds us that many fields of research and disciplines came into being in the course of the history of the sciences, and that their boundaries are neither primarily founded in objects nor in theory, but are historical as well.

He conceives of interdisciplinarity as re-establishing a larger disciplinary orientation, or increasing cognitive interest within or beyond given fields or disciplines:

> interdisciplinarity properly understood does not commute between fields and disciplines, and it does not hover above them like an absolute spirit. Instead, it removes disciplinary impasses where these block the development of problems and the corresponding responses of research. Interdisciplinarity is in fact transdisciplinarity. (Mittelstraß 2002)

Certain problems, often central and complex ones, may only be solved through the joint efforts of many disciplinary competencies, for example in the study of the human mind. While such co-operation in general entails a readiness to co-operate in research, a difference between inter- and transdisciplinarity can be drawn in that the former refers to a concrete but limited co-operation for a definite period of time, while the latter means that such co-operation results in a lasting and systematic order that alters the disciplinary order itself (2002).

In line with Mittelstraß' argumentation, we conceive of *transdisciplinarity* in this volume as a scientific research principle that is active and most effective where merely disciplinary, field-specific definitions of problems and their solutions are impossible. It is a principle that guides research and scientific organization in that it removes disciplinary deadlocks that are the product of institutional traditions and habits rather than of scientific necessity.

We thus avoid the term *interdisciplinarity* for various reasons. On the one hand, it may suggest a uniformity of independent cognitive (sub-)disciplines, but in the sciences, individual disciplines, we believe, become harder and harder to define, as exemplified terminologically by fields of study like *neurosciences, life sciences*, or *information sciences*. On the other hand, it appears that one reason for the yet unclear or even unrealized disciplinary exchange between what may be called 'core' cognitive sciences, such as psychology or the neurosciences, and other disciplines is the largely uni-directional application of theories, concepts, and methods without substantial transfer in the other direction. The actual day-to-day practice of interdisciplinarity thus appears to thrive on one-directional borrowings. Freeman (2007) gives an example of such uni-directionality that is particularly relevant to this volume:

> One question that inevitably arises is what new insights Cognitive Linguistics provides in literary studies that literary criticism has not already discovered. The corollary, what literary criticism can contribute to Cognitive Linguistics, is almost always never asked. (Freeman 2007: 1176)

Similarly, in the course of one of the most dominant paradigm shifts in the social sciences and the humanities, the 'cognitive turn', many disciplines have more or less selectively borrowed theories, concepts, and methods from the core cognitive sciences (see also Stefanowitsch, this volume). These borrowing disciplines are related to the study of the human mind insofar as their research objects are often products of the human mind, for example, language and literature.

Against this background, our goal is not to merge disciplines or to "dress up" literary studies or linguistics as cognitive disciplines in order to dissolve disciplinary boundaries, but rather to see how theories, concepts, and methods from one particular field within a cognitive, linguistic, literary, or pedagogical (sub-)discipline can influence or even transform a field into another field – whereby our interest lies also in the way in which this transformation itself transforms the field from which the methodology, knowledge, or approach was borrowed in the first place (a process that may betoken a kind of inadvertent rather than conscious bi-directionality). What we are interested in, then, are mutual transformations of smaller entities (potentially with repercussions for a discipline as such), that is, the movement and dynamics between such fields of study.

The chapters in this volume explore – and where appropriate and possible – establish bi-directional approaches that go beyond the uni-directional borrowing practice characteristic of existing interdisciplinary approaches. The terminological repercussion is that we use the term *transdisciplinarity* for *actual* interdisciplinarity, that is, where both disciplines are borrowing and lending.

While *transdisciplinarity* is thus a concept that ultimately works towards the conversion of disciplines, cementing bi-directionality in a way that ultimately dissolves distinct disciplinary entities into a new entity, bi-directionality as most of the contributors understand it, has no such goal. Rather, we aim to recuperate the middle ground once occupied by the term *interdisciplinarity* with its emphasis on a dialogue between two or more disciplines; after all, disciplines like linguistics and literary scholarship often co-exist as more or less discrete sub-disciplines under the umbrella of English Studies. Another advantage of bi-directionality lies in its independence from disciplinary definitions and – implicitly – the ideological and economical stakes of disciplinarity.

Just as the term *interdisciplinarity* describes a disciplinary two-way traffic that seems natural, but remains a one-way street in actual daily practice in the humanities, bi-directionality seems like a process inherent in the work of many humanities disciplines involved in the study of cognitive processes. And yet, the recent upsurge of work in Cognitive Linguistics and Cognitive Poetics is fraught with many an obstacle.

In what follows, we sketch several major obstacles to 'genuine' bi-directionality. This sketch then serves to motivate our suggestions as to how these obstacles may be overcome, outlining how the chapters in this volume – focusing on cognitive approaches in linguistics and literary studies – point to avenues, but also discuss challenges to and limitations for bi-directionality in both topics and methods.

2. Obstacles to bi-directionality

Linguistics is usually considered one of the 'founding' sub-disciplines of the cognitive sciences (Stillings et al. 1995: 13), and Cognitive Linguistics (CL) in particular has emerged as a school of linguistics that understands language creation, learning, and usage as best explained by reference to human cognition in general, assuming a close connection between socio-physical experience, the mind, and human language. Interest in cognitive approaches to literature has only emerged in the last decade or so, but has considerably accelerated in the last few years (Freeman 2007).

Cognitive linguistic theories have become sufficiently sophisticated and detailed in order to begin making predictions that are testable using the broad range of converging methods from the cognitive sciences (Gonzalez-Marquez et al. 2007), and CL witnesses a growing appeal of empirical methods that do not solely rely on intuition or introspection but are grounded in objective observation, for instance corpus linguistics, experimentation, and neurological modeling (Geeraerts and Cuyckens 2007: 16f.). Its theories, concepts, and findings have been applied to a wide range of neighboring disciplines such as (foreign) language acquisition, pedagogy, and teaching (Pütz, Niemeier, and Dirven 2001; De Knop and De Rycker 2008; Boers and Lindstromberg 2008), cultural studies (Sharifian, Yu, Dirven, and Niemeier 2008), and literary studies (Tsur 1992; Semino and Culpeper 2002; Stockwell 2002; Brône and Vandaele 2009; see also the chapters on applied and interdisciplinary perspectives of CL in Geeraerts and Cuyckens [eds.] 2007).

Insights and developments from CL are also becoming accepted ways of analyzing literary texts. Cognitive Poetics/Stylistics has become an important part of modern stylistics (Semino and Culpeper 2002; Stockwell 2002; Gavins and Steen 2003; Stockwell 2009), located at the interface of linguistics, literary studies, and the cognitive sciences. Cognitive Poetics is firmly grounded in modern CL, perhaps most prominently through the application of Conceptual Metaphor Theory or Blending Theory to the interpretation of literary texts, that is "explaining how interpretations are arrived at, rather than proposing new interpretations of texts" (Semino and Culpeper 2002: x). Despite this grounding, however, it is not simply an "offshoot" of CL, but a new brand of poetics, CL being only one of many cognitive influences on literary studies (Gavins and Steen 2002: 5, 11).

As far as scholarly exchange in these two disciplines and their positioning towards the cognitive sciences is concerned, there has traditionally existed a certain tension in that literary criticism is often said to contribute nothing to linguistic enquiry, and vice versa, a seemingly irresolvable conflict also manifest in the methodological split between scientific method and literary criticism. The connection between linguistics and literary studies has become undermined by the diversification of literary and linguistic research in the last sixty years or so. Moreover, in the last decades, both literary research and linguistics have undergone a series of theory shifts that have driven these two disciplines even further apart. While dominant paradigm shifts in linguistics have been the pragmatic and cognitive turns, literary studies have moved from a largely biographical and text-centered enterprise to an increasingly theorized discipline with a variety of different approaches such as Feminism, Queer Theory, Marxism, and New Historicism, often adopting models from other, non-literary fields of study such as psychoanalysis, philosophy, history, sociology, and political sciences.

It thus appears that one of the major obstacles to bi-directional exchange is the partially diverse epistemological goals, for example the tension between literary criticism, critical theory, and artistic sensibility on the one hand, and scientific rationalism paired with empirical research in the cognitive sciences such as psychology or linguistics on the other hand: "Literary critics focus on the emotional and aesthetic effects of literary works, cognitive linguists on accounting for the way language characterizes meaning" (Freeman 2007: 1179). This diversity is also apparent in diverse theoretical outlooks, methodological approaches, and terminological ambiguity and incompatibility, which is also due to the fact that the cognitive sciences are a relatively new field of study, located at the interface of several disciplines.

Important methodological issues and tensions do not only exist between disciplines as diverse as literary studies and linguistics, however, but also within CL, which conceives of itself as being at the heart of the "cognitive enterprise". Despite an intensification of the appeal of empirical research methods within CL – essentially a usage-based form of linguistic investigation – the application of experimental techniques that are well-established in other cognitive sciences (such as experimental psychology or neuroscience) is not (yet) the norm. It appears that empirical aspects of usage-based linguistics often remain programmatic, and much more methodological sophistication is needed (Geeraerts 2006; Geeraerts and Cuyckens 2007). In fact,

> there seems to exist a tension, in other words, between a broad methodological tendency in Cognitive Linguistics that considers introspection the most or perhaps the only appropriate method for studying meaning and a marginal but increasing tendency to apply empirical methods that are customary in the other cognitive sciences. Resolving that tension is likely to be on the agenda of Cognitive Linguistics in the near future. (Geeraerts and Cuyckens 2007: 18)

These obstacles to bi-directionality notwithstanding, several essays in this volume propose ways in which these hurdles can be overcome.

3. Overcoming the obstacles: Avenues, challenges, and limitations

The fifteen chapters in this volume are arranged not according to traditional disciplinary categories (linguistics, literary studies, or the like) but – in line with the volume's transdisciplinary perspective – in view of their position vis-à-vis bi-directionality. Part 1 "Avenues for bi-directionality" includes those contributions that identify and discuss avenues and crossroads, while the chapters in Part 2 "Challenges to and limitations on bi-directionality" address challenges and limitations in both topics and methods. The contributions discuss three major questions:

1. How do the disciplines that conceive of themselves as part of the 'cognitive enterprise' take position with regard to the cognitive sciences in general in terms of epistemological goals, research topics, and methods?
2. What are the obstacles to bi-directionality in the cognitive sciences?

3. How can these obstacles be overcome, and what goals, topics, and methods are promising to promote the desired cross-fertilization?

What, then, are the avenues, challenges, and limitations for "true" bi-directionality in cognitive approaches to linguistics and literary studies, and how are these addressed in the contributions to this volume?

Scholars working at the interface of Cognitive Poetics, Cognitive Stylistics and Cognitive Linguistics have argued that CL needs to be complemented by insights from other disciplines, for instance from the psychology of reading and text processing (Semino and Culpeper 2002: xi; and Steen 2002). Similarly, **Gerard Steen**'s "Genre between the humanities and the sciences" points to the need and potential of CL to integrate insights from cognitive psychology. Steen first outlines a number of obstacles that have slowed down or prevented the development of an integrated interdisciplinary framework for the study of genre across the humanities and the cognitive and social sciences. He finds that genre has not attracted much interest in CL, the closest relation between CL and genre possibly being the application of CL in Cognitive Stylistics and Cognitive Poetics. To overcome or at least alleviate these problems, Steen aims to pull together the insights of a wide range of diverse theoretical and empirical proposals to construct a unified, consistent, and explicit theoretical framework for genre that is based in cognitive psychology.

He defines genre as a type of complex knowledge schema that individual language users have at their disposal to engage in discourse and sketches his cognitive-psychological model of genre based on a multifeature-multidimensional model to discourse representation that includes four complementary mental models related to the three dimensions of discourse, i.e. communication, cognition, and language. In his view, the proposed model represents a truly bi-directional perspective on genre that can advance the study of genre and includes both the humanities, the cognitive, and the social sciences.

In accordance with the importance of (linguistic) context in CL, which also involves society and culture, it has been suggested that CL needs to strengthen the recently emerging interest in language as a social phenomenon and social aspects of language use (Geeraerts and Cuyckens 2007: 15f.). In fact, CL is now seeing studies that explore social cognition and sociovariational phenomena, for example by looking into cultural models (Dirven, Wolf, and Polzenhagen 2007) and the social nature of language in terms of a Cognitive Sociolinguistics (Kristiansen and Dirven 2008; Wolf and Polzenhagen 2009; Geeraerts, Kristiansen, and Peirsman 2010). Two contributions discuss how these issues are avenues for a bi-directional exchange of CL with other disciplines.

Susanne Niemeier's "Culture-specific concepts of emotionality and rationality" explores the bi-directionality of Applied Cognitive Linguistics, a culture-based linguistic approach that sees language as inseparably intertwined with culture, and Cultural Studies, a discipline that *inter alia* devotes itself to the search for cultural models that may have channeled metaphorical and metonymic conceptualizations. The study focuses on cultural models of emotionality and rationality as they are visible in metaphorical and metonymic expressions in a variety of languages. Metaphors are considered to be rooted in cultural models instead of either constituting them or reflecting them.

Niemeier examines the ways in which internal body organs are used in different languages to conceptualize human experiences such as emotions and/or the workings of the mind, and delineates the cultural models that account for the observed similarities and differences of the various conceptualizations. Across the cultures examined, these cultural models fall into three different global categories, independent of language families: abdominocentrism, cardiocentrism, and cerebrocentrism, the labels referring to those body organs that play a central role in conceptualization.

In "Widening the goalposts of cognitive metaphor research", **Marcus Callies** argues that research on linguistic and conceptual variation in metaphor to date has predominantly examined cross-cultural differences in metaphorical mappings, while language-internal variation along sociolinguistic parameters that may affect metaphor usage have remained an under-researched topic. Assuming that SPORTS is available as a generic-level source domain in metaphorical mappings across varieties of English, he shows how the types and aspects of (cross-cultural) variation can be applied to different varieties of one language. Taking into account within-culture variation in conceptual metaphor, Callies proposes, will give cognitive metaphor research a sociolinguistic spin, helping to better understand the historical, social, and cultural foundations of conceptual motivation and figurative language use, and at the same time providing a promising research agenda to promote cross-fertilization in cognitive studies.

The next chapters reflect the application of theories, concepts, and findings developed in CL to literary studies, also proposing an openness of cognitive approaches in the study of literary texts to a plurality of analytical methods. Ultimately, the question arises whether Cognitive Poetics will want to be conceived of as a cognitive "science" in its own right (as manifest in epistemological goals and research methods) or as applying the models and findings developed in the cognitive sciences. At the same time, however, it is not impossible to envision cognitive literary scholarship as resisting the drive to scientific empiricism in order to contribute to a plurality of methodologies and theories that can be mutually tested against one another. As such, Cognitive Poetics could function as a counter-balance to scientific inquiry.

This might be of particular service in studying the similarities and differences in the literary depiction of implicit and explicit reading processes, including the kind of literary critical and scholarly work (poetology, metacriticism, metafiction, etc.) that many literary texts are implicitly engaged in.

In "How novels feel: Emotional and rational reading processes in contemporary fiction", **Wolfram R. Keller**, **Astrid Lohöfer**, and **Christine Ott** investigate literary representations of interpretive processes, particularly the role of emotion in the interpretation of literary texts, focusing on two autobiographical novels, Serge Doubrovsky's *Fils* (1977) and David Williams's *Eye of the Father* (1985). Differences notwithstanding, both novels illustrate that spontaneous, emotional responses to literary texts often result in radically new perspectives on the studied texts. Rather than being an obstacle to textual interpretation, emotional reading fruitfully transgresses established cognitive schemata, a process that likewise holds, the authors argue, for the interpretation of scientific data. The interpretation of literary texts within the novels suggests ultimately that the bridge between cognitive science and traditional literary scholarship has to be crossed both ways: traditional literary critics need to be more open to empirical research methodologies, as much as cognitive scientists would most likely profit from literary representations of cognitive, interpretive processes that often appear to anticipate what empirical research explores only at a much later point in time.

Stephan Freissmann likewise pleads for the relevance of literature and its representation of cognitive processes and the formation of knowledge for cognitive sciences. In his "Cognitive Poetics and the negotiation of knowledge" he investigates the cognitive processes involved in the understanding of fictions of cognition, that is David Lodge's *Thinks* and Richard Powers's novel *Galatea 2.2*. On the one hand, this chapter contributes to the notion of knowledge that goes beyond empirically or scientifically testable hard facts, pointing up the alternative of a hermeneutics of knowledge formation. On the other hand, Freissmann argues for the validity of literature's power as an epistemic force for the generation of knowledge about cognition. Ultimately, he emphasizes alternative methods for knowledge formation about cognition, aside from scientific methods based on experiments and empiricism – underscoring the relevance of literatures of cognition for the cognitive sciences.

Returning to issues of methodology, several contributions document the increasing use of corpus-linguistic and (quasi-)experimental methods in literary studies. It has been suggested that cognitive approaches to the study of literary texts may benefit from an increased awareness of and openness to "scientific" empirical approaches and the testing of hypotheses (Steen 2002; see also Stefanowitsch, this volume). For example, Semino and Culpeper (2002) discuss a plurality of analytical methods that range from introspection based on the authors' own readings of texts or the reactions of literary critics to informant work for the testing of hypotheses. Such an openness towards the application of falsifiable empirical approaches may be facilitated by a sharpening of epistemological goals in view of an alignment towards the cognitive sciences.

This aspect is taken up by **Beatrix Busse** who modestly renders her contribution "WRITING IS MEDICINE: Blending cognitive and corpus stylistics" as an "attempt" to illustrate the potential bi-directionality between textual connections in Paul Auster's fictional universe and the effects these may have on the reader's mental processes. She shows how a cognitive stylistic approach – applying Fauconnier and Turner's theory of blending – can be married with text-based corpus stylistic methods to empirically investigate, measure, and describe in detail the lexico-grammatical character of two of Auster's novels. Busse explores the main characters' mental lives as well as the readers' inference processes, analyzing how the blend WRITING IS MEDICINE structures the two novels and how it addresses the readers' cognitive processing during reading and interpretation. She thus proposes to augment a cognitive approach by an informed stylistic analysis, suggesting a fruitful cooperation between textual investigation and cognition.

The first chapters in the second part of this volume also propose and document the necessity to use a plurality of analytical methods, but they point to challenges and limitations for bi-directionality. Focusing on Cognitive Poetics and the aesthetic effects of reading processes, these studies are oriented towards both, the literary text and its effects on the mind of the reader. All of the authors grant that the application of methodologies and findings from the core cognitive sciences to literary texts – and more specifically, Cognitive Poetics – have benefited literary studies. While some of these contributions break new ground for Cognitive Poetics, the main issue raised is that of the bi-directionality of such literary readings: To what extent does the literary-critical application and appropriation of the methodologies and approaches in the cognitive sciences feed back into the sphere of cognitive science research itself?

Alexandra Kleeman's "Collective aesthetics and the Mere Exposure Effect" examines a psychological phenomenon linked to implicit learning, conditioning, and aesthetic experience that describes a correlation between the frequency of exposure to a stimulus and an increase in the attractiveness of that particular stimulus. Discussing the implications of the Mere Exposure Effect for the nature of the artistic canon, Kleeman notes that its relevance appears to be more readily identifiable to literary scholars and sociologists exploring the reception of literature and collective aesthetics. She proposes, however, that experimental (modernist) texts that define their own aesthetics and shape the perception of their readers have something to offer to the empirical study of the mind in psychology and linguistics: they can serve as experimental stimuli for researchers working in the domain of language acquisition and processing since they create unique and identifiable effects in readers. Attending to the structure and dynamics of such texts provides new perspectives on current experimental practices and methodologies. Arguing that human understanding of how "unusual" texts are handled within the cognitive system is vague, the Mere Exposure Effect poses a challenge to cognitive scientists who appear to have neglected the aesthetic and affective dimensions of language.

In "Embodied mind and cross-cultural narrative patterns" **Yehong Zhang** analyzes cross-cultural narrative patterns and reading processes in Grimm's fairy tales with regard to Lakoff's concept of the "embodied mind" that challenges the assumption of a purely rational, self-conscious, and culturally determined subject embodied cognition. She advocates an empirical approach to literary studies based on developmental and social psychology and reports on the results of a reader-response-questionnaire showing that despite many cultural and educational differences, Chinese and German children respond to the tales in a very similar way. The tales' many references to the physical world and the characters' experiences make them culturally transferable. Zhang's chapter identifies two major challenges for future research. First, to substantiate hypotheses about embodied cognition and culture-independent reading responses, literary studies must integrate experimental methods borrowed from other disciplines to make its findings falsifiable and productive for other disciplines. This path towards bi-directionality – or rather, a path to an efficient one-directionality – thus requires the development and refinement of tools for reading research. Second, assumption-building in the comparative study of cross-cultural literary comprehension is impeded by the absence of any refined theoretical models.

In "The mind and the text / the mind in the text" **Dirk Vanderbeke** analyses the role of gaps in literary texts and what the latter reveal about human cognition. He first offers several excerpts from James Joyce's fiction to illustrate that reading is generally an exercise of filling *Leerstellen*, of homogenizing contradictory information and sensory input, which is naturally not confined to literary discourse. Vanderbeke then explores the relationship between gap-filling and the production of aesthetic pleasure in two recent novels that deal with neuro-scientific topics, Ian McEwan's *Enduring Love* (1998) and Richard Powers's *The Echo Maker* (2007), with respect to how their representation of (failed) cognition affects the readers' quest to construct coherence. The meta-narrative representations of how minds work – or do not work – Vanderbeke argues, may aid our understanding of cognition and mental mapping of the world. He considers literary studies an important partner in the *scientific* unraveling of the mysteries of cognition, and thus proposes an avenue towards a bi-directional exchange between literature and the cognitive sciences – as also anticipated by Freissmann and Keller, Lohöfer, and Ott.

In "Verbal irony in Shakespeare's dramatic works" **Wolfgang G. Müller** focuses on irony, a fruitful field for further study from the angle of Cognitive Poetics, which has hitherto ignored this trope that remains widely studied in CL. Moreover, he wonders if CL in turn might benefit from the "literary" insights his contribution has to offer. Müller approaches Shakespeare's *The Taming of the Shrew* and *Julius Caesar* by means of both, the echoic mention theory and the pretence theory of irony, demonstrating how the application of these enhances an audience's understanding of the intricate workings of irony. While this approach clearly provides insights into literary texts beyond what traditional literary approaches have thus far substantiated, the question vis-à-vis the bi-directional potential can only be tentatively answered. Müller concludes, however, that cognitive linguists may benefit from such analyses insofar as they bring to the fore further complexities of ironic utterances in the representation of the production and comprehension in literary text that may also attend to the latter in everyday communication. Moreover, borderline cases, where irony overlaps with deception, might be grasped more clearly.

The next three chapters address the limits of bi-directionality from a cognitive-linguistic point of view. **Liane Ströbel**'s "Invisible, visible, grammaticalization" examines the rise of a new category of empty verb constructions in typologically distinct languages, drawing on Lakoff's experientialist approach in which the human body and its parts play an important role in our conceptualization of the world. The study of these constructions is an attempt to bring together various guiding principles of Cognitive Linguistics into a unified case study of the concept "closeness to the body of the speaker" and its linguistic realizations. Ströbel argues that the motivation and mental images behind the rise of grammaticalization products can be reduced to the following formula: an abstract action or concept that is "invisible" in the discourse world is made linguistically "visible" with the help of a universal strategy. The potential for bi-directional exchange of Ströbel's cognitive-linguistic approach to the study of grammaticalization is obviously limited, though she argues that it has implications vis-à-vis neighboring disciplines such as philosophy of mind and neurolinguistics.

Gary Thoms and **Stefano Versace**'s "How does the mind do literary work?" analyses the structural properties of a number of examples of literary forms (meter and "deviant" syntactic structures in avant-garde literature) that they interpret as mismatches between these forms and language, and the way the mind processes them. The authors use generalizations from the study of literary forms to propose a specific model of how the mind works. They argue that while formal literary work has typically borrowed tools and findings from linguistic theory to perform its own analyses, such studies have rarely produced feedback for linguistic theory, an essentially uni-directional transfer of knowledge between these disciplines. Their study, however, provides an example of formal literary work that more widely feeds back into linguistic theory and the cognitive sciences. Arguing against traditional positions in literary and linguistic criticism, they demonstrate the advantages of divorcing functional and structural relations between language and literary structures. They suggest that the complex relations between literary form and language are best accounted for by building on a modular approach to cognition, thus developing a more theoretically and empirically coherent theory of literary forms.

Bringing together insights from cognitive schema theories and studies on vocabulary acquisition in foreign and second language learning, **Alexander Ziem**'s "Cognitive science meets language pedagogy" addresses the importance of semantic schemata as didactic tools for teaching and learning word meaning. He draws on recent findings in cognitive approaches to language pedagogy and vocabulary teaching that have highlighted the important role of conceptual motivation to help enhance the process of foreign language learning because it facilitates the understanding and retention of lexical meanings and seemingly opaque figurative expressions. Ziem argues in favor of "rich instruction", a teacher-centered method to support the process of vocabulary acquisition, shows how schema-based illustrations can help explain sense relations between words and the emergence of new meanings in complex constructions such as metaphorical expressions and compound words, and finally points out four advantages of applying semantic schemata in the classroom. Ziem's chapter thus clearly shows that while language pedagogy could considerably benefit from theoretical concepts such as "schema", "frame", and "blending", the didactic application of schemas and frames does not have discernable repercussions on cognitive schema theories, particularly because these theories have not been designed to provide didactic tools in the realm of language teaching and acquisition. According to Ziem, using cognitive schemata as heuristic devices for teaching and learning vocabulary may thus be considered a prime example for the uni-directional applicability of cognitive-linguistic theories and concepts to the domain of language pedagogy.

The final two chapters return to questions of bi-directionality in research topics and methods. The recent orientation towards social variation in CL addressed earlier in this introduction has in turn consequences for research methodology:

> The importance of socially and culturally determined variation for Cognitive Linguistics has an obvious methodological correlate: these socially oriented forms of Cognitive Linguistics would naturally support the adoption of the quantitative, empirical methodology that is dominant in sociolinguistic research at large. (Geeraerts 2006: 31)

While many cognitive linguists typically conduct their scholarship by relying upon their own intuition, there is a growing awareness in CL that introspection is often limited if not inaccurate, and that even trained experts often fail to recognize the real reasons for their beliefs and actions because our human ability to introspect upon many cognitive processes is extremely limited. This is not to say that introspection should be abandoned completely, but to point out the distinctive contribution and complementary character of the diverse methodologies since

> each methodology can be seen as having certain capacities and limitations that accord it a particular perspective on the nature of conceptual organization in language. In this respect, no single methodology is privileged over others or considered the gold standard of investigation. (Talmy 2007: XI)

The complementary character of diverse methodologies is addressed by **Juliana Goschler** who combines qualitative motif analysis with a quantitative corpus study in "The conceptualization of personality: Converging and diverging evidence". Focusing on people's beliefs about the location of the "core" of a person, she provides corpus-based evidence for the fact that metaphors pointing to a conventionalized idea of the heart as the seat of emotions are not only frequent in literature, but also very productive in everyday language, although they are not in line with the scientific consensus that cognition and emotion are processed by the brain. The results of a questionnaire study on folk belief, however, contradict the facts inferred from the examination of language use: people localize personality neither in the heart nor in the brain. Therefore, Goschler stresses the necessity to include empirical methods from psychology and sociology to analyze cognitive phenomena at the nexus of literature, language, and folk models, exemplifying the potential of a transdisciplinary approach to phenomena that have both a cultural and a cognitive dimension. Goschler thus demonstrates how theories and methods from one discipline can be borrowed and applied to another to enhance knowledge, for example by confirming or rejecting seemingly long-standing assumptions made solely on the basis of literary or linguistic data. Goschler points out, however, that while a combination of disciplines and methods is a useful extension, it is not bi-directional but rather an expansion into one direction. In view of some fundamental epistemological differences as to how linguistics and literary scholarship conceive of language as the object of study, Goschler remains skeptical whether this will lead to a transformation of one discipline, let alone a merging of two currently distinct disciplines.

Recently, a number of researchers have highlighted the importance of findings resulting from cross-disciplinary empirical and experimental methods, and have called for a commitment to empirical research methods that will place CL more firmly within the broader experimentally-oriented community of the cognitive sciences (Gibbs 2006, 2007; Rohrer 2006; Tummers et al. 2005; Geeraerts 2005, 2006):

> The specific way in which Cognitive Linguistics conceives of its 'cognitive' character would indeed seem to favor a more empirical approach to linguistic methodology. In generative grammar, the ultimate foundation for an introspective methodology is the autonomy of language and linguistics, whereas Cognitive Linguistics conceives of language as integrated with the other cognitive capacities of man [sic!]: giving up the autonomy assumption motivates a broader range of methodological techniques, in line with the practice used by the cognitive sciences. (Geeraerts 2006: 29)

The potential for such a methodological cross-fertilization is apparent in the broad range of methods applied in the papers collected by Gonzalez-Marquez et al. (2007). For example, Gibbs (2007) sketches how cognitive linguists and cognitive psychologists can more fruitfully explore their respective claims about the interaction of thought, language, and the body in their empirical work. Cognitive linguists need to alter their methods and their theories to better meet the demands of a scientific discipline. In particular, Gibbs argues for greater reliability in cognitive linguistic methods of linguistic analyses, and emphasizes the importance of falsification in cognitive linguistic theory construction.

This challenge for CL and other disciplines is the focus of the final contribution "Cognitive Linguistics as a cognitive science" by **Anatol Stefanowitsch**, who outlines three main ways in which a discipline can relate itself to the study of the human mind. He suggests that true bi-directionality in the cognitive sciences can only be achieved if researchers in disciplines that have a more indirect relationship to the study of cognition (for example the social sciences and humanities, in that their principal objects of research are products of the human mind) adopt strictly empirical methodologies. Focusing on CL, Stefanowitsch argues that to avoid a simple one-way transfer of models and theories from the core cognitive sciences (psychology and neurology), CL needs to rigorously commit itself to empirical, corpus-based linguistic analysis. He thus concludes that CL must become more linguistically accurate (focusing on detailed linguistic investigation), more scientific (overcoming introspection-based argumentation and generalization), more methodologically aware (adopting strict observational and experimental protocols and state-of-the-art statistical methods), and hence less theoretically ambitious, restricting itself to linguistic constructs that can be operationalized and research hypotheses that can be falsified.

Acknowledgements

We would like to thank a number of people who made this book possible. Our first thanks is to our volume authors for their contributions, their commitment to the project, and for their patience. We are further indebted to an anonymous reader who provided constructive criticism and encouraging feedback, which, we believe, resulted in a stronger volume. We would also like to thank Hanneke Bruintjes and Patricia Leplae who took care of our manuscript and worked hard to accomplish a swift continuation of the Human Cognitive Processing book series under new editorship. Needless to say, we are grateful to the current series editors, Linda L. Thornburg and Klaus-Uwe Panther, for their interest in including this collection into the series. All of the contributions in this book were originally presented at the conference "Bi-directional Perspectives in the Cognitive Sciences" held from February 26–28, 2009, at the historic city hall of Marburg, Germany. The conference was made possible by a generous grant by the Volkswagen Foundation. Additional funding came from the English Department at Humboldt University Berlin, the University of Freiburg, and the Graduate Center for Humanities and Social Sciences at the University of Marburg. We would like to thank these institutions for the generosity and support. Lastly, several individuals assisted in the organization of the conference. We would like to particularly thank Elena Borisova and Else Engelhardt at Humboldt University Berlin, Christian Pauls, Richard Wiese, Rüdiger Zimmermann, and Alexander Stein.

References

Boers, Frank and Seth Lindstromberg (eds.). 2008. *Cognitive Linguistic Approaches to Teaching Vocabulary and Phraseology*. Berlin: Mouton de Gruyter.

Brône, Geert and Jeroen Vandaele (eds.). 2009. *Cognitive Poetics*. Berlin: Mouton de Gruyter.

De Knop, Sabine and Teun De Rycker (eds.). 2008. *Cognitive Approaches to Pedagogical Grammar*. Berlin: Mouton de Gruyter.

Dirven, René, Hans-Georg Wolf and Frank Polzenhagen. 2007. Cognitive linguistics and cultural studies. In *The Oxford Handbook of Cognitive Linguistics*, Dirk Geeraerts and Hubert Cuyckens (eds.), 1203–1221. Oxford: Oxford University Press.

Evans, Vyvyan, Benjamin K. Bergen and Jörg Zinken. 2007. The Cognitive Linguistics enterprise: An overview. In *The Cognitive Linguistics Reader*, Vyvyan Evans, Benjamin K. Bergen and Jörg Zinken (eds.), 1–36. London: Equinox.

Freeman, Margaret H. 2007. Cognitive linguistic approaches to literary studies: State of the art in Cognitive Poetics. In *The Oxford Handbook of Cognitive Linguistics*, Dirk Geeraerts and Hubert Cuyckens (eds.), 1175–1202. Oxford: Oxford University Press.

Gavins, Joanna and Gerard Steen (eds.). 2003. *Cognitive Poetics in Practice*. London: Routledge.

Geeraerts, Dirk. 2005. Lectal variation and empirical data in Cognitive Linguistics. In *Cognitive Linguistics: Internal Dynamics and Interdisciplinary Interaction*, Francisco J. Ruiz de Mendoza Ibanez and M. Sandra Pena Cervel (eds.), 163–189. Berlin: Mouton de Gruyter.

Geeraerts, Dirk. 2006. Methodology in Cognitive Linguistics. In *Cognitive Linguistics: Current Applications, Future Orientations*, Gitte Kristiansen, Michel Achard, René Dirven and Francisco J. Ruiz de Mendoza Ibanez (eds.), 21–49. Berlin: Mouton de Gruyter.

Geeraerts, Dirk and Hubert Cuyckens (eds.). 2007. *The Oxford Handbook of Cognitive Linguistics*. Oxford: Oxford University Press.

Geeraerts, Dirk and Hubert Cuyckens. 2007. Introducing Cognitive Linguistics. In *The Oxford Handbook of Cognitive Linguistics*, Dirk Geeraerts and Hubert Cuyckens (eds.), 3–21. Oxford: Oxford University Press.

Geeraerts, Dirk, Gitte Kristiansen and Yves Peirsman (eds.). 2010. *Advances in Cognitive Sociolinguistics*. Berlin: de Gruyter Mouton.

Gibbs, Raymond W. Jr. 2006. Introspection and cognitive linguistics: Should we trust our own intuitions? *Annual Review of Cognitive Linguistics* 4: 135–151.

Gibbs, Raymond W. Jr. 2007. Why cognitive linguists should care more about empirical methods. In *Methods in Cognitive Linguistics*, Monica Gonzalez-Marquez, Irene Mittelberg, Seana Coulson and Michael J. Spivey (eds.), 2–18. Amsterdam: Benjamins.

Goldberg, Adele E. 1995. *Constructions: A Construction Grammar Approach to Argument Structure*. Chicago: The University of Chicago Press.

Gonzalez-Marquez, Monica, Irene Mittelberg, Seana Coulson and Michael J. Spivey (eds.). 2007. *Methods in Cognitive Linguistics*. Amsterdam: Benjamins.

Kristiansen, Gitte and René Dirven (eds.). 2007. *Cognitive Sociolinguistics*. Berlin: Mouton de Gruyter.

Mittelstraß, Jürgen. 2002. Transdisciplinarity – New structures in science. In *Innovative Structures in Basic Research (Ringberg-Symposium October 4–7, 2000)*, Uwe Opolka and Helen Schoop (eds.), 43–54. München: Max-Planck-Gesellschaft. <http://xserve02.mpiwg-berlin.mpg.de/ringberg/Talks/mittels%20-%20CHECKOUT/Mittelstrass.html> (last accessed March 13th, 2011).

Pütz, Martin, Susanne Niemeier and René Dirven (eds.). 2001. *Applied Cognitive Linguistics I: Theory and Language Acquisition. II: Language Pedagogy*. Berlin: Mouton de Gruyter.

Rohrer, Tim. 2006. Three dogmas of embodiment: Cognitive linguistics as a cognitive science. In *Cognitive Linguistics: Current Applications, Future Orientations*, Gitte Kristiansen, Michel Achard, René Dirven and Francisco J. Ruiz de Mendoza (eds.), 119–146. Berlin: Mouton de Gruyter.

Semino, Elena and Jonathan Culpeper (eds.). 2002. *Cognitive Stylistics: Language and Cognition in Text Analysis*. Amsterdam: Benjamins.

Sharifian, Farzad, Ning Yu, René Dirven and Susanne Niemeier (eds.). 2008. *Culture, Body, and Language: Conceptualizations of Internal Body Organs across Cultures and Languages*. Berlin: Mouton de Gruyter.

Steen, Gerard. 2002. Metaphor in Bob Dylan's 'Hurricane': Genre, language and style. In *Cognitive Stylistics: Language and Cognition in Text Analysis*, Elena Semino and Jonathan Culpeper (eds.), 183–209. Amsterdam: Benjamins.

Steen, Gerard and Joanna Gavins. 2003. Contextualising Cognitive Poetics. In *Cognitive Poetics in Practice*, Joanna Gavins and Gerard Steen (eds.), 1–12. London: Routledge.

Stillings, Neil A., Steven E. Weisler, Christopher H. Chase, Mark H. Feinstein, Edwina L. Rissland and Jay L. Garfield. 1995. *Cognitive Science: An Introduction* (2nd ed.). Cambridge, MA: MIT Press.

Stockwell, Peter. 2002. *Cognitive Poetics: An Introduction*. London: Routledge.

Stockwell, Peter. 2009. *Texture: A Cognitive Aesthetics of Reading*. Edinburgh: Edinburgh University Press.

Talmy, Leonard. 2007. Foreword. In *Methods in Cognitive Linguistics*, Monica Gonzalez-Marquez, Irene Mittelberg, Seana Coulson and Michael J. Spivey (eds.), XI–XXI. Amsterdam: Benjamins.

Tsur, Reuven. 1992. *Toward a Theory of Cognitive Poetics*. Amsterdam: Elsevier (North Holland).

Tummers, José, Kris Heylen and Dirk Geeraerts. 2005. Usage-based approaches in Cognitive Linguistics. A technical state of the art. *Corpus Linguistics and Linguistic Theory* 1: 225–261.

PART I

Avenues for bi-directionality

Genre between the humanities and the sciences

Gerard Steen
Vrije Universiteit, Amsterdam

This chapter outlines an approach to genre as the driving force behind language, cognition, and communication. Genre can and has been studied in the humanities as well as the cognitive and social sciences, but the present chapter presents a theoretical integration of these approaches and offers a model that can be empirically investigated by various methods across the board of these disciplines. The model itself is based in cognitive psychology, and a number of possibilities and implications for text analysis are illustrated. Both the role of language itself as well as the connection with literature are included in the discussion in order to suggest how a complete picture can be achieved of the bi-directional traffic between cognitive science and the humanities.

Keywords: empirical approach; relationship of language, literature, and cognition; text analysis

1. Why genre?

When you send an e-mail message to a colleague and do not receive a reply within 48 hours, you may begin to wonder whether something went wrong in digital space, whether your colleague has fallen ill, or whether she has forgotten to switch on an out-of-office message. You may also start worrying whether you said something wrong. Similarly, when you address an audience at an academic conference and you get strange or no responses, you may wonder about your audience, the form, content, or language of your paper, or the goal that you set yourself in preparing the talk in the way you did. Yet another example of the complexities of doing discourse that may cause puzzlement or concern, and eventually miscommunication, is the rhetorical variation between requests for reprints exhibited between on the one hand countries in the Middle East, which used to have elaborate personal, hand-written letters that begin with an enquiry into the health of yourself and your family, and on the other hand nations like the US, which used to have pre-printed message cards with one sentence: Could you please send me a copy of article so and so? Discourse is a risky business

This is the reason why we need genres. They provide socio-cultural, psychological, and formal (linguistic or multimodal) orientation as well as order to people in the complex universe of media and communication, even in the academe. Language users depend on this genre-driven orientation if they wish to function effectively as communicators (and stay healthy and sane in the process). This also applies to those genres where communication is not oriented towards some practical ulterior purpose, as in conversation and its digital equivalent, chat; in those genres, too, language, cognition, and communication are deemed to be intentional (in the sense used in pragmatics) and goal-directed (in the sense used in discourse psychology) and dependent on a frame of reference for 'success': conversations and chat, too, can derail and be experienced as failed. The encompassing framework for monitoring and judging such experiences is afforded by genre.

Indeed, the domain within which the notion of genre is perhaps most important, literature, the arts, and the media, has often been argued to have no ulterior purpose at all. And yet, as has been shown for two millennia, it is precisely these classes of discourse that have well-elaborated genre structures. All events of discourse consequently manifest diverse processes that are eventually aimed at complying with the expectations of genre: these processes can be more or less effective in comparison with those expectations. There is hardly any discourse without genre.

Genre is a notion that can be conceptualized in different ways, both in the humanities and in the cognitive and social sciences. The following phenomena can be seen as potentially constituting distinct genres: video games, films, talk shows, news shows on the radio or television, front page news articles, editorials, advertisements, instruction manuals, political rallies, the inauguration speech, board meetings, lectures, textbooks, and research articles, poems, plays, and novels, blogs, chat sessions, e-mail messages, doctor-patient meetings, interrogations in the police station and in court, and so on and so forth. Some of these genres are labeled by the document used for communicative purposes (e.g. video game, film, instruction manual, textbook), whereas others are labeled by the situation of interaction (e.g. talk show, political rally, doctor-patient meeting). Is there any order in this bewildering array of phenomena?

Yes, there is. All of these genres are manifested by concrete and situated *discourse events*, either comprising relatively freestanding discourse production or reception, or their combination in interaction. These discourse events are organized along three dimensions: they typically include goal-directed processes of *communication* between people on the basis of aligned processes of individual *cognition* by means of individual processes of *language* use (see Clark 1996). Major variation between genres occurs for instance when people interact via telephones or computers, or when some other modality of expression than language is used, but this variation can be taken into account. More specific aspects of genre-driven expectations about language, cognition, and communication include the appropriate *communicative* norms and conventions (contrast, for instance, the norms and conventions in the domains of science and the arts), the presupposed *cognitive* structures, opinions, and values (contrast, for instance, the discourse of experts and novices), and the adequate *linguistic* register and style that are to be used (contrast the language of law with the language of newspaper reporting on legal matters). Whether in discourse production, reception, or interaction, people have tacit assumptions about a whole range of linguistic, cognitive, and communicative genre properties, and they rely on these assumptions to make their participation in discourse more appropriate, effective, and comfortable.

This is also why these discourse events can be studied from three complementary angles: the social sciences, which begin with discourse as communication between people; the cognitive sciences, which begin with discourse as cognition within people; and linguistics, which begins with discourse as form in semiotic structures. In all of these approaches, however, aspects of the other two dimensions of discourse cannot be neglected: the social sciences, for instance, look at discourse as communication but cannot ignore the role of cognition and language; they give a social-scientific account of discourse in which the dimension of communication somewhat constrains how the dimensions of cognition and language are examined; and so on. This complex interaction between dimensions of discourse on the one hand and approaches to discourse on the other can be addressed in systematic and innovative fashion by examining the role of genre.

A general version of the insight that discourse is expectation-driven has been around for a long time in many disciplines, both in the humanities and in the cognitive and social sciences (e.g. Hamilton, Schiffrin, and Tannen 2001). Yet it is less generally accepted that these patterns may be attributable to one specific level of discourse organization, genre. This has prevented many discourse analysts from taking an exhaustive view of all of the relevant parameters of events of discourse. Thus, some discourse-analytical approaches focus on the formal structure and function of discourse and have difficulties in making empirical connections with the on-going linguistic, cognitive, and communicative processes in which genres manifest themselves in actual fact. Other approaches focus on the social dimension at the expense of the psychological one, as typically happens in Conversation Analysis. Yet others, when they are psychological, are too skeletal and/or selective in their model. That is why genre has remained dangling between the humanities and the cognitive and social sciences in a way that has prevented making full use of the theoretical and empirical potential that has in fact been there for some time.

Currently there is no generally accepted genre framework across the humanities and the cognitive and social sciences. It is true, though, that there are new interdisciplinary interests in the nature, function, and use of genres in communication science, social psychology, cognitive psychology, various brands of linguistics and literary studies, and computer science. In linguistics and discourse analysis, a new area labeled 'genre analysis' has been emerging since the early 1990s. In computer science, automated text classification on the web has now adopted the notion of genre as a leading concept for developing new tools. And in the study of the mass media as well as in poetics, the notion of genre has been fundamental from the beginning. But what is lacking is an integrated interdisciplinary framework for the study of genre.

What is more, there are on-going technological developments in the real world – outside academia – of modern media and communication that provide exciting leads to turn to genre as an obvious hole in the market. People are often clearly at a loss or frustrated in the world of the new media, and this may be largely due to a lack of appropriate genre knowledge and skills. Moreover, genres in the new media themselves are forever evolving, mixing, and being transformed, which does not help the media user (see Biber and Kurjian 2007). As Lemke (2005: 45) writes: "Genres are not what they used to be".

It is the aim of this chapter to pull the insights of a wide range of diverse theoretical and empirical proposals together and argue that it is possible to construct a unified, consistent, and explicit theoretical framework for genre that is based in cognitive psychology. The main reason of my preference for a psychological approach over a linguistic or communication-scientific approach to genre is that psychological schemas for genre can be acquired, trained, monitored, improved, and transformed by individual language users. Such a framework will hence facilitate the identification of important areas in language, cognition, and communication for innovative interdisciplinary research that is both fundamental and applied.

The next section begins by outlining a number of basic issues that have slowed down or prevented the development of a full-blown theory of genre across the humanities and the cognitive and social sciences. In the third section I turn to the basis of the present proposal, which has to do with the notion of a genre event and the knowledge that people have of genres in the form of a genre repertoire. Section 4 offers an outline of the structure of the genre theory pursued in the present chapter. The fifth section illustrates some of the possibilities of this theory, and is followed by some brief concluding comments.

2. Basic issues

There are several long-standing challenges to achieving a full grasp of the structure, function and processes of genre across the broad range of disciplines that should be concerned with it:

1. Genre has not been ubiquitously recognized as important to the study of language use and, more broadly, communication (or 'discourse', in some circles).
2. When it has been studied, genre has been conceptualized and researched in varying ways in distinct disciplines, causing discrepancies between distinct theoretical and terminological traditions.
3. Genre itself may be seen to include a wide range of disparate aspects of discourse (linguistic, cognitive, and communicative), raising issues of grouping, relevance, and proportion.

The present proposal does not aim to produce an unbiased response to all of these challenges, but adopts one well-defined position in the field in which the strategy of fostering bi-directional perspectives between the humanities and the sciences is prominent. This will enable a consistent approach to each of the issues, while awareness is maintained that the adopted approach is one (strong) candidate out of a range of possible alternatives that may be developed along comparable lines.

The theoretical grounding of the present approach lies in cognition, not in language or communication: genre is here defined as a type of complex knowledge schema that individual language users have at their disposal to engage in discourse. This position is emergent from and compatible with many psychological approaches to discourse (e.g. Paltridge 1995; Graesser, Millis, and Zwaan 1997; Slater 1997; Steen 1999; Slater and Rouner 2002; Hall 2003; Bilandzic and Rössler 2004; Knobloch et al. 2004; Sherry 2004; Donovan and Smolkin 2002; Van Dijk 2008). Some of this work has also led to related studies in the empirical study of literature (e.g. Fishelov 1993). There is also a rapidly growing number of publications on genre in applied linguistics, especially in English for Specific Purposes, which includes a large number of cognitively oriented studies of language use in various genres of discourse (e.g. Sengupta et al. 1999; Ramanathan and Kaplan 2000; Fernández Toledo 2005; Kirkpatrick and Klein in press), with an initial impact on education (Hyland 2003, 2007). Moreover, some of these applied linguistic studies pay explicit attention to cross-cultural variation and interaction in particular genres, mainly from the perspective of Contrastive Rhetoric (e.g. Vergaro 2002; Connor 2002, 2004; Baumgarten 2007). In adopting this cognitive-psychological perspective on genre and its use for special purposes across cultures, I am also giving a positive response to the first issue listed above, that genre is important, indeed fundamental, to the study of discourse conceived as events of interaction involving language, cognition, and communication.

It is interesting to note that there is not much interest in genre in cognitive linguistics, the brand of linguistics that has most affinity with the role of language in usage when that is studied as language, cognition, and communication. Cognitive linguistics itself seems most focused on the small-scale details of the use of language as semiotic code, in interaction with knowledge, typically at utterance level. The closest relation between cognitive linguistics and genre, in my opinion, may therefore be found in the application of cognitive linguistics found in cognitive stylistics and cognitive poetics. For instance, there is a clear interest in genre in Semino and Culpepper (2003) as well as in Stockwell (2002) and Gavins and Steen (2003). There is still a wide gap between linguistics and genre in many traditions, which may have to be closed by an approach from literature and style.

The cognitive-psychological starting point adopted above does not deny that there are other theoretical approaches to genre (see Muntigl and Gruber 2005; Johns et al. 2006). These have defined the object of study from the perspective of for instance language instead of cognition, that is, as formal, linguistic objects, as has happened in Systemic-Functional Grammar (e.g. Eggins and Martin 1997; Martin and Rose 2005) and in computational linguistics (e.g. Dillon 2007). Still other approaches have defined genres as social action (e.g. Winsor 1999; Dunmire 2000; Jordens and Little 2004) or as communication in the form of socially conventionalized interaction patterns (e.g. Swales 1990; Bhatia 1993; Freedman and Medway 1994; Berkenkotter and Huckin 1995). These alternative approaches to genre via either language or communication instead of cognition have produced questions and answers that are relevant to our understanding of genre, and they can be translated into cognitive-psychological terms. In doing so, I am also defining my position regarding the second issue raised above, that it is possible to order the wide range of heterogeneous issues that have to do with modeling genre from one single perspective, that of cognitive psychology.

The cognitive-psychological standpoint also offers a theoretical and empirical basis for selecting, grouping, weighing, and ordering the most important variables that need to be modeled. There are two problems here. Firstly, particular features of discourse that are labeled in particular ways, such as narrative/story or argumentation/argument, are regarded as genres by some whereas they are seen as aspects of genres, such as the novel or a political speech, by others. Secondly, variation between classes of discourse can be modeled in more than one way, with variables and categories receiving different names, functions, and weights in different models. It is my claim that both problems can be alleviated by modeling genre as involving a number of complementary mental models that language users are supposed to construct during discourse processing (which itself is an encompassing notion referring to production, reception, and interaction).

The cognitive-psychological model of genre to be presented in a moment will be based on the multi-level approach to discourse representation advanced by Van Dijk and Kintsch (1983; Kintsch 1998; Van Dijk 2008). This includes a number of complementary mental models that turn out to be related to the three dimensions of discourse labeled as communication, cognition, and language distinguished at the beginning of this chapter:

- a *context model* (which may be interpreted as an individual's mental representation of discourse as 'communication', or interaction between people);
- a *situation model* and *text base* (which may be interpreted as an individual's mental representations of discourse as the conceptual structure of discourse, or the series of thoughts a discourse is supposed to contain and convey; in brief, discourse as 'cognition');
- and a *surface text* (representing discourse as 'language', or the linguistic and multimodal meanings of the semiotic elements of text and talk).

Each of these mental models accommodates different aspects of discourse that have been shown to be relevant to the psychological processes and products involved in discourse processing (cf. Clark 1996). In this contribution I propose to model these distinct mental representations and their components in a more specific fashion: they need to be oriented towards the presumed functions and effects of the organizing notion of genre. This aspect of my theoretical approach resolves the third challenge mentioned above, how to identify and order the many variables and issues in the field from a well-motivated angle.

In sum, genre has been an intractable object because of its many dimensions and the way these have been highlighted or ignored in the many disciplines that have taken an interest in the study of discourse. The adoption of one single and explicitly formulated standpoint in cognitive psychology is not meant to downgrade the dimensions of language or communication, or to ignore some of the relevant alternative approaches. Moreover, it is meant as a further development of a more general approach to the psychology of discourse that is relatively independently motivated. This will lead to a theoretical consideration of the distinction, weights, and relations of the several variables that have been discussed in the various theories and research projects.

Evidence for postulating and characterizing distinct dimensions, variables and categories is eventually a matter for meta-analysis. It can be based on a wide range of materials: theoretical models in the humanities and the sciences, reference works such as encyclopaedias of language and linguistics, manuals and advice literature in communication, distinctions and labels used in discourse communities, processing patterns and effects revealed in psycholinguistic and sociolinguistic research, and statistical patterns revealed in corpus work. When these materials point to potential cognitive parameters in the use of genres, they can be included in the research. The perspective on this theory of genre is therefore truly bi-directional and includes both the humanities and the cognitive and social sciences.

3. Genre events and genre knowledge

As a basis for comparison between all of these variations of discourse I should like to advance the new notion of the *genre event*. When language users orient their psychological processes towards discourse by employing the notion of genre, they do so in situated events of genre use. They do this when they have conversations, are in meetings, read the newspaper, watch the news, write and read email messages, text messages, or interact on the telephone or via Skype. In each of these situations, we can speak of a genre event in which aspects of the context model, situation model and text base, and the surface text have to be constructed, maintained, updated, aligned, monitored, and so on, by the language users that are involved in the genre event.

Genre events are a new variation on the more familiar theme in linguistics of speech event, usage event, discourse event, and so on. The new notion shifts the emphasis from such events as primarily grounded in the use of language (*speech* or *usage* event) to their being grounded in genres as individual action patterns that involve not just language but also, and crucially, cognition and communication. When people engage in genre events, they need to muster three types of knowledge: about the expectations regarding appropriate language use, use of world knowledge, and conventions of communication. This is what they also need to be taught when they learn how to interact in a second language, as when businessmen acquire the communicative conventions for negotiation across cultures. This is also what students have to learn when they acquire professional skills in order to design and execute an academic article, a news report, a home page, etc. Genre knowledge is the type of knowledge that people need to have about language, cognition, and communication, and which may be abstracted by both participants as well as discourse analysts from the observations of thousands if not millions of genre events.

The notion of a genre event highlights the action-oriented character of discourse in which individual people interact with each other. It suggests that some genre labels, such as 'the novel' or 'the newspaper article', are in fact metonymic:

from a psychological discourse perspective, they typically stand for 'reading a novel' or 'reading a newspaper article', a bias in favor of the large number of receivers as opposed to the small number of producers, while the genre label itself focuses on the document because of its relative prominence in the genre event. This interpretation of document-oriented genre labels as metonymic becomes all the more relevant when we turn to some potential new genre labels such as texting, chatting, emailing, skyping, browsing the web, and so on (see Bateman 2008). In general, the action-oriented perspective on the mental representation of genres allows for a motivated theoretical integration of approaches to genres that focus on participants and their (inter)action on the one hand (people need to represent conversations as conversations), with approaches that focus on documents and their meaning on the other (people need to represent education materials as messages, which captures the document perspective, but also as products of their own reception process, which captures the interaction perspective).

In addition, other genres such as conversations are typically labeled at the level of the interaction between participants. This allows for another distinction. First of all, for this class of genres there are the contexts models, situation models, text bases and surface texts that are presumably *shared* between participants – this is the typical interpretation of terms like 'conversation', 'discussion', and 'negotiation'. These shared models clearly have to be mentally represented as such by individual discourse participants in their various mental models ("it takes two to tango"). But, for the same class of genres, there are also the *private* contributions to as well as interpretations of context models, situation models, text bases, and surface texts, for either senders or receivers or even by-standers in conversations, discussions, and negotiations ("what is my plan and which are my moves going to be?"). These private mental models of say a conversation may sometimes be at odds with the presumably shared models of the same conversation. Both models need to be monitored by individuals when they represent their participation in a genre event. This leads to a dual perspective on verbal interaction that has been successfully laid out for all discourse by the psychologist Herb Clark (1996), for instance. The difference between that type of proposal, which is typical of many studies of discourse as discourse, and the present proposal is that the present approach constrains and specifies the broad notion of 'discourse' by a behavior-regulating notion of genre knowledge.

It may take some effort to step away from genre as either a document-related phenomenon (novel, film, letter, and so on) or an interaction-related phenomenon (conversation, meeting, interview, and so on), for these approaches represent the traditionally most popular ones. But that is precisely what is being proposed here. In order to study language use as individual behavior in context, we need to adopt a cognitive-psychological standpoint and see both genre-as-document and genre-as-interaction as intimately connected to our knowledge schemas for genre. These are both long-term knowledge schemas that have been acquired over lifetime and in formal training, as well as short-term knowledge schemas that are constructed and monitored in production and reception as we engage in the moment-to-moment processes of a specific genre event. These knowledge schemas need to be studied as the mental models that are deemed to be constructed during discourse processing in general models of discourse, such as van Dijk and Kintsch (1983). They have suggested, as we saw above, that we need a number of complementary models, which, it now turns out, can account for the idea that some genres are most naturally represented as interaction (see 'communication model') whereas other genres are most naturally represented as documents (see 'surface text' and 'text base').

The inclusion of both classes of genre within one psychological approach that is less familiar opens up new avenues of theory and research. To give one example, the distinction between the various mental models may raise the question whether these mental models have different values for the two classes of genre distinguished above. That is, could it be possible that genres typically labeled as documents are processed with more attention to the surface text and text base models, and with less attention to the context model, as opposed to genres that are labeled as interaction-types are processed with more attention to the context model and with less attention to the surface text and text-base models? And is this one reason why oral presentations may benefit from visual support by means of PowerPoint presentations, so that addressees have to (or can) concentrate on constructing both a context model (prompted by the genre labeled as interaction-type) and a surface text and text base model (prompted by the genre labeled as document). This is just one possible consequence of the present theoretical approach, and it can be followed up in both theoretical and empirical research.

A specific concern of the present approach is the role of cognitive-psychological theories of categorization, including exemplar theory and prototype theory (e.g. Murphy 2002). A particular genre event can be a central or marginal case for the category it exemplifies in that it can display better or worse characteristics of the genre it belongs to. Any historical novel, for instance, is (a) a novel but (b) less typical since it is not completely fictive. This perspective therefore allows for the inclusion of a particular genre event within the class of a genre as more or less typical, or even as a hybrid between two genres, without undermining the complete system. This is in fact how many language users operate with genres that are in a stage of transition or that have ended up on the border between two well-defined but mutually exclusive categories.

In all of these questions, the orientation of the project remains firmly cognitive-psychological, in that it aims to model what language users in fact do in terms of discourse processing. Different classes of genre events may highlight different aspects of the various mental models (context model, situation model, text base, surface text) in different ways. They may even lie at the basis of different routes of genre processing. Yet the overarching idea of discourse suggests that there should be a common ground to all of these genre events that affords effective communication.

4. Towards a cognitive-psychological model for genre

The main purpose of this line of thinking is to develop one encompassing model of the structure, function, and process for various groups of genres across the media. In order to do so, we need to examine theories and research on linguistic, cognitive, and communicative aspects of discourse in relation to genre. When we adopt a cognitive-psychological standpoint, these findings can be linked to the following mental models in discourse analysis: (a) the surface text, (b) the text base and situation model, and (c) the context model (Van Dijk and Kintsch 1983; Kintsch 1998; Van Dijk 2008). These mental models are distinct types of cognitive representation that give natural prominence to diverging aspects of genres, which adds to the model's potential for explanatory adequacy.

When we consider the structure and function of each of the mental models, and the way they have been researched in discourse psychology, it may be useful to simplify the model somewhat, and to go back to an old distinction that has been used in various manifestations throughout the humanities and the sciences, the one of code, text, and context. 'Text' refers to the message that is being exchanged or co-constructed between participants; 'code' refers to the semiotic means comprising the text; and 'context' refers to all relevant structures and entities outside the text that are involved in the genre event. The mental representation of code yields Van Dijk and Kintsch's surface text, the representation of text produces both a text base and a situation model, and the representation of context leads to a context model. Descriptions of code, text, and context therefore provide input for the reconstruction of multi-level production and reception processes in genre events.

The following sketch constitutes a provisional ordering of the resulting field of research and presents tentative ascriptions of specific discourse and genre phenomena to distinct variables and submodels:

a. **Context model**:
represents the *interaction between language users* in various types of discourse, including the major classes of talk and text; the question is which aspects of

context have to be included in the envisaged multi-level mental model for genres. Obvious candidates are the following variables:

i. participants and their relations and roles; there is a wide range of issues that has to do with identities, power relations, and conventions of interaction that has been studied in many disciplines. Moreover, this variable also includes numbers of participants and the question whether they are known or anonymous or something in between (mass communication versus other types of communication).
ii. goals and functions of discourse; there is a vast literature on the informative, persuasive, and instructive functions of discourse, which can lead the way to developing a more refined taxonomy.
iii. situations and settings and their properties; special attention needs to be given here to spatio-temporal co-presence of participants versus other variants.
iv. domains and their norms and values; macro-social divisions between culturally acknowledged and sanctioned environments of discourse include distinctions between religion and law, arts and entertainment, science and education, government and politics, business and everyday affairs, and so on.
v. medium, including devices and channels of communication; this is one traditional area of the communication sciences, with attention to the printed press, telecommunication via telephone, radio, television, and now the internet; included in this variable is therefore the role of text versus talk, where sound and gesture are an integral part of talk while layout is an integral part of text.

b. **Text model:**
represents various *aspects of the text* (text base and situation model) in various types of discourse, including talk and text; the question here is which aspects of **texts** have to be included in the envisaged multi-level mental model for genres. Obvious candidates here are the following variables:

i. content of the text, with various types of studies of topics, thematics, issues, and so on in various domains (see context model) as well as media (see context model).
ii. type of the text, with the vast literatures on narration and argumentation leading the way.
iii. form of the text, with formal superstructures that are more general across genres (e.g. Problem-Solution) or are more genre-specific (e.g. installment structure for news reports; Introduction-Methods-Results-Discussion structure for research reports).

iv. structure of the text, with the vast literatures on referential and relational coherence leading the way.

c. **Code model:**
represents various aspects of the *language and other means of expression* of messages ('surface text'), for talk and text; which aspects of **code** have to be included in the envisaged multi-level mental model for genres? Obvious candidates here are:
   i. modality, which has to do with language versus other sign systems, such as visuals and sound that may interact with the structure, function and effect of the linguistic expressions used.
   ii. language, in that choice of a specific language in different genres may be essential for achieving or hindering effective communication (e.g. English as lingua franca in the domains of politics, business, science).
   iii. register, in that diverging genres have varying degrees of latitude for using an appropriate, generally recognized language variety, or register (legalese, journalese, academese, and so on), and that there may be a general tendency for leveling this variation in the western world partly as a result of mass media discourse (conversationalization).
   iv. style, in that diverging genres display varying degrees of appreciation for allowing for distinct personal styles, or group-oriented styles of a particular kind (macho versus feminine, and so on).
   v. rhetoric, in that the deliberate manipulation of language for particular communicative ends may play a role between sender and receiver.

The overall approach hence makes use of a multifeature-multidimensional model, quite comparable to what Douglas Biber (1988) has developed for the study of register by itself. The allocation of particular variables to the three major areas of context, text, and code is tentative. Thus, it is immediately clear that visuals in the dimension of code are strongly related to channels and devices of communication in the media variable (in the dimension of context). It will be one exciting challenge to keep an eye on this initial ordering and evaluate its adequacy.

The leading research question that can be addressed by adopting this framework is the following:

> Which variables are to be postulated for each of the three assumed dimensions (for context, text, and code), and how are they to be modeled in terms of their number and types of categories per variable, as well as their interrelations within and between the three dimensions?

That is, what are the criteria for distinguishing between, for instance, distinct *domains* of discourse, distinct *types* of discourse, and so on, as related to their constitutive function for diverging genres? Consequently, and to give another illustration, are there only four categories for type, *argumentation, narration, exposition, and description*? And how are the distinct variables, of for instance domain and type, related to each other across diverging genres?

Even though many of these issues arise at least implicitly in contemporary genre analysis, they have not been placed on the agenda in this fashion before. The present approach is different in that it adopts a multidimensional model within which one dimension includes a set of variables that all need to be taken into account. Moreover, questions can be raised about the distinct categories per variable from the perspective of psychological validity and functionality, which is another new element in the discussion. These questions can be addressed by utilizing experimental methods in psychological studies, but input for these experiments in terms of expectations about code, text, and context is needed from small- and large-scale observational studies in discourse analysis, which can include both linguistic as well as social and communicative findings. Only a bi-directional perspective on genre can advance this type of investigation, of which both linguistic and literary studies will eventually be able to benefit.

5. Possibilities for research

The full range of variables and their values of the genre model sketched just now represents a set of possible configurations of genre properties that can characterize specific genres. Thus a lecture typically exhibits a number of co-present participants who have distinct roles, who communicate with or without additional media such as a computer, in a setting that includes a lecture room and a number of props such as blackboards, whiteboards, a computer and a data projector, in the domain of science and education; the lecture's content has to do with a topic in an academic discipline, its type may be descriptive or expository or narrative or argumentative, its form may be loose or bound by certain conventions, and its structure will be organized by relatively short speech paragraphs; its language may be in the native tongue of the participants, or in a lingua franca such as English, in a register that is appropriate to science, in a style that may be more or less idiosyncratic depending on the speaker, who may make more or less extensive use of rhetorical ploys and devices, with concomitant features in gesture and body language. For more information about a similar exercise, but from a sociolinguistic perspective, see Biber (1994).

Other genres can be described in the same way as consisting of different combinations of values of the three sets of variables. Producers and addressees have a repertoire of this type of genre knowledge that is more or less developed, and in diverging ways, in accordance with their communicative needs. This repertoire of knowledge and the way it drives human cognition and verbal communication is the focus of the model that is aimed for. One interesting variable in that model is the interaction between professional expertise and genre knowledge as well as its use.

There are many ways in which this approach to genre can be used in empirical research. I have given one illustration in my discussion of 'love stories' (Steen 2003), which deals with the cognitive scenario that may be assumed as a basis for all love stories, whether these are poems, novels, Hollywood films, celebrity reports in popular magazines, doctor-patient interaction in counseling sessions, or law court sessions related to obtaining a divorce. I have argued that the cognitive scenario related to instances of these diverse genres can be expected to be relatively constant, involving three distinct stages: (1) the people involved in a love story either do or do not want each other, (2) do or do not get each other, and (3) do or do not keep each other. Moreover, in instances of all of the above classes of love stories there are interesting causes and consequences of (not) wanting, (not) getting, and (not) keeping each other. *Pride and Prejudice* is a novel (and several films) entitled by the two major causes preventing the two potential lovers from even wanting each other, at least for a while. *Romeo and Juliet* is a play in which the two protagonists do want each other but are prevented from getting each other by external impediments, the long-lasting feud between the two lovers' families. And *Fatal Attraction* is a film about the dangers of keeping each other as presented by attractive competitors outside the love relationship.

All of this involves well-known world knowledge about love relationships that people use when they are in genre events that have to do with love stories. This world knowledge drives or influences their treatment of the variable of content. Indeed, personal ads and speed dating sessions are two distinct genres that are completely geared towards making the relevant content explicit for the addressee so that he or she can decide whether they wish to engage in a process of wanting the sender. That this is not as strange as it sounds is shown by the parallel with the genre of the job interview, where the content of both the letter, the CV, and the meeting are all tailored to comparable ends. Indeed, any sales talk for any object could be subjected to a similar type of analysis.

To show some of the potential of the multi-dimensional/multi-feature approach adumbrated above, there are several aspects that can now be attended to. These have to do with the relations between the cognitive scenario on the one hand and some of the other variables on the other hand. I now present one concrete illustration of this potential (see Steen 2003 for more details).

Let us first turn to the relation between content and some variables of code. Consider the following poem by Sir Henry Wotton:

(1) *Upon the death of Sir Albert Morton's wife*

He first deceased; she for a little tried
To live without him, liked it not, and died.

The fact that we need a love scenario for successfully negotiating the content of this genre event as readers is first and most explicitly signaled by the last word of the title of the poem. Marriage is one specific variant of the love scenario. Moreover, what the poem says about this particular love relationship is expressed extremely sparsely, which is possible because the scenario is so well-known: the lovers do not keep each other, because one of them dies, which, in turn, is the cause of so much grief in his partner that she also dies. In terms of the code variables of the genre model, these observations all have to do with the register of (love) poetry, as these are content elements that are expressed by verbal means that may be expected.

Two additions need to be made to this initial description to throw more aspects into relief. One has to do with the way in which this poem expresses its content in linguistic forms. I should like to see that as a matter of the poem's rhetoric, which has to do with its linguistic design as a function of the sender's intentions upon the receiver. Even though the verbal means are rather conventional as such, their rhetorical deployment is not comparable to the more regular type of expression of this instantiation of the scenario. That would be more like Harry Nilsson's line "I can't live if living is without you". By contrast, in this poem there is a special quality to the use of the vocabulary "for a little" and "liked", and to the way in which the cohesion between the three clauses after the semi-colon is presented as merely temporally ordered. It is not debunking, but it does have a humorous effect, which raises questions about the intentions of the author and the way we as readers are supposed to take this in. This is where the variables of content and rhetoric ineluctably tie in with the variables of participants and goal/function. One possibility that is raised by this conjunction of genre information is that the rhetoric is meant to be a reflection of the character of Sir Albert Morton's wife, perhaps even of her view of the perennial wisdom that lies at the basis of this poem, that it is simply impossible to live on without your lover.

The second aspect has to do with the way in which this text gives shape to the content by presenting it as a narrative text type. This is what adds to its causal, dynamic character. In order to demonstrate that not all love stories are stories in the sense of narrative text types, consider the following love poem:

(2) *A marriage ring*

The ring so worn as you behold,
So thin, so pale, is yet of gold:
The passion such it was to prove;
Worn with life's cares, love yet was love.

This poem evokes a love story in terms of its content, which can be described with reference to the third stage of the lovers keeping each other, in spite of 'life's cares'. The language signals that we have to do with a marriage version of the love scenario again, and the rhetoric is quite different from the previous poem. Most important, however, is the fact that we are here dealing with a descriptive text type, not a narrative one. The text presents aspects of the marriage ring as symbolic of the marriage, as opposed to Sir Henry Wotton's poem, which presents a series of actions that are causally connected.

The multi-dimensional/multi-feature approach to genre hence facilitates precise description of the similarities and differences between two instantiations of the same genre. We are here dealing with two love poems whose content can be related in different ways to the same scenario, whose language looks like belonging to the same register, but whose rhetoric is fairly different, and the text type of which is different in absolute terms. Moreover, the relation of these aspects with the contextual variables of participants and goal is intriguing for the first poem but seems much more mainstream for the second. The specific nature of distinct genre events like the reception of these two poems can in principle be explained by this approach, provided we agree on the number of dimensions, variables, categories, and their precise nature.

The latter is not a simple task. I used the variable of rhetoric when I discussed the way in which the content was expressed in the first poem, whereas another candidate might have been the notion of style. In my opinion, style has more to do with the personal stamp put on a text by a particular author (or even group of authors) in comparison with the more regular patterns of a register. These stylistic patterns are typically fairly unconscious and have little or no interactive function – at most, they can be called expressive. Rhetoric, by contrast, has to do with the relatively deliberate manipulation (design) of language by the sender in order to optimize the effect of the genre event as a whole. Rhetoric can hence be more or less prominent, subtle, and eventually successful in terms of audience design.

This is a conceptualization of rhetoric and style that is closely related to the study of persuasion in social psychology, where experimental work on text design and its effects has a rich and long-standing tradition (e.g. O'Keefe 2002). As long as there is no generally accepted agreement about this sort of assumption about the difference between terms (and concepts) like 'style' and 'rhetoric', and as long as scholars do not avail themselves of the bi-directional perspectives between the humanities and the sciences, studies of discourse will not be completely comparable to each other. Separate attention to the theoretical underpinnings of the elements and relations of a model of genre like the one proposed here may be helpful to advance this type of discussion.

6. Concluding comment

I have presented an approach to genre with a sketch of a provisional model. In principle each variable represents a dimension of variation in the language user's knowledge system for capturing distinct aspects of genres. For instance, the variable 'type' may be postulated to have four categories: narrative genres, argumentative genres, descriptive genres, and expository genres. These categories are hence expected to exhibit particular cognitive qualities that differentiate them from each other in systematic ways. These categories and this variable (and all other ones) need to be described and explained in their own right (for instance *as* 'type'), as well as in relation to their position and function in the overall model (for instance, as 'type' in relation to 'content', 'form', 'domain', and so on), in order to increase the descriptive and explanatory adequacy of the model.

When this type of model has been tested and further refined, it will facilitate research on the role of distinct categories of variables within distinct genres. To illustrate, does the cognitive love scenario exemplified above work in the same way for each genre that can intuitively be called a love story? Are the same aspects of narrative text types important to all genres that are deemed to be narrative? Can the special nature of unique texts, including literary ones or historical speeches and so on, be more systematically accounted for by obtaining more empirical knowledge about the general patterns of such texts regarding all details of genre? These are just some of the questions that are triggered by reflecting on the position of genre between the humanities and the cognitive and social sciences.

There is one final aspect that needs to be accommodated, and this is cross-cultural variation. It is assumed that the need for various mental models and their variables as such is constant across cultures: for any genre event, people will always have to build a partial mental representation of context, of text, and of code, and each of these is expected to exhibit properties regarding participants, domain, content, type, language, register, and so on. That these properties may be somewhat different for some variables between cultures is natural. This variation has been discussed in some discourse theories and research in ways that can throw into relief important issues for the various genre variables. However, at the moment the main aim of the project is to produce a genre model that may be operationalized in empirical research, which may then be used for more fine-grained follow-up work on cross-cultural and intercultural communication in various genres. If and how findings will affect the model proposed here is a question that can be turned to at a later stage.

References

Bateman, John. 2008. *Multimodality and Genre: A Foundation for the Systematic Analysis of Multimodal Documents.* London: Palgrave Macmillan.

Baumgarten, Nicole. 2007. Converging conventions? Macrosyntactic conjunction with English *and* and German *und. Text & Talk* 27 (2): 139–170.

Berkenkotter, Carol and Thomas N. Huckin. 1995. *Genre Knowledge in Disciplinary Communication.* Hove: Lawrence Erlbaum.

Bhatia, Vijay K. 1993. *Analysing Genre: Language Use in Professional Settings.* Harlow: Longman.

Biber, Douglas. 1988. *Variation Across Speech and Writing.* Cambridge: Cambridge University Press.

Biber, Douglas. 1994. An analytical framework for register studies. In *Sociolinguistic Perspectives on Register Variation,* Douglas Biber and Edward Finegan (eds.), 31–56. Oxford: Oxford University Press.

Biber, Douglas and Jerry Kurjian. 2007. Towards a taxonomy of web registers and text types: A multi-dimensional analysis. In *Corpus Linguistics and the Web,* Marianne Hundt, Nadja Nesselhauf and Carolin Biewer (eds.), 109–132. Amsterdam: Rodopi.

Bilandzic, Helena and Patrick Rössler. 2004. Life according to television: Implications of genre-specific cultivation effects – The gratification/cultivation model. *Communications* 29: 295–326.

Clark, Herbert H. 1996. *Using Language.* Cambridge: Cambridge University Press.

Connor, Ulla. 2002. New directions in contrastive rhetoric. *TESOL Quarterly* 36 (4): 493–510.

Connor, Ulla. 2004. Intercultural rhetoric research: Beyond texts. *Journal of English for Academic Purpose* 3: 291–304.

Dillon, George L. 2007. The genres speak: Using large corpora to profile generic registers. *Journal of Literary Semantics* 36: 159–187.

Donovan, Carol A. and Laura B. Smolkin. 2002. Children's genre knowledge: An examination of K-5 students' performance on multiple tasks providing different levels of scaffolding. *Reading Research Quarterly* 37 (4): 428–465.

Dunmire, Patricia L. 2000. Genre as temporally situated social action: A study of temporality and genre activity. *Written Communication* 17 (1): 93–138.

Eggins, Susan and James Martin. 1997. Genres and registers of discourse. In *Discourse Studies: A Multidisciplinary Introduction.* Vol. 1: *Discourse as Structure and Process,* Teun A. Van Dijk (ed.), 230–256. London: Sage.

Fernández Toledo, Piedad. 2005. Genre analysis and reading of English as a foreign language: Genre schemata beyond text typologies. *Journal of Pragmatics* 37: 1059–1079.

Fishelov, David. 1993. *Metaphors of Genre: The Role of Analogies in Genre Theory.* Penn State Press: University Park, PA.

Freedman, Aviva and Peter Medway (eds.). 1994. *Genre and the New Rhetoric.* Basingstoke: Burgess Science Press.

Gavins, Joanna and Gerard J. Steen (eds.). 2003. *Cognitive Poetics in Practice.* London: Routledge.

Graesser, Arthur, Keith Millis and Rolf Zwaan. 1997. Discourse comprehension. *Annual Review of Psychology* 48: 163–189.

Hall, Alice. 2003. Reading realism: Audiences' evaluations of the reality of media texts. *Journal of Communication* 53: 624–641.
Hamilton, Heidi E., Deborah Schiffrin and Deborah Tannen (eds.). 2001. *The Handbook of Discourse Analysis.* Oxford: Blackwell.
Hyland, Ken. 2003. Genre-based pedagogies: A social response to process. *Journal of Second Language Writing* 12: 17–29.
Hyland, Ken. 2007. Genre pedagogy: Language, literacy and L2 writing instruction. *Journal of Second Language Writing* 16: 148–164.
Johns, Ann M., Anis Bawarshi, Richard M. Coe, Ken Hyland, Brian Paltridge, Mary Jo Reiff and Christine Tardy. 2006. Crossing the boundaries of genre studies: Commentaries by experts. *Journal of Second Language Writing* 15: 234–249.
Jordens, Christopher F. C. and Miles Little. 2004. In this scenario, I do this, for these reasons': narrative, genre and ethical reasoning in the clinic. *Social Science & Medicine* 58: 1635–1645.
Kintsch, Walter. 1998. *Comprehension: A Paradigm for Cognition.* Cambridge: Cambridge University Press.
Kirkpatrick, Lori C. and Perry D. Klein. In press. Planning text structure as a way to improve students' writing from sources in the compare-contrast genre. *Learning and Instruction.*
Knobloch, Sylvia, Grit Patzig, Anna-Maria Mende and Matthias Hastall. 2004. Affective news: Effects of discourse structure in narratives on suspense, curiosity and enjoyment while reading news and novels. *Communication Research* 31 (3): 259–287.
Lemke, Jay L. 2005. Multimedia genres and traversals. *Folia Linguistica* XXXIX (1–2): 45–56.
Martin, James R. and David Rose. 2005. *Genre Relations: Mapping culture.* London: Equinox.
Muntigl, Peter and Helmut Gruber. 2005. Introduction: Approaches to genre. *Folia Linguistica* XXXIX (1–2): 1–18.
Murphy, Gregory. 2002. *The Big Book of Concepts.* Cambridge, MA and London: MIT Press.
O'Keefe, Daniel J. 2002. *Persuasion: Theory and Research* (2nd ed.). Thousand Oaks, CA: Sage.
Paltridge, Brian. 1995. Working with genre: A pragmatic perspective. *Journal of Pragmatics* 24: 393–406.
Ramanathan, Vai and Robert B. Kaplan. 2000. Genres, authors, discourse communities: Theories and application for (L1 and) L2 writing instructors. *Journal of Second Language Writing* 9 (2): 171–191.
Semino, Elena and Jonathan Culpepper (eds.). 2003. *Cognitive Stylistics: Language and Cognition in Text Analysis.* Amsterdam: John Benjamins.
Sengupta, Sima, Gail Forey and Liz Hamp-Lyons. 1999. Supporting effective English communication with the context of teaching and research in a tertiary institute: Developing a genre model for consciousness raising. *English for Specific Purposes* 18: 7–22.
Sherry, John L. 2004. Flow and media enjoyment. *Communication Theory* 14 (4): 328–347.
Slater, Michael D. 1997. Persuasion processes across receiver goals and message genres. *Communication Theory* 7 (2): 125–148.
Slater, Michael D. and Donna L. Rouner. 2002. Entertainment-Education and elaboration likelihood: Understanding the processing of narrative persuasion. *Communication Theory* 12 (2): 173–191.
Steen, Gerard J. 1999. Genres of discourse and the definition of literature. *Discourse Processes* 28 (2): 109–120.

Steen, Gerard J. 2003. 'Love stories': Cognitive scenarios in love poetry. In *Cognitive Poetics in Practice,* Joanna Gavins and Gerard J. Steen (eds.), 67–82. London: Routledge.
Stockwell, Peter. 2002. *Cognitive Poetics: An Introduction.* London: Routledge.
Swales, John. 1990. *Genre Analysis.* Cambridge: Cambridge University Press.
Van Dijk, Teun A. 2008. *Discourse and Context: A Sociocognitive Approach.* Cambridge: Cambridge University Press.
Van Dijk, Teun A. and Walter Kintsch. 1983. *Strategies of Discourse Comprehension.* New York: Academic Press.
Vergaro, Carla. 2002. "Dear Sirs, what would you do if you were in my position?" Discourse strategies in Italian and English money chasing letters. *Journal of Pragmatics* 34: 1211–1233.
Winsor, Dorothy A. 1999. Genre and activity systems: The role of documentation in maintaining and changing engineering activity systems. *Written Communication* 16 (2): 200–224.

Culture-specific concepts of emotionality and rationality*

Susanne Niemeier
Universität Koblenz-Landau

The cultural models of emotionality and rationality showing up in figurative language use vary extensively on a worldwide scale. Such variation is largely due to different cultural backgrounds, for instance, concerning underlying ethnomedical or philosophical beliefs. The present chapter reports on a research project involving various, partly unrelated languages and the different cultural models that underlie these cultures' conceptualizations of emotions and of the mind. Three general categories of conceptualization (namely abdominocentrism, cardiocentrism, and cerebrocentrism) are distinguished and discussed. As an example from the category of cerebrocentrism, metaphorical and metonymic expressions referring to emotionality and rationality concepts in present-day English are analyzed in depth.

Keywords: abdominocentrism, cardiocentrism, cerebrocentrism, figurative language use, metaphor, metonomy

1. Introduction

The general aim of this contribution is to explore the bi-directionality of Applied Cognitive Linguistics (ACL) on the one hand and Cultural Studies on the other hand. ACL as a culture-based linguistic approach sees language as inseparably intertwined with culture. Accordingly, among other research agendas, ACL –

* This contribution owes a lot to co-researchers who have worked on emotion concepts in various languages and who have joined the discussion about emotionality and/or rationality models in a symposium convened during the 9th International Cognitive Linguistics Conference in Seoul in 2005. The majority of the papers presented during this symposium were subsequently published in Sharifian, Dirven, Yu, and Niemeier, eds. (2008). My sincere thanks go to my colleague Fred Thompson (Universität Koblenz-Landau) for having checked my English. All remaining flaws are, as usual, my own responsibility.

especially in its search for the impact of figurative thought – aims at exploring the cultural underpinnings and connections of language-related phenomena. Cultural Studies, on the other hand, devotes itself (again, among other research agendas) to the search for the cultural models that may have channeled metaphorical as well as metonymic conceptualizations.

In bringing these two viewpoints together, this contribution focuses on cultural models of emotionality and rationality as they show up in metaphorical and metonymic expressions in a variety of languages. Both scientific perspectives mentioned above are co-activated in the analysis of the culture-specific models under discussion.

More specifically, this contribution wants to pursue three different aims: (i) to explore the ways in which internal body organs are used in different languages to conceptualize human experiences such as emotions and/or the workings of the mind; (ii) to explore the cultural models that account for the observed similarities and differences of the various conceptualizations; and (iii) to exemplarily explore the head/heart-dichotomy in present-day English.

After having provided some background information, the findings on the emotionality and rationality concepts across a variety of cultures under investigation is summarized and discussed, and in a final section the focus is narrowed down to the cultural models of emotionality and rationality, i.e. the cultural models of the heart and of the head, in present-day English.

2. Background

The central question dealt with in my contribution is how feeling, thinking, and knowing are related to internal body organs in different cultures as they are reflected in different languages. The chapter investigates the different roles of the heart as well as at those of the rational mind in various languages/cultures, from a diachronic as well as from a synchronic perspective. The reference point is present-day English in which the heart is conceptualized as the seat of emotion and not as that of reason, which is seen as located in the brain. This view of course contradicts modern scientific theory – we find a nearly complete dichotomy between the head and the heart in the English language (see Section 4 of this contribution, Niemeier 2000, 2008), which modern neuroscience disagrees with.

Admittedly, the search for the conceptualization of the mind and of body organs is not a new topic but has been undertaken by many scholars before (e.g. Jahoda 1993; Palmer et al. 2003; Shweder 1991; Wierzbicka 1989). What the current project is adding to already existing research is that it is looking at the possible impact of metaphorical thought as well as of the different cultural models that may have guided and brought forth such conceptualizations. 'Metaphorical thought' is used here in a very broad sense, including metonymical thoughts and expressions, tying in with the hypothesis that potentially all metaphors developed from earlier metonymies (cf. Barcelona (ed.) 2000, Niemeier 2000). Cultural models, on the other hand, are defined here in Quinn's and Holland's sense as

> (...) presupposed, taken-for-granted models of the world that are widely shared (although not necessarily to the exclusion of other, alternative models) by the members of a society and that play an enormous role in their understanding of the world and their behavior in it. (Quinn and Holland 1987: 4)

Expanding this definition slightly, cultural models can be seen as complex conceptual systems which are used by a cultural group in order to construe their "cultural cognition" (see Sharifian 2008a) and which lend themselves as templates for understanding certain aspects related to human inner and outer experiences, especially referring to abstract, hard-to-grasp notions such as emotions and mental experiences. As a working hypothesis in our project, we see metaphors as rooted in cultural models (see also Sharifian et al. 2008: 12) instead of either constituting them (as Lakoff and Kövecses 1987 claim) or reflecting them (as Quinn 1991 claims). Although metaphors are indeed very often grounded in immediate bodily experience, they can also go beyond these.

Mere physiological motivation does not explain many of the conceptualizations of internal body organs, as otherwise we would find many more similarities across languages and cultures, given the fact that our bodies work in identical ways all over the world. Many cultural models have arisen in times when knowledge about the human body (including the brain/mind) was scarce (or – from a present-day-perspective – simply wrong) and have persisted until modern times, when at least some more facts about the human body and brain/mind are better understood. Therefore, we find outdated beliefs in the currently used cultural models which once may have counted as "physiological state-of-the-art knowledge" but which rather relied on other sources of explanation. Such expansions of body organ conceptualizations frequently happened on the basis of cultural models from, for instance, ethnomedicine, philosophy, or religion. This becomes obvious when metaphors from Aboriginal cultures are analyzed, in which embodiment plays only a minor role as many Aboriginal cultures favor non-bodily models rooted in the Aboriginal worldviews.

It has to be stressed here that we often find various cultural models for the same experience – these models may complement each other, may simply coexist, but can also be contradictory. Generally, though, the major models are identifiable by the sheer quantity of metaphors and metonymies in the languages of the cultures observed.

3. Locating cultural models of emotions and the mind on the world map

The task to create order in chaos, i.e. categorizing the various contributions to our project from the various languages, was not an easy one. Still, we discovered that the cultural models discussed in these contributions seem to fall into three different global categories of how emotions and the workings of the mind are conceptualized across the cultures examined. Presumably, other cultures fit into this tripartite model as well; there is at present just no evidence, i.e. no data, for fitting them in.

As is generally the case for all categories – at least in the Cognitive Linguistic view on categorization, cf. Taylor (2003) – there are prototypical instances where the fit between the language in question and the category is perfect and there are also more marginal members of the categories where an overlap with one of the other categories is evident.

The three general types of conceptualization of the mind and also of emotions have been labeled abdominocentrism (i), cardiocentrism (ii), and cerebrocentrism (iii) (see Sharifian et al. 2008), the labels referring to those body organs that play a central role in conceptualization in the category in question. These categories are independent of language families and are described in the following paragraphs. Cerebrocentrism is dealt with in more detail than the two other approaches, as it paves the way for the exemplary analysis of present-day English in Parts 4 and 5.

3.1 Abdominocentrism

In the abdominocentric approach, the mind is located in or around the abdomen. Abdominocentrism is the predominant approach in Southern Asia, Polynesia, and other disparate cultures, including – as a marginal category member – the Basque culture, which is bordering on cardiocentrism. In the abdominocentric cultures, feelings are located in the belly, in the liver, or in the kidney, and rationality is located in the liver, in the ear, or – partially – in the "gogo" (see below). Mostly, we are dealing with cultural models of animistic spirituality.

Not only in Malay but also in Indonesian, for example, the liver (*hati*) is seen as the seat of emotional as well as mental activities (see Goddard 2008, Siahaan 2008), which imposes a comparison with the role of the liver in Ancient Greek philosophy, where it was seen as hosting one part of the three-part soul, the "natural soul" controlling nutrition (Erickson 1997:4). The importance of the liver stems from ancient, now forbidden, liver divination rituals (Siahaan 2008:47), in which specially ordained priests interpreted the livers of sacrificed animals as containing messages from the gods or divine spirits for the humans. The liver was thus seen as the central organ by which spiritual beings communicated with the human species.

In Kuuk Thaayorre, an Australian aboriginal language, the belly is associated with emotions (see Gaby 2008). This is mainly due to the Aboriginal worldview in which food is of significant importance in determining physical and spiritual health and therefore the place for the food – the belly – is also seen as connected to physical and spiritual health. Rationality and the mind are seen as connected to the ear, which is not really astonishing as Kuuk Thaayorre is an oral language and therefore all information enters the body via the ears.

Basque, as mentioned above, is a borderline case. It relates emotions to a diversity of inner organs (heart, liver, stomach, intestines), relying on a variety of cultural models for a variety of emotions. For example, positive feelings are associated with the heart, relying on a Cartesian model (see also Section 3.3), whereas negative feelings are associated with the liver, relying on the theory of the Four Humors. The stomach and the intestines are only marginal, linguistically rather unproductive organs in the realm of emotions (see Ibarretxe Antuñano 2008). What is special about Basque is its concept of a "gogo", i.e. an inner entity for intellectual reasoning processes based on intuition and emotion, relying on a cultural model of an (animistic-) spiritual cosmovision.

3.2 Cardiocentrism

In the cardiocentric approach, emotions as well as the mind are located in the heart region. Cardiocentrism is the traditional view of China, Korea, and Japan, based on philosophical and medical cultural models of holistic, heart-centering conceptualizations with a remarkable degree of similarity between them.

In China, the heart is seen as the ruler of the body (Yu 2008). It is not only connected to emotions, but also seen as the central faculty of cognition and as the physiological centre of the body. In a cosmic view, the heart can be seen as the "mirror of the universe". This view relies on cultural models from traditional Chinese medicine, from the yin-yang concept as well as on the five elements (wood, fire, earth, metal, water) in traditional Chinese philosophy. The brain is only given scant attention in traditional Chinese medicine.

Also in Korea, intellectual and emotional activities are localized in and around the heart (Yoon 2008). Korean has three different expressions roughly corresponding to the English concept of "heart": one expression referring to the actual body organ, the second expression referring to a highly polysemous concept including the body part as well as the psychological faculties of a person, and the third expression referring to psychological and mental faculties. The latter one is frequently used as a metonymy and/or a metaphor for a person.

In Japanese, we also find three different expressions for the heart, similar to Korean. One of these is a Chinese loan word for the body organ itself, the second a highly polysemous native word for the workings of the heart, referring to emotions and reason and the third is the English loan word *haato* for emotions connected to the heart, e.g. romantic love. The increasing use of *haato* is an obvious case of the influence of English in Japan (Occhi 2008), as this term was nonexistent in ancient Japanese. Parallel to this development, in present-day Japanese we find some references to thought processing in the brain; however, so far these only play a minor role (Ikegami 2008).

3.3 Cerebrocentrism

In the cerebrocentric approach, the mind is located in the head or more precisely the brain region. This view is held by Greek-based West Asian, European, and North-African cultures, prototypical examples being the major Indo-European languages. This approach relies on philosophical, medical, and/or religious cultural models. We find a dualism between the mind and the body and thus between the head/brain as the seat of intellect and the heart as the seat of emotions. This head-heart dualism grew out of the earlier interaction between cardiocentrism and cerebrocentrism in ancient Greek philosophy and medicine (Foolen 2008). Modern English, Modern Dutch and Modern German, for example, represent prototypical examples in which we find a nearly complete dichotomy between *head-* and *heart-*expressions.

In Ancient Greece, the question whether the heart or the brain is the seat of the soul and thus the central organ of thought and sensation was discussed controversially. Hippocratic writers (5th cent. BC) claimed that the "heart is a strong muscle with a large hot chamber housing the intelligent ruling principle of the soul" (Erickson 1997: 199), and also Aristotle (4th cent. BC) agreed that the "heart is the chief organ of the body and the center for thought and the appreciation of sensation" (Clarke and O'Malley 1968: 6f.), whereas Plato argued that the soul was made up of three parts and he assigned the most important part – the rational part – to the head/brain. At that time, this was a minority opinion; however, it was supported by Galen, a very influential and important medical doctor and anatomist (1st cent. AD; cf. also Clarke and O'Malley 1968: 15).

In the medieval period, cardiocentrism once more gained primacy, mainly due to Avicenna, a Persian physician and philosopher (10th/11th cent. AD, Stevens 1997). Also the Bible can be seen as an important authority representing

Christianity's cardiocentric view. After the Middle Ages (16th/17th cent. AC), however, religious and scientific views started to drift away from each other: whereas the Catholic Church continued to believe in and to propagate cardiocentrism (visible for example in the Sacred Heart cult), medical scientists (such as Andreas Vesalius or William Harvey) discovered the heart's real function as a blood pump (Foolen 2008; Niemeier 2008).

The influence of René Descartes (1596–1650) in all of Western Europe and beyond marked the birth of the fully outspoken dualism. We see a shift from monocentrism/cardiocentrism to dualism/cerebrocentrism linked to the developments in medicine and philosophy in the 16th and 17th centuries, followed by the development of the Cartesian dualistic belief in the "disembodied rational mind", i.e. in the mind/brain split and thus also in the mind/body split, where emotions were seen as located in the heart and reason as located in the head (Stevens 1997; Damasio 1994).

Following this, we can witness the replacement of *heart*-expressions by *head*-expressions in Persian, Neo-Aramaic, English, and Dutch – and presumably in other languages. The Cartesian model has even persisted until today, although it has been seriously questioned in the 20th century in philosophy by phenomenology and in linguistics by experientialism or embodied realism (Geeraerts 1985, 1987; Johnson 1987; Lakoff 1987; Lakoff and Johnson 1999).

In languages beyond Western Europe, we find the influence of competing cultural models, such as for example Ancient Syriac thought on ethnomedicine and early Christian mythical thought in Neo-Aramaic, the Koran in Tunisian Arabic, or Sufism in Persian.

Although Neo-Aramaic has many similarities with English, the dualistic concepts of these languages are not completely alike. In Neo-Aramaic, on the one hand, the heart concept (*libbā*) also includes the stomach, and on the other hand we find traces of the Galenic Four Humors theory: although the organs liver and spleen hardly play a role, a differentiation of wet vs. dry and of hot vs. cold has survived. Furthermore, we find the influence of a cultural model of morality (Wolk 2008).

Tunisian Arabic also shows the Cartesian split, but is additionally influenced by a cultural model based on the Koran (Maalej 2008). Similar to Christianity, the Koran teaches a conflated model of body and mind; however, this view has progressively yielded to the split between the two concepts (mainly owing to the translations of Greek philosophy into Arabic).

Modern Persian is a marginal member of the category of cerebrocentrism, as the notion of *del* ('heart') may also encompass aspects of cognition. We find a correspondence between conceptualizations of *del* especially in Sufi literature, in Avicenna's version of the theory of the Four Humors and the metaphorical expressions referring to the actual body organ (see Sharifian 2008b). *Del* actually describes more than just the heart, because it refers to the area of the upper chest down to the stomach and also accounts for inner drives (somewhat similar to the Basque *gogo*). Thought processes in Persian are seen as located in the *zehn*, which can be roughly glossed with 'mind', being located in a person's brain.

4. The cultural model of the heart in English

Although in present-day, post-Cartesian English the dichotomy between the head and the heart is nearly complete, Old English still adhered to cardiocentrism and saw the heart (or rather the mood) as connected to mental life, emotions, and volition, according to the Galenic humoral doctrine (see Geeraerts and Gevaert 2008). Due to the Cartesian influence, the cultural model gradually changed on the conceptual and then also on the lexical level to cerebrocentrism and – as already mentioned before – today we find a nearly perfect dichotomy that sees the heart as the seat of emotions and the head as the seat of reason.

Such concepts and such language usage of course contradict scientific knowledge that argues that not only cognition, but also emotions are generated in the mind. However, the latter is not yet reflected in the language as language change is quite slow. Still, it is relatively predictable that a new cultural model of the mind as the seat of emotions may be a next step in language change.

The head-heart dichotomy of present-day English comes to the fore when we look at metaphors and metonymies involving these two expressions, where the "heart" nearly always refers to emotion concepts, whereas the "head" refers to rationality concepts, making English a perfect and prototypical example for the cerebrocentric category. Each subcategory can, of course, be analyzed further, as none is internally homogeneous but incorporates expressions with varying degrees of generality and focusing on different facets of the rather broad categories "emotionality" and "rationality".

The following two sub-sections therefore first offer a condensed version of an analysis of the heart expressions in present-day English (for a more detailed analysis see Niemeier 2000, 2008), followed in the next section by a brief analysis of head expressions in present-day English (again, for more details, see Niemeier 2008). Both analyses are based on extensive corpora of heart expressions resp. head expressions compiled from general dictionaries, collocation dictionaries, and thesauri. Through analyzing a vast number of heart expressions in English, four different categories with an increasing degree of generality could be extrapolated: (i) the heart as a metonymy for a person's feelings; (ii) the heart as a metonymy for a person as a whole; (iii) the heart as an object of value and (iv) the heart as a container. Each of these categories comprises several submodels – some of them more prototypical and some less so.

The expressions of the first category, i.e. the heart as a metonymy for a person's feelings, refer to emotions, courage, attitudes, etc., the most prototypical emotion being that of romantic love, which is easily symbolizable (for a history of the heart symbol see Niemeier 2008), but we also find more negative emotions such as *heart of stone* for an emotionless attitude or *half-hearted,* referring to a lack of courage. The heart as the most salient body part in the cultural model of emotions is seen here as a part of a person.

In the second category, i.e. heart as a metonymy for a person as a whole, the heart is seen in a more general perspective as referring to the complete person, not only to feelings. Here, the heart acts as a focalized centre of a larger entity, e.g. perceiving a person as a *sweetheart* or in expressions such as *all hearts throb for her* or *two hearts that beat as one.*[1] The expressions in this category refer mostly to failed love where negative emotions within oneself are described as causing damage to the heart (*my heart is bleeding, broken-hearted*).

The third category consists of expressions referring to the heart as an object of value. The heart is seen in its totality as a treasure chest. Emotions are involved, but not named (*my heart belongs to you, to lose one's heart*). In this category, at an even more generalized level of conceptualization, the location for romantic love has to be shared with other feelings such as courage (*to put one's heart into sth.*), resentment (*to take sth. to heart*), or discontent (*to dishearten sb.*). We are presumably more aware of this third category in our everyday language usage than we are of the first two categories because it is closer to the metaphoric pole of the metonymy-metaphor continuum (cf. Niemeier 2000, 2008).

The fourth category, the heart as a container, is the most general one. We are dealing here with a more universal type of schema, focusing on the contents rather than on the container. Its submodels highlight various aspects from the outer qualities of the container (*sth. is locked up in one's heart*) including its handleability (*to pour out one's heart to sb.*) via its inner qualities (*from the bottom of one's heart*) to its internal structure (*to have a place in every heart*). The container model is obviously not unique to the cultural model of the heart, as it is used in a plethora of other models, among them models of the human body (head, womb, chest, etc.). This category is dependent on the existence of the other three categories that provide its metonymic basis (see Niemeier 2008: 356) and is more "metaphtonymically" structured (Goossens 1995).

1. This expression was even turned into a hit in the 1980s by the Irish pop band U2 – my thanks to Fred Thompson who provided me with this information.

As mentioned previously, the dichotomy between *heart*- and *head*-expressions is nearly complete, only two exceptions could be detected in the corpus of *heart*-expressions, namely *to learn sth. by heart* and *to take sth. to heart*. The first expression refers to memorizing and the second to thinking about something very deeply, i.e. in both of these cases the heart is connected to mental faculties. These two expressions may well be leftover reminiscences of Old English cardiocentrism.

5. The cultural model of the head in English

In English, the head is primarily seen as the locus of rational thought, which in the cerebrocentric group of languages is a culturally shared model. The basic reference of the expression "head" is to the actual body part, but we also find frequent references to its presumed contents, i.e. the brain, the mind, human ratio, intelligence. Furthermore, the head can be seen as the location of abilities such as, amongst others, perception, spatial imagination, creativity, and logic.

As mentioned in Section 4, the dichotomy between head expressions and heart expressions is nearly complete. Although Damasio (1984) argues that mere intellectual reasoning is an impossibility and that human thinking is always supported and strengthened by emotions (basically turning around the famous saying by Descartes into "I am therefore I think", i.e. the body pre-exists thought), has not (yet) shown up in the English language.

We can find three different categories of head expressions, again showing increasing degrees of generality (see also Niemeier 2008: 357ff.). The first category refers to the head as the locus of reasoning, the second to the head as a control centre and the third to the head as a container.

The most specific level of meaning is to be found in the category of the head as a locus of reasoning (e.g. *We need some good heads to solve the problem*). We find a clear metonymic basis here, as the head stands for the brain (the first whole-for-part metonymy) and the brain then stands for the process of rational thinking that is taking place in it (the second whole-for-part metonymy).

The second category, referring to the head as a control centre, is more general than the previous one. Here, the head is seen as the main site for decision-making. The focus is not on mental processes, but rather on the general processes that we assume to take place in the head, and on the necessity of balance in our heads/minds (as a precondition for reasoning). This is in keeping with Geeraerts (2002), who identified the metaphor THE HEAD IS THE LOCUS OF ONE'S SELF-CONTROL. The examples in this category are mostly negative and the majority refer to a lack of or loss of balance (e.g. *the success went right to his head; to lose one's head*). As in any normal situation people are supposed to possess a balance of mind, these examples show features of markedness, i.e. refer to a non-prototypical, marginal situation.

The third, most general category sees the head as a container. Here, we find more (metonymy-based) metaphors than pure metonymies. In the previous categories, containment is of course also implied, as both reasoning and the maintenance of control or balance happen *inside* the head, but the containment in these cases is only tacitly assumed, whereas in the third category, it is explicitly mentioned.

The expressions in this third category focus either on the static fullness/emptiness state of the head (e.g. *empty-headed, to have a head like a sieve*) or on the dynamic process of filling/emptying it (e.g. *Put these weird thoughts out of your head; it never entered my head that he might be lying*), with the latter sub-model being more salient.

Again, as in the *heart* analysis above, there were only two potential exceptions in the whole corpus of *head*-expressions. The first exception is *heady* (roughly: extremely exciting) and the second one is *headstrong* (roughly: obstinate), both referring to emotional attitudes. However, none of these two exceptions refers to a prototypical emotion: in *heady*, something is weakening or eliminating the control centre and in *headstrong*, the person thinks that their arguments are good enough not to give in – in that way, (distorted) ratio is involved after all and therefore in both cases the notion of "rationality" is at least marginally present, so these two cases may not be exceptions after all, but rather marginal exemplars of the category of head expressions rooted in a Cartesian model of rationality.

6. Conclusion

We find a nearly complete division of labor between the head and the heart in English and it has become obvious that the English language is still governed by a Cartesian cultural model, although modern neuroscience does not agree with it.

Whereas the conceptualization of the heart has in no meaningful way evolved from that of a mental faculty (emotions) to a mere organ (blood pump), the conceptualization of head as ratio/mind has fully found its way into the language. This may be due to the enormous symbolicity of the heart, deeply rooted in Western tradition, which is absent for the head. We may hypothesize that the next step might lead away from the present dualism to a complete cultural model of cerebrocentrism, where not only the mind but also the emotions are conceptualized as being located in the head.

Research results from modern neuroscience are, of course, also present and available in cultures adhering to abdominocentric and cardiocentric cultural models, especially in the age of worldwide communication. It is hard to predict whether – and if so when – changes may take place in these cultures as well. Apparently, Japanese is already slightly drifting away from cardiocentrism towards cerebrocentrism, presumably influenced by the English language. Cultural models are indeed dynamic and open to changes, as was discussed for the change from Old English cardiocentrism to present-day English cerebrocentrism, although such changes tend to happen in a very slow manner. Scientific knowledge may have a hard time to reach people still living with evidently different cultural models. As these models are largely located in our social and cultural common subconsciousness, i.e. our cultural cognition, they cannot be changed at will by individuals but may need centuries to change for the complete population of a culture. Time will tell.

References

Barcelona, Antonio (ed.). 2000. *Metaphor and Metonymy at the Crossroads. A Cognitive Perspective*. Berlin: Mouton de Gruyter.

Clarke, Edwin and Charles D. O'Malley. 1968. *The Human Brain and Spinal Cord: A Historical Study Illustrated by Writings from Antiquity to the Twentieth Century*. Berkeley: University of California Press.

Damasio, Antonio. 1994. *Descartes' Error: Emotion, Reason and the Human Brain*. New York: Avon.

Erickson, Robert A. 1997. *The Language of the Heart, 1600–1750*. Philadelphia: University of Pennsylvania Press.

Foolen, Ad. 2008. The heart as a source of semiosis: The case of Dutch. In *Culture, Body, and Language: Conceptualizations of Internal Body Organs across Cultures and Languages*, Farzad Sharifian, René Dirven, Ning Yu and Susanne Niemeier (eds.), 373–394. Berlin: Mouton de Gruyter.

Gaby, A. 2008. Gut feelings: Locating intellect, emotion and life force in the Thaayorre body. In *Culture, Body, and Language: Conceptualizations of Internal Body Organs across Cultures and Languages*, Farzad Sharifian, René Dirven, Ning Yu and Susanne Niemeier (eds.), 27–44. Berlin: Mouton de Gruyter.

Geeraerts, Dirk. 1985. *Paradigm and Paradox: Explorations into a Paradigmatic Theory of Meaning and its Epistemological Background*. Leuven: Universitaire Pers.

Geeraerts, Dirk. 1987. Cognitive linguistics and the history of philosophical epistemology. In *Conceptualisations and Mental Processing in Language*, Brygida Rudzka-Ostyn and Richard Geiger (eds.), 53–79. Berlin: Mouton de Gruyter.

Geeraerts, Dirk. 2002. The interaction of metaphor and metonymy in composite expressions. In *Metaphor and Metonymy in Comparison and Contrast*, René Dirven and Ralf Pörings (eds.), 153–180. Berlin: Mouton de Gruyter.

Geeraerts, Dirk and Caroline Gevaert. 2008. Hearts and (angry) minds in Old English. In *Culture, Body, and Language: Conceptualizations of Internal Body Organs across Cultures and Languages*, Farzad Sharifian, René Dirven, Ning Yu and Susanne Niemeier (eds.), 319–348. Berlin: Mouton de Gruyter.

Goddard, Cliff. 2008. Contrastive semantics and cultural psychology: English heart vs. Malay *hati* 'liver'. In *Culture, Body, and Language: Conceptualizations of Internal Body Organs across Cultures and Languages*, Farzad Sharifian, René Dirven, Ning Yu and Susanne Niemeier (eds.), 77–104. Berlin: Mouton de Gruyter.

Goossens, Louis. 1995. Metaphtonymy: The interaction of metaphor and metonymy in figurative expressions for linguistic action. In *By Word of Mouth: Metaphor, Metonymy and Linguistic Action in a Cognitive Perspective*, Louis Goossens, Paul Pauwels, Brygida Rudzka-Ostyn, Anne-Marie Simon-Vandenbergen and Johan Vanparys, 175–204. Amsterdam: Benjamins.

Ibarretxe Antuñano, Iraide. 2008. Guts, heart and liver: The conceptualization of internal organs in Basque. In *Culture, Body, and Language: Conceptualizations of Internal Body Organs across Cultures and Languages*, Farzad Sharifian, René Dirven, Ning Yu and Susanne Niemeier (eds.), 105–130. Berlin: Mouton de Gruyter.

Ikegami, Yoshihiko. 2008. The Heart. What it means to the Japanese speakers. In *Culture, Body, and Language: Conceptualizations of Internal Body Organs across Cultures and Languages*, Farzad Sharifian, René Dirven, Ning Yu and Susanne Niemeier (eds.), 171–188. Berlin: Mouton de Gruyter.

Jahoda, Gustav. 1993. *Crossroads between Culture and Mind: Continuities and Change in Theories of Human Nature*. New York: Harvester Wheatsheaf.

Johnson, Mark. 1987. *The Body in the Mind. The Bodily Basis of Meaning, Imagination, and Reason*. Chicago: The University of Chicago Press.

Lakoff, George. 1987. *Women, Fire, and Dangerous Things. What Categories Reveal about the Mind*. Chicago: The University of Chicago Press.

Lakoff, George and Mark Johnson. 1999. *Philosophy in the Flesh. The Embodied Mind and its Challenge to Western Thought*. New York: Basic Books.

Lakoff, George and Zoltán Kövecses. 1987. The cognitive model of anger inherent in American English. In *Cultural Models in Language and Thought*, Dorothy Holland and Naomi Quinn (eds.), 195–221. Cambridge: Cambridge University Press.

Maalej, Zouhair. 2008. The heart and cultural embodiment in Tunisian Arabic. In *Culture, Body, and Language: Conceptualizations of Internal Body Organs across Cultures and Languages*, Farzad Sharifian, René Dirven, Ning Yu and Susanne Niemeier (eds.), 395–429. Berlin: Mouton de Gruyter.

Niemeier, Susanne. 2000. Straight from the heart – metonymic and metaphorical explorations. In *Metaphor and Metonymy at the Crossroads. A Cognitive Perspective*, Antonio Barcelona (ed.), 195–213. Berlin: Mouton de Gruyter.

Niemeier, Susanne. 2008. To be in control: kind-hearted and cool-headed. The head-heart dichotomy in English. In *Culture, Body, and Language: Conceptualizations of Internal Body Organs across Cultures and Languages*, Farzad Sharifian, René Dirven, Ning Yu and Susanne Niemeier (eds.), 349–372. Berlin: Mouton de Gruyter.

Occhi, Debra. 2008. How to have a heart in Japanese. In *Culture, Body, and Language: Conceptualizations of Internal Body Organs across Cultures and Languages*, Farzad Sharifian, René Dirven, Ning Yu and Susanne Niemeier (eds.), 189–212. Berlin: Mouton de Gruyter.

Palmer, Gary, Cliff Goddard and Penny Lee (eds.). 2003. Talking about thinking across languages. Special Issue of *Cognitive Linguistics* 14 (2–3).

Quinn, Naomi. 1991. The cultural basis of metaphor. In *Beyond Metaphor: The Theory of Tropes in Anthropology*, James W. Fernandez (ed.), 56–93. Stanford: Stanford University Press.

Quinn, Naomi and Dorothy Holland. 1987. Culture and cognition. In *Cultural Models in Language and Thought*, Dorothy Holland and Naomi Quinn (eds.), 3–40. Cambridge: Cambridge University Press.

Sharifian, Farzad. 2008a. Distributed, emergent cultural cognition, conceptualisation, and language. In *Body, Language, and Mind.* Vol. 2: *Sociocultural Situatedness*, Roslyn M. Frank, René Dirven, Tom Ziemke and Enrique Bernárdez (eds.), 109–136. Berlin: Mouton de Gruyter.

Sharifian, Farzad. 2008b. Conceptualizations of *del* 'heart-stomach' in Persian. In *Culture, Body, and Language: Conceptualizations of Internal Body Organs across Cultures and Languages*, Farzad Sharifian, René Dirven, Ning Yu and Susanne Niemeier (eds.), 247–266. Berlin: Mouton de Gruyter.

Sharifian, Farzad, René Dirven, Ning Yu and Susanne Niemeier (eds.). 2008. *Culture, Body, and Language: Conceptualizations of Internal Body Organs across Cultures and Languages.* Berlin: Mouton de Gruyter.

Sharifian, Farzad, René Dirven, Ning Yu and Susanne Niemeier. 2008. Culture and language: Looking for the "mind" inside the body. In *Culture, Body, and Language: Conceptualizations of Internal Body Organs across Cultures and Languages*, Farzad Sharifian, René Dirven, Ning Yu and Susanne Niemeier (eds.), 3–23. Berlin: Mouton de Gruyter.

Shweder, Richard A. 1991. *Thinking Through Cultures.* Cambridge: Harvard University Press.

Siahaan, Poppy. 2008. Did he break your *heart* or your *liver*? A contrastive study on metaphorical concepts from the source domain ORGAN in English and in Indonesian. In *Culture, Body, and Language: Conceptualizations of Internal Body Organs across Cultures and Languages*, Farzad Sharifian, René Dirven, Ning Yu and Susanne Niemeier (eds.), 45–76. Berlin: Mouton de Gruyter.

Stevens, Scott M. 1997. Sacred heart and secular brain. In *The Body in Parts: Fantasies of Corporeality in Early Modern Europe*, David Hillman and Carla Mazzio (eds.), 263–282. New York: Routledge.

Taylor, John. 2003. *Linguistic Categorization* (3rd ed.). Oxford: OUP.

Wierzbicka, Anna. 1989. Soul and mind. Linguistic evidence for ethnopsychology and cultural history. *American Anthropologist* 91 (1): 41–58.

Wolk, Daniel P. 2008. Expressions concerning the heart (Libbā) in Northeastern Neo-Aramaic in relation to a classical Syriac model of the temperaments. In *Culture, Body, and Language: Conceptualizations of Internal Body Organs across Cultures and Languages*, Farzad Sharifian, René Dirven, Ning Yu and Susanne Niemeier (eds.), 267–318. Berlin: Mouton de Gruyter.

Yoon, Kyung Joo. 2008. The Korean conceptualization of *heart*: An indigenous perspective. In *Culture, Body, and Language: Conceptualizations of Internal Body Organs across Cultures and Languages*, Farzad Sharifian, René Dirven, Ning Yu and Susanne Niemeier (eds.), 213–243. Berlin: Mouton de Gruyter.

Yu, Ning. 2008. The Chinese heart as the central faculty of cognition. In *Culture, Body, and Language: Conceptualizations of Internal Body Organs across Cultures and Languages*, Farzad Sharifian, René Dirven, Ning Yu and Susanne Niemeier (eds.), 133–170. Berlin: Mouton de Gruyter.

Widening the goalposts of cognitive metaphor research*

Marcus Callies
Johannes-Gutenberg Universität, Mainz

This chapter discusses patterns of use and variation in selected sports-related idiomatic expressions based on conceptual metaphors in five national varieties of English. Previous studies have mostly focused on the pervasiveness of sports metaphors and their linguistic realization in American English, while the use of such metaphorical expressions in other varieties of English has been much less studied. In general, cognitive-linguistic research on linguistic and conceptual variation to date has predominantly examined cross-cultural differences in metaphorical mappings, while language-internal variation along sociolinguistic parameters that may affect metaphor usage have remained an under-researched topic. The present study shows how the types and aspects of (cross-cultural) metaphor variation can be fruitfully applied to different varieties of only one language. It investigates differences in the prominence of specific source domains across the varieties under study by taking into account metaphorical transparency and idiom variation based on a cognitive linguistic-model of idiom representation.

Keywords: metaphorical mapping, within-culture variation, sports-related idiomatic expressions, metaphorical transparency, idiom variation, Cognitive Sociolinguistics

1. Introduction: Why sports metaphors?

David Crystal (2003) in his book *English as a Global Language* reports an incident that vividly illustrates the variability and culture-specific nature of figurative language, but also the confusion and mutual incomprehensibility that the use of sports-related expressions may cause:

---

* This paper is dedicated to Bernd Kortmann on the occasion of his 50th birthday.

> Some time ago, I was attending an international seminar at a European university. Around the table were representatives of some 20 countries. (…) The lingua franca of the meeting was English, and everyone seemed to be using the language competently – even the native speakers. We were well into the discussion period following a paper which had generated a lively buzz of comment and counter-comment. Someone then made a telling remark. There was a silence round the table, which was broken by one of the US delegates observing: "That came from out in left field". There was another silence and I could see some of the delegates turning to their neighbors in a surreptitious way, as one does when one does not understand what on earth is going on, and wants to check that one is not alone. But they were not pondering the telling remark. They were asking each other what 'from out in left field' meant. My neighbor asked me: as a native speaker, he felt confident I would know. I did not know. Baseball at that time was a closed book to me – and still is, very largely. One of the braver of the delegates spoke up: "out where?", he asked. It took the US delegate by surprise, as plainly he had never had that idiom questioned before; but he managed to explain that it was a figure of speech from baseball, a ball coming from an unusual direction, and what he had meant was that the remark was surprising, unexpected. There were nods of relief from around the table. Then one of the UK delegates chipped in: "You played that with a straight bat", he said. "Huh?", said the American. "Oh, I say, that's not cricket", I added, parodically. "Isn't it?", asked a delegate from Asia, now totally confused. The next few minutes of the meeting were somewhat chaotic. The original theme was quite forgotten, as people energetically debated the meaning of cricket and baseball idioms with their neighbours.
> (Crystal 2003: 186f.)

Sports is an important part of a nation's popular culture, and numerous studies have shown the importance of sports and sports metaphors in the construction and reinforcement of national identity, e.g. in political and media discourse (Blain, Boyle, and O'Donnell 1993; Bairner 2001, Charteris-Black 2004: Chapter 6). Certain types of sports are even said to embody the character of a nation or country: football in England,[1] rugby union in Wales and New Zealand, or baseball in the United States. For over a century, baseball has been a distinctive feature of American culture (McGimpsey 2000), and in modern American fiction, baseball provides tropes of exile and homecoming, metaphors of the American Dream, and ludic models for writing and fictionality (Lauricella 1999).

It is tempting to conclude that in a language (or variety of that language) and its culture a certain type of sports contributes to shaping its identity that

1. In this paper, the term 'football' is used to refer to the type of sport that is popular in Europe and South America, called 'soccer' in the United States. Football as played in the United States is referred to as American football.

may also be reflected in the types and the salience of sports metaphors found in that language:

> the high frequency and diversity of a particular metaphor can sometimes be taken as a reflection of a country's history (e.g. the comparatively high number of sailing metaphors in British English) or even its national stereotypes (e.g. the relatively high number of gardening metaphors in British English; Boers 1999). In such cases, awareness of metaphor might even serve as a window onto a community's "culture". (Boers 2003: 235)

However, examining ice hockey which has often been portrayed as having an enduring link to the idea of 'Canadianness', Gruneau and Whitson (1993) show that ideas about hockey representing a unified Canadian identity are mythical: Hockey has been shaped by the Americanization and globalization of the entertainment industry and becomes increasingly subjugated to the imperatives of the global entertainment market, and hence de-Canadianised. Thus, it becomes more and more difficult to sustain the myth that it epitomizes a distinctive Canadian identity. The authors suggest that hockey may eventually resemble soccer, where national sentiments co-exist in an international framework.

Shore (1996) considers sporting events as a kind of ludic model that engages serious cultural issues in the very rules and structures constituting the field of play:

> baseball, as our most venerable national sport, has something to tell us about the American mind. More specifically, in its complex representations of time, space, and action, baseball models important tensions in mainstream American culture between communitarian and individualistic values. And, in turn, baseball provides a source domain for modeling other areas of life where analogous tensions are experienced. (Shore 1996: 75f.)

Assuming that figurative language can be considered a reflection of a community's conventional patterns of thought or world views, sports metaphors are thus an ideal testing ground for the interaction of studies on language, culture, history, and even national identity, hence a topic that lends itself to interdisciplinary research. What makes cognitive linguistics fundamentally interdisciplinary in nature is its 'cognitive commitment': models of language and linguistic organization, and principles of linguistic structure should reflect what is known about the human mind and human cognition from other cognitive sciences. Taking into account within-culture variation in conceptual metaphor along, for example, regional, social, ethnic, and diachronic dimensions (thereby giving cognitive metaphor research a sociolinguistic spin) can help to better understand the historical, social, and cultural foundations of conceptual motivation and figurative language use, and at the same time provides a highly promising research agenda to promote cross-fertilization in cognitive studies. For example, it opens up new perspectives for cultural studies in terms of a cognitive cultural studies (see, e.g. Kövecses 2005: Chapter 8) and, in accordance with the importance of context in cognitive linguistics (involving linguistic, social and cultural dimensions), strengthens the recently emerging interest in sociovariational phenomena in terms of a Cognitive Sociolinguistics (Kristiansen and Dirven 2008; Wolf and Polzenhagen 2009; Geeraerts, Kristiansen, and Peirsman 2010). Ultimately, such an approach may help to delineate and sharpen the interfaces of cognitive linguistics with cognitive approaches in other research fields that conceive of themselves as part of the cognitive enterprise.

In general, it seems fair to say that cognitive-linguistic research on variation in metaphor to date has predominantly examined cross-cultural differences in metaphorical mappings (with a focus on body parts and the conceptualization of emotions, see, e.g. Kövecses 2000a; Sharifian et al. 2008), while language-internal variation along typical sociolinguistic parameters (e.g. region, social class, ethnicity, and gender) that are likely to affect metaphor usage have remained an under-researched topic (Kövecses 2005: 88f.; Geeraerts and Cuyckens 2007: 16). Thus, it is not surprising that studies on the impact of (geographic) region on within-culture variation in metaphor are sparse and only slowly emerging, see for example the study on Dutch and Afrikaans by Dirven (1994) and recent corpus-based studies on African Englishes from a cultural-linguistic perspective (Wolf and Polzenhagen 2007a, b).

The present chapter examines patterns of use and variation in selected sports-related idiomatic expressions based on conceptual metaphors in five national varieties of English, taking into account metaphorical transparency and idiom variation based on a cognitive linguistic-model of idiom representation. One of the central aims of the study is to show how the types and aspects of variation that are involved in cross-cultural variation of metaphors can also fruitfully be applied to different varieties of only one language, taking into account regional variation (and potentially other sociolinguistic parameters). Assuming that SPORTS is available as a generic-level source domain in metaphorical mappings across varieties of English, I investigate if there are differences in the prominence of specific source domains across the varieties under study when SPORTS is instantiated as a specific-level, culturally-embedded conceptual metaphorical mapping, and if these differences are reflected by the frequency of use and the extent of lexico-grammatical variability of sports-related idiomatic expressions. Finally, I discuss the question if usage data may evidence a convergence or even erosion of differences in metaphor usage across World Englishes as a result of ongoing economic and cultural globalization.

2. Sports metaphors in English

Previous studies on sports metaphors in English (Hardaway 1976, Howe 1988, Ching 1993, Liu and Farha 1996) have largely focused on their special role, salience and pervasiveness in American culture due to the importance of sports and competition in US American society:

> Nobody would argue the place of sports in American life; they are big business. And they are big business because they fit philosophically with the widely accepted American dream of open competition in a free market economy. Americans believe in competition, foster it, and encourage it. They live by its rules. No wonder the language of athletic competition has found its way as metaphor into every aspect of American life. (Hardaway 1976: 78)

Ching (1993) identifies LIFE IS PLAYING A GAME as an underlying metaphor for life in the United States and points out that

> playing results in attaining the American middle-class ideal of investing our lives in competitive, risky activity that might reap rewards of enjoyment, self-esteem, and survival. We prove our worth and gain our identity through winning games. To Americans, "TO LIVE IS TO PLAY". (Ching 1993: 43, 46; emphasis in original)

Kövecses (2005) goes even one step further by assuming a foundational metaphor in American culture that is "LIFE IS A SHOW OR SPECTACLE, or more generally, ENTERTAINMENT" (Kövecses 2005: 184ff., emphasis in original).

Liu and Farha (1996: 38) identify the major driving forces for the increasing popularity and pervasiveness of sports metaphors in American culture in particular: a growing intensity of competition in almost all aspects of public and private life, as well as the increasing commercialization of sports, accompanied by an ever-growing coverage and accessibility of sports events in the media. It is important to point out that COMPETITION can be considered a general source domain for common concepts like success, progress, and failure that are used to comprehend and evaluate everyday situations and activities (Langlotz 2006: 143f.). Thus "the claim here is not that only Americans have the game and sports metaphors, but that they have them for a more extensive range of target concepts than other nations" (Kövecses 2005: 244). Sports metaphors have also become increasingly popular in Chinese in the last two decades when socio-political reform has raised China's level of competition in both economy and sports (Yu 1998: 44).

Many different sports abound as source domains in metaphorical mappings, but the types of sports that have been shown to feature most prominently as source domains in American English are baseball, American football, and boxing (Howe 1988; Liu and Farha 1996; Kövecses 2000b; Tottie 2002). The most prominent target domains that have been identified are political discourse (Howe 1988), law, warfare, business/economics, work, personal relationships, love, and sex (Liu and Farha 1996; Kövecses 2000b: 295f.).

By contrast, the use of sports-related metaphorical expressions in and across other varieties of English, for example those based on source domains such as FOOTBALL or CRICKET, have been much less studied systematically in an empirical way, with Charteris-Black (2004: Chapter 6), who examines the use of metaphors based on football, rugby, and cricket in the British press, being one major exception. However, it is interesting to note that there appears to be a growing awareness among speakers of other varieties of English that certain types of "American" expressions are spreading due to linguistic and cultural contact. Consider the following quotes from *The Guardian* and *The Sydney Morning Herald*, respectively:

> Last week, the defence secretary, John Hutton, said that it was time for our European allies "to step up to the plate" and send more of their troops to Afghanistan. How distressing to see a man who shares a surname with one of England's greatest cricketers using a baseball metaphor so lazily when our own national games offer so many richer ones. People seem to be "stepping up to the plate" at an ever increasing rate these days. Guests on the Today programme, pundits on Newsnight – everyone wants to do it. Last year, there were 341 references in the national media to this term, an increase on the previous year and well up from 307 in 2006. (Campbell 2009)

> Baseball has invaded the Australian language without most of the population ever having seen a game. Only Australians over 60 continue to derive imagery from our own sports: 'sticky wicket', 'playing a straight bat', 'hit for six', 'a free kick', 'dropping a catch', and at funerals: "Well, he had a good innings". The rest of us are more in tune with politicians promoting a "three strikes you're out" crime policy; businessmen talking about "ballpark figures" or "a whole new ball game"; surprises coming "out of left field"; and acquaintances wanting to "touch base" or asking for a "rain check" if they can't. [...] So let's dump cricket and make baseball our national summer sport. Then the way we live will catch up with the way we talk. (Dale 2004)

Similarly, Moon (1998) in her corpus-based study of idioms and fixed expressions maintains that

> there are a few cases where British and American English have parallel idioms, with similar meanings, usages, and even source domains for the metaphors, but different lexis altogether [...]. While many of these distinctions are well established, the situation in general is complex. The influence of American culture and media in Britain means that Americanisms and American variations become established in British English, or at least established in certain registers or genres of British English use. (Moon 1998: 133f.)

Kövecses (2005: 94) also argues that American English seems to have had a greater impact on British English than British English on American English and that there is a large number of metaphorical expressions used in British English that originate in American English. Such observations may in fact be the first signs of a more encompassing trend. Boers (2003) even hypothesizes that "as a result of ongoing economic and cultural globalisation, some cross-cultural differences in metaphor usage will eventually be eroded" (2003: 236). It will thus be interesting to see if usage data in fact evidence a leveling of differences in metaphor usage across varieties of English and how this might be explained. Liu and Farha (1996) propose several candidates for what they call "stay-at-home" and "exported" metaphorical expressions: they assume *out in left field*, *switch hitters*, *slam dunk* to have remained predominantly American, while *touch base with someone* and *off base* are said to have been adopted abroad (Liu and Farha 1996: 40).

3. Dimensions of variation in metaphor

Kövecses (2005) broadly distinguishes between cross-cultural variation and within-culture variation in conceptual metaphor. Conceptual metaphors are systematic and recurring source-target mappings in natural language, and it is generally assumed that certain conceptual metaphors are potentially universal or can be near universal on a generic level. In particular, these metaphors are "simple" or "primary" metaphors, based on universal human experiences. Often, primary metaphors map image-schemas onto abstract experience (e.g. GOOD IS UP, BAD IS DOWN), motivated by correlations in the domain of general physical experience (embodiment). Complex conceptual metaphors combine (or compound) different primary metaphors (e.g. LIFE IS A JOURNEY) and are more likely to be susceptible to culture-specific influences and thus, to cross-cultural variation, which arises when generic-level metaphors are instantiated in culture-specific ways at the specific level (see Figure 1).

Kövecses (2005: Chapter 4) discusses the various aspects that are involved in variation at the conceptual level: the source domain (languages may use different source domains for a certain target domain), the target domain (languages may use a certain source domain for the conceptualization of a different target domain), the range of the target domain (the number of source domains languages have available for the conceptualization of a certain target domain), and the scope of the source domain (the set of target domains to which a particular source domain can apply). Differences can also arise as to preferential conceptualization (two languages have a set of the same conceptual metaphors for a given target domain, but speakers of the languages prefer to use a different set of metaphors for

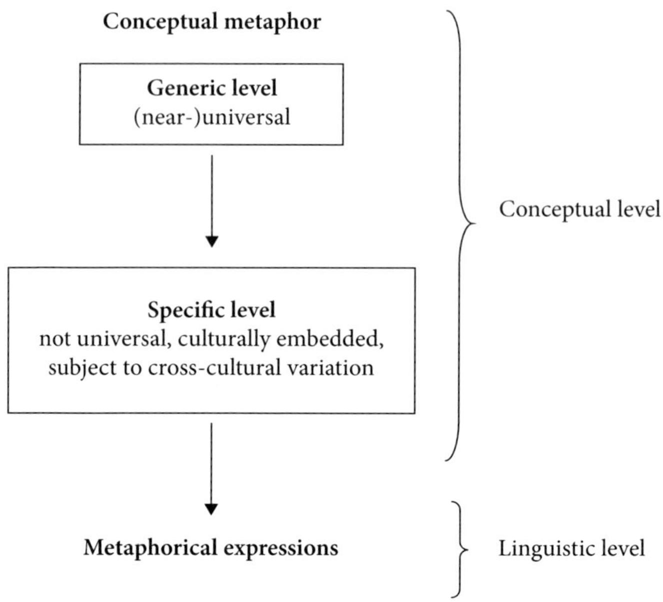

Figure 1. The locus of variation in conceptual metaphor

this target) and uniqueness (a metaphor that has both a culturally unique source and target domain). Boers (2003: 232) identifies three types of cross-cultural variation that are involved in actual usage of metaphors: differences with regard to the particular source-target mappings that have become conventional in the given cultures (similar to Kövecses' "preferential conceptualization"), differences with regard to value-judgments associated with the source or target domains of shared mappings, and differences with regard to the degree of pervasiveness of metaphor as such, as compared with other (rhetorical) figures. Of course, a great deal of variation also arises on the linguistic level, i.e. when the conceptual metaphors are realized by metaphorical expressions in the respective languages.

To exemplify some of these patterns of variation, consider first the metaphor POLITICS IS SPORT in American English and Chinese. Both English and Chinese have this generic-level metaphor (same source-target mapping). However, the two cultures differ as to the kinds of sports that are popular, and this results in variation when the generic-level metaphor is instantiated at the specific level. In American English the specific source domain is typically BASEBALL, AMERICAN FOOTBALL, or BOXING, but Chinese uses VOLLEYBALL, FOOTBALL, and PING-PONG as common source domains for sports metaphors in political and everyday discourse (Yu 1998: 44).

An example of cross-cultural variation where two languages use different source domains for a certain target domain is provided by Liu (2002). He proposes that Americans view their life as sports and therefore tend to think about life and act in such a way as if they were in a game (LIFE IS A GAME). Thus politics, business, and personal relationships are viewed as a game in which one has to win, and thus they serve as target domains to which the source domain SPORTS (with its various instantiations) applies. The Chinese, however, do not view their world as a game, and so in this context SPORTS is not considered a dominant metaphor in China. Liu claims that what is more central to Chinese life is family and eating, and accordingly, these occur as source domains for the conceptualization of the target domain LIFE.

Abadi and Sacerdoti (2001) give an example of preferential conceptualization (or differences in the degree of conventionalization of a source-target mapping) in two different cultures. They carried out a study of the source domains of metaphors in Israeli and American political discourse. The main source domains that emerged were WAR, certain branches of sports, and two games, all of which share the feature of competition. Their analysis showed that the source domain WAR is more wide-ranging in Israel than in the US, whereas the source domain SPORTS is more varied in the US than in Israel. The reasons for this difference are on the one hand the pervasive presence of the army in Israeli life, and on the other hand the greater variety of sports that are popular in the US.

Importantly, metaphors can vary not only cross-culturally but also within cultures and languages. Kövecses points out that

> an obvious place to look for variation in metaphor are the social, cultural, stylistic, individual, and so on, dialects and varieties that have been identified by sociolinguists, linguistic anthropologists, and other researchers of language variation in a social and cultural context. (Kövecses 2005: 88)

Similar to cross-cultural variation, within-culture variation is subject to the "social and cultural divisions that are likely to produce metaphor variation due to people's divergent experiences in social and cultural life" (Kövecses 2005: 89). Thus, he identifies a social, ethnic, regional, stylistic, diachronic, developmental, and individual dimension of variation.

The types and aspects of variation outlined earlier can also apply to different varieties of only one language. The kinds of variation for the metaphor POLITICS IS SPORT or LIFE IS A GAME in American English and Chinese sketched above, as well as the differences in preferential conceptualization in Israeli and US-American culture, are likely to be found in (sports) metaphors across varieties of the same language, e.g. English. Football (soccer) is clearly more popular in Britain than in the US, and the two main varieties are likely to differ in the degree of productivity and conventionality of football-based metaphorical expressions.

Charteris-Black (2004: 129) in his analysis of the use of metaphors in sports reporting in the British press discusses a conceptual map that also illustrates the type of variation and preferential conceptualization that is expected to emerge across the different varieties of English that all have their 'national' sport. He argues that there is a generic-level conceptual metaphor SPORT IS WAR with specific-level instantiations FOOTBALL IS WAR, RUGBY IS WAR, CRICKET IS WAR, etc. Obviously, how this generic-level metaphor is realized both conceptually and lexically depends on the type of sport that is prevalent in the respective variety of English, e.g. American football and baseball in the US, hence the greater likelihood of BASEBALL IS WAR as a conceptual metaphor in American English than for example in Australian English. In addition, Charteris-Black identifies a higher order conceptual key SPORT IS A STRUGGLE FOR SURVIVAL that accounts for the interrelatedness of sport, conflict, and survival, and enables SPORT to be "incorporated into a biological schema that is also found in other competitive domains such as politics and business" (2004: 129f.).

Apart from regional variation, there are several other dimensions of variation in metaphor usage that can, however, not be addressed in detail in this chapter. For example, Kövecses (2005: 183) highlights the close correlation between personal history and individual use of metaphors. He investigated the metaphors used by US-American politicians in the course of their election campaigns in 1996 (on the basis of *Time*). All the candidates running for office used sports metaphors, but clear differences in preferences were discernible.

Variation in the use and preferences of sports metaphors may also be due to different attitudes towards competitiveness in general. Several psychological studies suggest that men are more competitive than women in the selection and preference of (non-)competitive games and activities in childhood, a tendency that increases through puberty, so that by adulthood more men than women describe themselves as competitive (Campbell 2002). This may also be reflected in the workplace:

> Perhaps what really bothers me about 'straight bats' and 'sprints for the finishing line' is the fear that they might reveal something deeper about work than the fact that gossiping about sport is the best way to show you're a good bloke. Perhaps the reason all this jocular sports chat is so popular is that men have a tendency to see life as a contest; perhaps the reason it tends to irritate their female colleagues so much is that women don't. It sounds like sexist psychobabble, but increasing evidence is emerging that men and women approach competitions, sporting and otherwise, somewhat differently. (Stewart 2007)

Niederle and Vesterlund (2007) investigated whether men and women of the same ability differ in their preferences for competition with respect to the type of compensation scheme they prefer to receive for their work when holding other job characteristics constant. Participants in a laboratory experiment had to solve a task where no gender differences in performance were expected. Participants first performed the task under a non-competitive piece-rate compensation and then under a competitive tournament. Having experienced both compensation schemes, participants then were to choose which of the two schemes they wanted to apply to their performance of the next task. Although there were no gender differences in performance, men selected the tournament twice as much as women when choosing their compensation scheme for the next performance (73 vs. 35 percent). The authors argue that this "gender gap" in tournament entry is driven by men being more overconfident and by gender differences in preferences for performing in a competition: women shy away from competition and men embrace it.

Archer and Cohen (1997) studied the effect of sports metaphors in legal jargon on women's participation in the field of law by examining 22 baseball-, boxing-, and football-based metaphorical expressions of the conceptual metaphor LAW IS SPORT in the *LexisNexis* database to search for evidence of their presence in all federal and state court cases from the late 1800s until 1995. The expressions they selected were considered semantically opaque to non-sports fans and not easily understood nor part of popular American English. The authors hypothesized that such metaphors have the potential to exclude or confuse anyone not familiar with sports, and that women in particular, who may not have equal access to sports, would be negatively impacted by the use of sports metaphors. They concluded that the use of such metaphors hindered women from fully participating in the legal arena.

Finally, the popularity of a specific type of sports in a culture and thus, the use of idiomatic expressions related to this type of sports, is often closely connected to national history and identity, see for example the somewhat provocative statement by Tharoor:

> Americans have about as much use for cricket as Lapps have for beachwear. The fact that elsewhere in the civilized world grown men dress up like poor relations of Gatsby and venture hopefully into the drizzle clutching their bats invariably mystifies my American friends. And the notion that anyone would watch a game that, in its highest form, could take five days and still end in a draw provokes widespread disbelief among results-oriented Americans. (…) In any event, nothing about cricket seems suited to the American national character: its rich complexity, the infinite possibilities that could occur with each delivery of the ball, the dozen different ways of getting out, are all patterned for a society of endless forms and varieties, not of a homogenized McWorld. They are rather like Indian classical music, in which the basic laws are laid down but the performer then improvises gloriously, unshackled by anything so mundane as a written score. (Tharoor 2007)

Why did cricket and baseball become national passions in two of the world's biggest erstwhile colonies? Majumdar and Brown (2007) examined the question why cricket flourished in countries like Pakistan and India, but has not become popular in the United States with its heritage and "special relationship" with Britain. In both cases, the common reference point remains the Empire, but with an inverse invocation of nationalism. India, where the Empire lasted far longer, saw an appropriation and subsequent indigenization of British sports for purposes of resistance. The nationalist movement from the close of the 19th century made cricket imperative to be taken up as a non-violent means to compete with the ruling British. By contrast, in the US, where independence was achieved more than a century earlier, the reaction to Empire sport was a dissociation of American sport from British sport to emphasize an independent American identity and there was no need to retort to imperial games.

4. Methodology and data

Due to the lack of comparable and sufficiently large corpora for the diverse varieties of English under study, I used selected national newspapers as proxies of current usage of sports-related idiomatic expressions in the press, accessed through the *LexisNexis* database. While it is clear that the choice of newspaper for each variety of English may have an impact on the type of language represented (consider factors such as tabloid vs. broadsheet newspapers, and the influence of a certain in-house writing style, copy editing policies, and personal preferences), this influence is hard to estimate and seems impossible to measure. I searched all issues of a number of major national newspapers published between February 2007 and February 2009, representing five main varieties of English (see Table 1). The newspapers were searched for frequencies of occurrence of 15 idiomatic expressions from three different types of ball games as source domains (BASEBALL, FOOTBALL, and CRICKET, see Table 2).

Table 1. National newspapers examined (February 2007 – February 2009)

variety	newspaper	publication details
American English (AmE)	*New York Times*	daily, Monday – Sunday
British English (BrE)	*The Guardian* *The Observer*	daily, Monday – Saturday weekly, Sunday
Australian English (AusE)	*Sydney Morning Herald*	daily, Monday – Sunday
Indian English (IndE)	*Hindustan Times*	daily, Monday – Friday; occasionally Saturday
South African English (SAE)	*The Star* *The Sunday Times*	daily, Monday – Saturday weekly, Sunday

Table 2. Sports-related idiomatic expressions extracted from national newspapers

type of sport	expression
baseball	– (V$_{[throw]}$) *a* ***curve*** (*ball*); 'unexpected, surprising'; 'deceptive' – *step* ***up to the plate*** 'take on or accept a challenge or responsibility' – *be* (*way*) ***off base*** 'be completely wrong' – ***out*** P$_{[from, in, of]}$ ***left field***; with *in*: 'eccentric, odd' with *of/from*: 'a source of the unexpected or illogical' – *have* NUM$_{[one, two, three]}$ ***strikes against*** 'condition or situation that makes it extremely difficult for someone or something to be successful'
football	– ***play it safe*** 'act carefully, avoid risks' – V$_{[get, set, start, keep, nudge]}$ ***the ball rolling*** 'start something happening' – V$_{[change, move, shift, narrow, widen, raise, set]}$ *the* ***goalposts*** 'change the rules, limits etc. for something while someone is trying to do something, making it more difficult for them' – *score an* ***own goal*** 'create a problem for oneself', 'an action or remark that has the opposite effect from what one intended' – V$_{[hit, find]}$ *the* ***back of the net*** 'reach one's goal, be successful'
cricket	– (*be on a*) ***sticky wicket*** 'difficult or tricky situation' – (*just*) ***not cricket*** 'unfair or unjust' – V$_{[reply, respond, offer, play]}$ (*with*) *a* ***straight bat*** 'offer a non-committal or evasive answer to a question' – (*have*) *a* ***good innings*** 'having spent a long time doing something, having had a long, successful life' – ***hit for six*** 'score a big success', 'being astonished or amazed by something; 'being devastated'

In view of the considerable lexico-grammatical variability of these expressions that will be discussed further below, the most effective way to ensure a relatively high recall rate was to retrieve them automatically using lexical anchors (indicated in bold print in Table 2; optional elements are given in round brackets). These are essentially lexical key elements present in many figurative expressions that activate their idiomatic configuration. Irrelevant and literal uses were filtered out and separated from the metaphorical ones, which were in turn analyzed manually.

The idiomatic expressions listed in Table 2 would traditionally not be considered metaphorical. Idioms have long been thought to be frozen phraseological units of language that fall outside of metaphorical thought and linguistic creativity, and whose meaning is opaque since it is semantically non-compositional. Cognitive-linguistic research on idioms, however, has shown that a large number of them is in fact conceptually motivated and thus reflects linguistic creativity, compositionality, and metaphorical thought, thus falsifying the classical treatment of idioms as semantically non-compositional lexical units. In particular, the recent monograph by Langlotz (2006) shows how "idioms can be motivated and structured by well-entrenched patterns of semantic extension, particularly conceptual metaphors" (Langlotz 2006: 143). Langlotz provides an explanatory framework for systematic idiom variation based on a cognitive linguistic-model of idiom representation. In this model, idioms are considered "complex linguistic and conceptual activation networks, which can be unfolded variably" (Langlotz 2006: 178). Consequently, idioms are not fixed or even formulaic, but subject to idiom creativity and variation:

> many idioms can be attributed an intrinsically creative semantic structure. [...] the degree to which this structure can be (re)motivated by a cogniser in relation to a set of entrenched conceptual metaphors, metonymies and emblems correlates with their potential for systematic lexicogrammatical variation. (…) Idiom variation is the process by which an idiomatic construction is adapted to the usage-context to fulfil its cognitive modelling function. (Langlotz 2006: 287f.)

As will be shown in the subsequent analysis, the sports-related idiomatic expressions studied in this chapter do have an underlying conceptual motivation that is also reflected in a considerable degree of lexico-grammatical variability. Obviously, this has some implications for the present study. Assuming that LIFE IS PLAYING A GAME (and the underlying conceptual key SPORT IS A STRUGGLE FOR SURVIVAL) is a conventional and entrenched generic metaphor that motivates the expressions listed in Table 2, this should correlate with, or at least enable, the potential for idiom creativity and variation of sports-based idioms in different Englishes in relation to the culture-specific preferential elaborations of the source domain SPORTS (e.g. FOOTBALL, BASEBALL). In other words, the more prominent a specific source domain is in a certain variety of a language, the higher the potential for lexico-grammatical variability of the idiomatic expressions that are based on that conceptual metaphor.

5. Results and discussion

A total of 2,372 instances of the 15 target items were extracted from the database, and 1,572 tokens (66.3%) proved to be relevant, metaphorically used expressions. These fall into the three ball game categories as follows: 465 (29.6%%) are baseball expressions, 884 (56.2%) from the domain of football, and 223 (14.2%) from cricket. Table 3 shows their distribution across the five varieties of English.[2]

2. For technical reasons, it was impossible to determine the number of words of the newspaper database that was searched for each variety of English. This leads to the methodological problem that normalized frequency counts for the 15 expressions under study cannot be reported.

Table 3. Distribution of expressions per source domain in five varieties of English

source domain	AmE	BrE	AusE	SAE	IndE
baseball	162 (58.9%)	133 (24.8%)	105 (25.6%)	41 (21.6%)	24 (15%)
football	108 (39.3%)	331 (61.6%)	217 (52.9%)	122 (64.2%)	106 (66.25%)
cricket	5 (1.8%)	73 (13.6%)	88 (21.5%)	27 (14.2%)	30 (18.75%)
total	275 (100%)	537 (100%)	410 (100%)	190 (100%)	160 (100%)

Interestingly, it is not American English that shows the highest amount of sports-related expressions, but British English followed by Australian English. The figures are lowest in South African and Indian English. However, and unsurprisingly, American English stands out in that baseball clearly emerges as the most frequent source domain. What is surprising, though, is that baseball expressions are not only used in American English, but also in the other varieties, and much more frequently than one might expect in view of the relative lack of popularity of this sport in English-speaking countries outside of the US. By contrast, cricket-based expressions are very rare in American English, but fairly common in all other varieties. While it may have been predictable that football is the prevalent source domain in British English, it is again striking that football expressions show a fairly high frequency of occurrence in all varieties, not only in British English. While for South African English this may be – at least partially – due to the increasing popularity of football in the build-up to the FIFA World Cup hosted in South Africa in 2010, this trend needs a more detailed explanation.

Looking at the individual frequencies of occurrence of the 15 expressions across the five varieties, it becomes clear that there are great differences in the popularity of individual expressions as evidenced by their usage in the respective national newspapers. Table 4 illustrates that – except for Indian English, which shows the lowest number of baseball idioms in general – the expressions V *up to the plate*, and V *a curve* (*ball*) occur frequently in all varieties. The use of *out* P *left field* is somewhat limited to American, British, and Australian English, while *off base* and NUM *strikes against* is hardly used except for in American English. In general, the data from Indian English have to be interpreted with great caution because of their somewhat limited value: many of the Indian newspaper texts represent newswire distributed by internationally operating news agencies, and also include a great deal of reported speech by speakers from other varieties of English.

However, the distribution of the three types of expressions across the five corpora is significantly different. A chi-square test of independence yielded the following results: the log-likelihood ratio (LL) is 174.207 for eight degrees of freedom, $p < 0.001$. In the chi-square test, proportional data such as normalized frequencies are not necessary and discrepancies in corpus size are unimportant because the test automatically compares the raw frequencies proportionally.

Table 4. Distribution of baseball expressions in five varieties of English

	V up to the plate	*V a curve ball*	*out P left field*	*off base*	NUM *strikes against*
AmE	76 (34.4%)	22 (19.1%)	20 (29.4%)	37 (78.7%)	7 (50%)
BrE	74 (33.5%)	47 (40.9%)	9 (13.2%)	1 (2.1%)	2 (14.3%)
AusE	40 (18.1%)	29 (25.2%)	34 (50%)	0 (0%)	2 (14.3%)
SAE	21 (9.5%)	15 (13%)	1 (1.5%)	3 (6.4%)	1 (7.1%)
IndE	10 (4.5%)	2 (1.7%)	4 (5.9%)	6 (12.8%)	2 (14.3%)
	221 (100%)	115 (100%)	68 (100%)	47 (100%)	14 (100%)

A similar picture emerges from Table 5. Football-related idioms that are used productively in all varieties appear to be V *the ball rolling*, V *the goalposts* and *play it safe*, while the remaining two, (V) *an own goal* and V *the back of the net* in particular, are hardly used (if at all) in varieties where football is not a very popular sport.

Table 5. Distribution of football expressions in five varieties of English

	V *the ball rolling*	V *the goalposts*	*play it safe*	(V) *an own goal*	V *the back of the net*
AmE	31 (10.1%)	19 (8.8%)	58 (30.7%)	0 (0%)	0 (0%)
BrE	74 (24.2%)	94 (43.7%)	37 (19.6%)	119 (71.7%)	7 (87.5%)
AusE	72 (23.5%)	52 (24.2%)	60 (31.7%)	32 (19.3%)	1 (12.5%)
SAE	64 (20.9%)	29 (13.5%)	17 (9%)	12 (7.2%)	0 (0%)
IndE	65 (21.2%)	21 (9.8%)	17 (9%)	3 (1.8%)	0 (0%)
	306 (100%)	215 (100%)	189 (100%)	166 (100%)	8 (100%)

Finally, Table 6 shows first that the number of cricket-based expressions is lowest in the data, and almost absent from American English. In fact, it is only in the British and Australian newspapers where all five selected expressions are used with moderate frequencies. It comes as a surprise that some cricket expressions are fairly infrequent in the South African and Indian newspapers despite the popularity of cricket in (parts of) these countries. Whether this can be interpreted as a sign of preferential conceptualization (cf. the fairly high number of football-related expressions), or a comparatively low prominence of SPORTS as a general source domain in these varieties (cf. Table 3) needs to be examined in future research.

How can these differences in the (non-)use of individual expressions in and across the five varieties be explained? At first sight, it may appear tempting to conclude that those varieties that can be assumed to have a more intensive contact with US-American media and popular culture in the form of music, movies, TV,

Table 6. Distribution of cricket expressions in five varieties of English

	not cricket	*sticky wicket*	V *a straight bat*	*hit for six*	*a good innings*
AmE	1 (1.4%)	4 (8.3%)	0 (0%)	0 (0%)	0 (0%)
BrE	16 (21.6%)	12 (25%)	19 (41.3%)	7 (23.3%)	19 (76%)
AusE	36 (48.6%)	10 (20.8%)	19 (41.3%)	19 (63.4%)	4 (16%)
SAE	14 (18.9%)	6 (12.5%)	2 (4.4%)	4 (13.3%)	1 (4%)
IndE	7 (9.5%)	16 (33.4%)	6 (13%)	0 (0%)	1 (4%)
	74 (100%)	48 (100%)	46 (100%)	30 (100%)	25 (100%)

and sports also exhibit a higher frequency of occurrence of baseball expressions. While this may at first hand be evidence for an impact of an ongoing economic and cultural globalization that may ultimately lead to an erosion of some cross-cultural differences in metaphor usage (Boers 2003: 236), I would like to propose a different line of argumentation here.

It is interesting to note that there are two types of expressions: those that are used across different varieties of English (the "exports") and those that are not (the "stay-homes"). Both groups share certain characteristics. Assuming that many idiomatic expressions are conceptually motivated, the extent to which a structure can be motivated in relation to one or a set of conceptual metaphors depends on a language user's ability to relate the words used in an expression to a certain domain. If this is not achieved, then the expression remains opaque. The "exports" (V *up to the plate*, V *a curve* (*ball*), *out* P *left field*, V *the ball rolling*, V *the goalposts* and *play it safe*) are expressions for which an underlying conceptual motivation without knowledge of sport-specific terminology or rules is available, and therefore these have the highest degree of metaphorical transparency. Metaphorical transparency refers to a constellation in which a metaphorical expression that belongs to a conceptual metaphor in one language (or variety of that language) is more or less transparent than another expression that belongs to the same conceptual metaphor in another language (or variety of that language). For example, the three football-based expressions that are used across all five varieties (V *the ball rolling*, V *the goalposts* and *play it safe*) are generally applicable to other ball games that involve balls and goals/goalposts, or similar devices that mark the field of play.

In the following, I briefly describe the conceptual basis that motivates "the exports". V *a curve*(*ball*) draws on an image schema employed in the orientational metaphor FORESEEABLE FUTURE EVENTS ARE AHEAD (Lakoff and Johnson 2003: 16). Straightforward movement along a linear path is considered unimpeded and predictable, whereas movement along a non-linear path that involves bends or curves is associated with unpredictable events, or even impediments and obstacles that one may have to face. A curve ball, the object that represents the issue to be dealt with or the event/object that is approaching, is a ball with a spin that causes its trajectory to curve. Although the curve is smooth, the batter sees the ball as if going straight, but suddenly changing direction, making it hard to hit, hence the association with an unpredictable outcome.

In *out* P *left field*, the 'left field' literally refers to the part of the baseball outfield to the left of the batter looking out from the plate, facing the pitcher. This idiom can also be motivated by referring to the orientational metaphor FORESEEABLE FUTURE EVENTS ARE AHEAD. While the precise origin of the expression is unclear and disputed (Ammer 1997), a plausible explanation seems to be that a runner who attempts to score at home has his back to the left field, thus a throw to the plate coming from left field can arrive as a surprise to the runner. Although this requires some sport-specific knowledge, it may suffice to know that the left side is sometimes generally associated with awkwardness and clumsiness as in the expression *having two left feet*, and questionable or doubtful status as in *left-handed compliment*. *Left* also has a negative connotation in various obsolete proverbial expressions (see the respective entry for "left" in the *Oxford English Dictionary*). The choice of preposition effects conceptual specification: *of* and *from* refer to the direction or source, *in* denotes position, compare the following two examples.

(1) Kerry Keady, a solicitor representing three leaseholders, said the revoking of leases "came completely **out of left field**".
(*Sydney Morning Herald*, February 15, 2008)

(2) "They won't have the time to read that they have now when they're on the outside", she said. "For us, we no longer feel like we're **out in left field**. We feel that we're in center field, that our culture will become the dominant culture of Washington". (*New York Times*, November 6, 2008)

Step up to the plate is motivated by the metaphor LIFE IS PLAYING A GAME and pictures a player (a human agent) who steps up (comes out to act) to the plate (the scene or the basis/context of activity).

The three football-based idioms can also be motivated by the metaphor LIFE IS PLAYING A GAME. In *play it safe*, *play* conceptualizes acting in life, *it* is the life situation to be dealt with, and *safe* conceptualizes the avoidance of risk-taking. In V *the ball rolling* the ball again conceptualizes the life situation to be dealt with, and *rolling* conveys the idea of dealing with or taking on this situation. Here, the choice of matrix verb that precedes these two constituents effects conceptual elaboration/specification: *set*, *get*, *start*, and *nudge* may be used to convey the initiation of activity, while *keep* denotes a continuation of the activity. Finally, in V *the goalposts*, the goalposts represent the limits or boundaries of the basis/context of action, and the choice of verb through lexical substitution determines these boundaries (again creating conceptual elaboration or specification): *set* simply denotes the implementation of boundaries, the use of the verbs *change*, *move*, and *shift* signals a change of boundaries, while *narrow*, *widen*, and *raise* specify the type of change.

It is clear that not all sports-related idiomatic expressions can be conceptually motivated in such a way because they require the language user to have some knowledge about the game. For example, *score an own goal* and V *the back of the net* are specific to ball games such as football where the goal has a net and own goals can be scored, something that is impossible or not featured in other ball games. Similarly, for a speaker-listener to motivate *off base* and NUM *strikes against* they need some background knowledge about baseball. In particular, the cricket expressions contain lexical key elements that are sport-specific and contain little else that would enable non-specialists to motivate them conceptually, despite the fact that for example *wicket* and *inning* do exist in both American and British English. *Wicket* has the general meaning of 'small gate or door; opening like a window'. *Inning*, a division of a cricket match, is even used in baseball ('a division of the game consisting of a turn at bat for each team', 'a team's turn at bat ending with the third out', or simply 'a player's turn'), but the idiom *have a good innings* is very rare in American English. Similarly, *bat* is also used to refer to the wooden implement used for hitting the ball in baseball, but V *a straight bat* is rare, while *off the bat* is common in American English for 'without delay, immediately' (note, however, that *off one's own bat* means 'through one's own efforts without being told to do so' in British English). It is possible to identify an underlying conceptual motivation, but difficult for non-experts to relate these words to a certain domain. Hence, the metaphorical transparency of idiomatic expressions in the domain of cricket is lowest which inhibits the use of cricket-based metaphorical expressions in socio-cultural contexts outside of those in which speakers are familiar with the terminology.

This approach, linking differences in frequency of occurrence and use to differences in the transparency and conceptual motivation of these idiomatic expressions, is in my view supported by two additional observations. First, the examples in (3)–(9) illustrate that the expressions under study show a considerable degree of lexico-grammatical variability in terms of constructional adaptation, i.e. regular morpho-syntactic adaptations of their base form (Langlotz 2006: 206–207). For instance, verbs can be passivized as in (4), verb phrases can be converted into nominalizations as in (5), or into complex adjectival (often hyphenated) adnominal modifiers, see Examples (8)–(9). Other principles of idiom variation are also at play, allowing conceptual elaboration or specification through lexical substitution or pre- and postmodification: verbs can be substituted as in Example (3) (see also Table 2 as to how the verb slot in some expressions can be filled), and simple adnominal modifiers can be freely added as in (4), (6), and (7).

(3) Perhaps it's time for a literateuse to **hobble up to the plate**. (*New York Times*, August 3, 2008)

(4) Round a small table with Brown, I saw him consider the warnings of Treasury officials and of Tony Blair who argued child poverty was proving so costly to tackle that **the statistical goalposts ought to be shifted**. (*The Guardian*, April 19, 2008)

(5) Now we can return to the wider debate about executive pay, which has little to do with the housebuilder's **outrageous movement of the goalposts** during its financial year. (*The Guardian*, January 17, 2009)

(6) Hunt-Davis, a tall, courtly grey-haired gent, **played the deadest of straight bats** to the questioning of how much he knew, though periodically, under cross examination by Michael Mansfield QC, Al Fayed's counsel, he looked as though he was thinking longingly of the Gurkhas' lethal kukri knife. (*The Guardian*, December 14, 2007)

(7) It's the most polarising hip-hop album of the year. Kanye West's sparse, blunt and shockingly earnest newly re-leased album, Heartbreak, is one of this year's **musical curveballs**. (*Sydney Morning Herald*, December 5, 2008)

(8) Germans tend to be the strait-laced, **play-it-safe types** in financial matters. (*New York Times*, October 4, 2008)

(9) The nominations, considered important Oscar tea leaves despite their occasional **out-of-left-field nature**, were fairly predictable in the acting categories and spread the love around, with no one film dominating as has happened in the past. (*New York Times*, December 12, 2008)

Both constructional adaptation and, more importantly, conceptual elaboration and specification are only possible

> if the speaker is able to establish a systematic, directly meaningful and functional relationship between the elaborated literal scene and the target conceptualisation. Moreover, this manipulation of the literal scene must be fully compatible with the idiomatic meaning. (Langlotz 2006: 256)

The second observation is the cross-functionality of the expressions under study, referring to the fact that they are also used in various other types of sports without causing ambiguity (see Examples (10)–(13)), which is one of five major constraints on idiom variation (Langlotz 2006: 222). I take this as further evidence that they have a more general, not-sport specific conceptual motivation.

(10) American Michael Phelps **got the ball rolling**, smashing Ian Thorpe's 200 m freestyle record to claim gold. (*Sydney Morning Herald*, March 28, 2007)

(11) Yet, it wasn't all Williams's work. Safina **scored some own goals**, starting with three double faults in her first service game. (*Sun Herald*, February 1, 2009)

(12) In order to advance from a heat, many surfers take what waves they can catch and **play it safe** for sure points. (*New York Times*, November 27, 2007)

(13) The sport's leaders will be hoping that Jamaica's Asafa Powell or America's Tyson Gay can **step up to the plate** in the 100m to help bring these championships alive. (*The Observer*, August 26, 2007)

6. Conclusion and outlook

The present chapter has studied patterns of use and variation in selected sports-related idiomatic expressions based on conceptual metaphors. On the basis of a cognitive linguistic-model of idiom representation I have argued that these expressions are conceptually motivated. I have sketched the underlying conceptual metaphors for some of them, and outlined both their lexico-grammatical variability in terms of constructional adaptation, as well as their conceptual elaboration and specification through lexical substitution and pre- or postmodification.

Assuming that SPORTS is available as a generic-level source domain in metaphorical mappings across varieties of English, I have applied the types and aspects of variation that have been shown to be involved in cross-cultural variation of metaphors to different varieties of only one language. When SPORTS is instantiated as a specific-level, culturally-embedded metaphorical mapping, there are indeed differences in the prominence of specific source domains across the five varieties under study. These are reflected in the frequency of use of sports-related idiomatic expressions (cf. the frequent use of baseball-related expressions in American English and cricket expressions in British and Australian English).

On account of the limited amount of usage data examined here it is difficult to answer the question if there is evidence for a convergence or even erosion of differences in metaphor usage across World Englishes as a result of ongoing economic and cultural globalization. It seems obvious that cross-cultural contact on a larger scale is a prerequisite to enable the use of a variety-specific expression in another variety. But exposure alone is not sufficient for productive use, and possibly functions as a catalyst once the contact becomes more intense. For these expressions to catch on, however, speakers must be in a position to recognize their underlying conceptual motivation by relating their words to a certain domain. It appears that some sports-related expressions allow such a motivation more readily, while others remain opaque to a majority of speakers who are unfamiliar with the specific type of sport.

It is clear that the findings presented here are subject to the methodological limitations mentioned earlier. Thus, in the following I would like to point out some issues that need to be tackled in future research. First, the hypothesis advanced earlier in the chapter, namely that the potential for creative use and variability (including formal reduction) of an idiomatic expression may correlate with the salience and entrenchment of its underlying conceptual metaphor in a certain language or variety of language needs to be tested empirically on a wider scale. Second, there is some preliminary evidence that there are differences between the varieties as to the discourse domains in which sports-related idiomatic expressions are preferably used. In the present data, V *a curve* (*ball*) is used very often in the context of arts (film, music, literature) and fashion in British English as shown in Examples (14)–(16). If there are indeed variety-specific preferences needs to be studied more closely.

(14) As the Oscars roll towards us, the Berlin film festival has **thrown the world a curveball** by picking as best film the most thoroughly anti-Hollywood offering you can imagine. (*The Guardian*, February 20, 2009)

(15) On this, their second album – which translates as With You We Are Four – they puncture their smooth mix of R&B and exotic techno with **sonic curveballs** and lyrical wit. (*The Guardian*, August 24, 2007)

(16) Kate Atkinson **threw her readers a curveball** a few books back. (*The Guardian*, August 16, 2008)

Finally, similar to the corpus-based approach adopted by Charteris-Black (2004), it is not only the use of idiomatic expression but also more general lexical items used in metaphorical expressions that need to studied (exemplified in (17)–(19)).

(17) The Reserve Bank has been an active **player** in the market in an attempt to reduce the volatility of the exchange rate, and Ray Block, economist with brokers Dominguez Barry Samuel Montag. (*Australian Corpus of English*)

(18) Among the reasons for public dissatisfaction with the American legal system was contentious procedure, which turned litigation from a search "for truth and justice" into a **game** or sport "that the parties should fight out in their own way without interference". (*Freiburg-Brown Corpus of American English*)

(19) By the end of his 21-year tenure it was no secret that Shaw's orchestra was well into the **major leagues**, and by common consent his Atlanta Symphony Orchestra Chorus was second to none. (*Freiburg-Brown Corpus of American English*)

These can be conceived of as forming a 'sports lexicon' (e.g. verbs like *play*, *score*, *beat*, *win*, and *lose*; nouns like *player*, *winner*, *loser*, *game*, *match*, *league*, *ball*, *field*). Such a study would be based ideally on large electronic corpora that allow more refined quantification and statistical comparison. This would also make possible a closer investigation of the assumed special role of sports metaphors in American English. If the concepts of competition and sports really are so salient and pervasive in US-American society, then this should have two effects for the use of SPORTS as a source domain in metaphorical conceptualization (reflected in the occurrence of sports-based metaphorical expressions):

- In comparison to other varieties of English, American English is characterized by a wider range of sports metaphors because the popularity and omnipresence of sports in general boosts the use of different types of sports as source domains.
- American English is also characterized by a wider scope of metaphor because sports-based source domains apply to a wider range of target domains.

References

Abadi, Adina and Yaakova Sacerdoti. 2001. Source domains of metaphors in political discourse. A cross-cultural study: Israel and the U.S.A. *RASK: Internationalt Tidsskrift for Sprog og Kommunikation* 15: 27–51.

Ammer, Christine. 1997. *The American Heritage Dictionary of Idioms*. Boston, MA: Houghton Mifflin.

Archer, Maureen and Ronnie Cohen. 1997. Leaving women out in left field: Sports metaphors, women, and legal discourse. *Women and Language* 20: 2–6.

Bairner, Alan. 2001. *Sport, Nationalism, and Globalization: European and North American Perspectives*. Albany, NY: State University of New York Press.

Blain, Neil, Raymond Boyle and Hugh O'Donnell. 1993. *Sport and National Identity in the European Media*. Leicester: Leicester University Press.

Boers, Frank. 2003. Applied linguistics perspectives on cross-cultural variation in conceptual metaphor. *Metaphor and Symbol* 18: 231–238.

Campbell, Anne. 2002. *A Mind of Her Own: The Evolutionary Psychology of Women*. Oxford: Oxford University Press.

Campbell, Duncan. 2009. Stumped by curveballs. Baseball metaphors are meaningless to most of us. It's time we hit them into the long grass. *The Guardian*, January 19, 2009.

Charteris-Black, Jonathan. 2004. *Corpus Approaches to Critical Metaphor Analysis*. Basingstoke: Palgrave Macmillan.

Ching, Marvin K. L. 1993. Games and play: Pervasive metaphors in American life. *Metaphor and Symbolic Activity* 8: 43–65.

Crystal, David. 2003. *English as a Global Language*. 2nd edition. Cambridge: Cambridge University Press.

Dale, David. 2004. Strike me lucky, it just isn't cricket. *The Sydney Morning Herald*, February 10, 2004.

Dirven, René. 1994. *Metaphor and Nation: Metaphors Afrikaners Live By*. Frankfurt/Main: Lang.

Geeraerts, Dirk and Hubert Cuyckens. 2007. Introducing Cognitive Linguistics. In *The Oxford Handbook of Cognitive Linguistics*, Dirk Geeraerts and Hubert Cuyckens (eds.), 3–21. Oxford: Oxford University Press.

Geeraerts, Dirk, Gitte Kristiansen and Yves Peirsman (eds.). 2010. *Advances in Cognitive Sociolinguistics*. Berlin: de Gruyter Mouton

Gruneau, Richard and David Whitson. 1993. *Hockey Night in Canada: Sport Identities and Cultural Politics*. Toronto: Garamond.

Hardaway, Francine. 1976. Foul play: Sports metaphors as public doublespeak, *College English* 38: 78–82.

Howe, Nicholas. 1988. Metaphor in contemporary American political discourse, *Metaphor and Symbolic Activity* 3: 87–104.

Kövecses, Zoltan. 2000a. *Metaphor and Emotion: Language, Culture, and Body in Human Feelings*. Cambridge: Cambridge University Press.

Kövecses, Zoltan. 2000b. *American English. An Introduction*. Peterborough, Ont.: Broadview.

Kövecses, Zoltan. 2005. *Metaphor in Culture. Universality and Variation*. Cambridge: Cambridge University Press.

Kristiansen, Gitte and René Dirven (eds.). 2008. *Cognitive Sociolinguistics. Language Variation, Cultural Models, Social Systems*. Berlin: Mouton de Gruyter.

Lakoff, George and Mark Johnson. 2003. *Metaphors We Live By* (2nd ed.). Chicago: The University of Chicago Press.

Langlotz, Andreas. 2006. *Idiomatic Creativity. A Cognitive-Linguistic Model of Idiom-Representation and Idiom-Variation in English*. Amsterdam: Benjamins.

Lauricella, John A. 1999. *Home Games: Essays on Baseball Fiction*. Jefferson, NC: McFarland & Company.

Left, adj.[1], *adv.*, and *n. The Oxford English Dictionary*. 2nd ed. 1989; online version November 2010. <http://www.oed.com:80/Entry/106982>; accessed 11 March 2011.

Liu, Dilin. 2002. *Metaphor, Culture and World View: The Case of American English and the Chinese Language*. Lanham, MD: University Press of America.

Liu, Dilin and Bryan Farha. 1996. Three strikes and you're out. A study of the use of football and baseball jargon in present-day American English. *English Today* 12: 36–40.

Majumdar, Boria and Sean Brown. 2007. Why baseball, why cricket? Differing nationalisms, differing challenges. *International Journal of the History of Sport* 24: 139–156.

McGimpsey, David. 2000. *Imagining Baseball: America's Pastime and Popular Culture*. Bloomington, IN: Indiana University Press.

Moon, Rosamund. 1998. *Fixed Expressions and Idioms in English. A Corpus-Based Approach*. Oxford: Clarendon Press.

Niederle, Muriel and Lise Vesterlund. 2007. Do women shy away from competition? Do men compete too much? *Quarterly Journal of Economics* 122: 1067–1101.

Sharifian, Farzad, Ning Yu, René Dirven and Susanne Niemeier (eds.). 2008. *Culture, Body, and Language: Conceptualizations of Internal Body Organs across Cultures and Languages*. Berlin: Mouton de Gruyter.

Shore, Bradd. 1996. *Culture in Mind: Cognition, Culture, and the Problem of Meaning*. New York: Oxford University Press.

Stewart, Heather. 2007. It's not cricket, but economics ought to be a game of two sexes. *The Observer*, August 19, 2007.

Tharoor, Shashi. 2007. Our cricket problem. *New York Times*, March 23, 2007.

Tottie, Gunnel. 2002. *An Introduction to American English*. Malden, MA: Blackwell.

Wolf, Hans-Georg and Frank Polzenhagen. 2007a. Fixed expressions as manifestations of cultural conceptualizations: Examples from African varieties of English. In *Phraseology and Culture in English*, Paul Skandera (ed.), 399–435. Berlin: Mouton de Gruyter.

Wolf, Hans-Georg and Frank Polzenhagen. 2007b. Culture-specific conceptualisations of corruption in African English: Linguistic analyses and pragmatic applications. In *Applied Cultural Linguistics*, Farzad Sharifian and Gary B. Palmer (eds.), 125–168. Amsterdam: Benjamins.

Wolf, Hans-Georg and Frank Polzenhagen (eds.). 2009. *World Englishes. A Cognitive Sociolinguistic Approach*. Berlin: Mouton de Gruyter.

Yu, Ning. 1998. *The Contemporary Theory of Metaphor. A Perspective from Chinese*. Amsterdam: Benjamins.

How novels feel

Emotional and rational reading processes in contemporary fiction

Wolfram R. Keller, Astrid Lohöfer, and Christine Ott
Humboldt-Universität zu Berlin / Philipps-Universität, Marburg / Philipps-Universität, Marburg and Harvard University

This chapter investigates the depiction of cognitive processes within two "autofictional" novels, Serge Doubrovsky's *Fils* and David Williams's *Eye of the Father*. Both works feature literary scholars as the main protagonists, who recognize their own emotional predicaments through their emotional and rational interpretations of literature. These texts speak in interesting ways to – and challenge – recent work in cognitive poetics, raising the question whether rational or emotional reading processes trigger epiphanies. Both novels suggest that the two different literary responses are interdependent, thus rendering suspect the possibility of purely rational, objective approaches to literature. Reading the novels at the nexus of emotion, "autofiction", and interpretation, we argue that literary scholarship needs to account for emotional-subjective and rational-objective approaches.

Keywords: autofiction, cognitive poetics, David Williams, literary depictions of cognitive processes, relationship of objective and subjective approaches to literature, Serge Doubrovsky

1. Introduction

In recent years, the increased attention to cognition in the humanities has also impacted literary studies, prompting a "cognitive turn", a veritable "cognitive revolution" (Richardson and Steen 2002: 1–8). One important point of intersection between cognition and literature is the concept of emotion, since literary communication is not only concerned with the transmission of rational meaning, but also affects the senses (Stockwell 2002: 171; for definitions of *emotion*, see further Izard 1977; Ulrich and Mayring 2003: esp. 51–57). Recent empirical studies in psychological aesthetics have widely explored emotional reading responses and their connection to cognition and rational interpretation. Most of these studies, however, remain restricted to the analysis of data from non-professional readers, which might be explained by the fact that among professional readers – that is, academics – emotional reading is still frowned upon. Thus, Peter Stockwell (2002: 172) remarks that in the past, the "emotional 'content' of a literary work has almost been regarded as a coincidental side-effect […]. The text is seen as responsible for emotional cues, but the actual emotions that are evoked have been regarded as within the bounds only of readerly whim".

In this essay, we examine two contemporary 'autofictions', written by authors who are also literary scholars: Serge Doubrovsky's *Fils* (1977) and David Williams's *Eye of the Father* (1985). Both novels, we argue, challenge the equation of professional reading with rational reading by portraying a scholar's emotional engagement with a literary work and the epiphanies it triggers regarding both textual understanding and real life experience. Concerning reading processes, these texts raise the following questions: Are purely rational approaches to literature possible and, above all, are they preferable to emotional approaches? Is the distinction between 'natural', emotional reading and professional, rational interpretation sustainable, or in itself fictitious? Reading texts at the nexus of emotion, autofiction, and interpretation, we wonder if literary scholarship as a discipline can continue to sidestep emotional-subjective approaches as has been largely the case throughout the last decades.

2. Emotional and rational reading

Reflections about the relation between emotion and rationality have a long tradition. Already in the eighth century BCE, Indian philosophy addresses this issue by conceiving of human life as universal suffering caused by avidity and lust, which needs to be suppressed by rigorously training the opposing domain of the mental, and in ancient Greece, Epicureans and Stoics stress the importance of *ataraxia*, the absence of emotional disturbances of the mind (Radhakrishnan 1953: 807–808; Edwards 1972: vol. 1, 171–173). The pre-Socratics, by contrast, consider emotions as the driving forces of all being, and thus as a basis and support of rational thinking; and Aristotle divides emotions into rationally controlled "excellences" and unintentional, irrational "affections" (Aristotle 2002: 115–116; Curd 2008: 444). The question of whether emotions and rationality are separate or mutually dependent phenomena still remains controversial in current psychological research. While earlier, so-called "affective" approaches primarily defined emotions as mere physiological reactions to pleasure or displeasure, recent studies are rather interested in the cognitive processes involved in the production of emotions (Ulrich and Mayring 2003: 10). That some kind of relation exists between emotional and cognitive processes has lately been substantiated by neuropsychological data, which point to the fact that both processes activate similar cerebral structures (Ulrich and Mayring 2003: 43). Since the popularization of the term "emotional intelligence" (Goleman 1995), the increasing conceptual conflation of emotionality and rationality also becomes apparent in recent philosophical studies (e.g. Nussbaum 2001).

In the following, we leave aside the Hindu, Buddhist, or Stoic ideas of the mind as an instrument for influencing or suppressing emotions, and focus instead on the potential impact of emotions on rational thinking. Numerous studies in the realms of psychology and empirical social research provide demonstrable evidence for the fact that emotions often act as aids to cognition. Antonio Damasio, for example, substantiates that subjects who have a reduced ability to experience emotions due to injury or illness are also hindered in making practical decisions (Damasio 1994; and see further Matthews and Wells 1994; Evans 2004). Other studies show, by contrast, that our rational assessment of past and future is systematically distorted by emotions. Being a frequent source of self-deception, emotions can also inhibit cognitive processes such as evaluation, judgment, or interpretation (Fingarette 1969; Mele 1987). This has, *inter alia*, increasingly fostered the necessity (in the realm of empirical social studies) to more rigorously reflect – and document – the actual research process (Steinke 2003; Behse-Bartels and Brand 2009).

The interpretation of literary texts represents a particularly complex relationship between emotion and rationality. One aspect of this relation is the Platonic idea that literature, as well as all other art, can excite emotions in a harmful way, even to the point of losing self-control:

> It is likewise true in the case of erotic desire, and anger, and all the cravings, pains and pleasures in the soul that we agree attach themselves to every action of ours, that such is the effect which poetic representation works on us. For it waters and nourishes these feelings, when they ought to be dried up, and it puts them in control of us, when *they* are the things that ought to be controlled if we are to become better and happier people, not worse and more miserable. (Plato 1988: 69)

This is a prevalent issue also in literary works, such as *Madame Bovary* or *Don Quixote*. On another view – thus, for example, Hegel – the understanding of literature requires a particularly high degree of rationality in comparison to other art forms. As opposed to visual arts and music, "Poetry is the *universal art of the mind* which has become free in its own nature, and which is not tied to find its realization in external sensuous matter" (Hegel 1993: 96, our italics). This, too, leads to negative judgments about reading processes based on identification and feeling. According to Vladimir Nabokov, for instance, Madame Bovary's fault lies in the fact "that she is a bad reader. She reads books emotionally, in a shallow juvenile manner, putting herself in this or that female character's place" (Nabokov 1980: 136f.). Madame Bovary's emotional engagement with the texts does not only hinder her rational understanding of what she is reading, but also has negative consequences for her life.

Similar opinions continue to be widespread in the field of what Eric Livingston calls "reading *cultura*", that is, the cultivated practices of professional reading (2006: 662). Academic readers and literary critics generally try to keep their personal experiences and emotions apart from the objects of their professional work, as Peter Stockwell points out:

> They ["natural" readers] read for the pleasure of it, and because they feel better for it, for engagement or escape, to illuminate and enhance the day. Almost all of these things are inadmissible in professional discourse, or are articulated at a meta-theoretical level that places them at several removes from the natural experience. (Stockwell 2005: 144)

German literary criticism, in particular, is skeptical about emotional reading – paradoxically, given that it is located in the tradition of Wilhelm Dilthey and Emil Staiger, whose interpretational approaches were based on the experience of "being moved" (Staiger 1957: 13; on German critics' stance on emotional reading, see further Anz and Huber 2007: 282–285). Nevertheless, Thomas Anz and Evelyne Keitel have recently criticized the prevalent "hostility to psychology" of literary criticism in Germany and its "speechlessness [...] vis-à-vis emotions involved in the reading process" (Anz 1998: 13; Keitel 1996: 23, our translations). In contrast to this, the American academic tradition appears to leave more room for emotions, subjectivity, and creativity, which is already reflected in the respective terminology: 'literary criticism' vs. 'Literaturwissenschaft'. The American scholar Mark William Roche claims, for instance, that "[a] good question to ask scholars and teachers of literature is which literary works grabbed them; with which works they had a moving reading experience; or which works changed their perception of the world. If the question draws a blank, one can be suspicious [...]" (Roche 2004: 85). E. M. Dadlez similarly argues that strongly emotional and identificational readings do not necessarily exclude a rational access to a literary work: "The paradox may lie not in our fearing and pitying fictions, but in the unquestioning acceptance of an alliance between the rational and the unimaginative" (Dadlez 1996: 302). And Donald Wesling argues for the acceptance of the fact that even academic writing is always partially based on emotions: "All our writing is spun out of our guts, whatever kind of writer we are" (Wesling

2007: 363; see further Kimmich and Schahadat 2009; Adorno 1998). While it would be interesting to contextualize such statements in statistical data, the rapidly growing body of empirical work on the interaction between emotions and literary production/reception is mainly concerned with non-professional readers (see, e.g. the studies on psychological aesthetics, such as Kneepkens and Zwaan 1994: 125–138; Cupchik, Leonard, Axelrad, and Kalin 1998: 825–847). As David S. Miall (2006a: 292) states, the readers studied in most empirical research are (first year) students, or drawn from the general population.

An early and influential example of an approach based on data from 'ordinary' readers is Keith Oatley's taxonomy of what he calls "emotions of literary response". Oately distinguishes two basic categories: "text-external", cognition-based emotions are stimulated by the reader's confrontation with the text's surface structure, for example, irritation or curiosity about the development of the plot. Text-internal emotions, on the contrary, arise when the reader enters the text-world, identifies with the characters, feels sympathy or antipathy for them, or compares their situation to his or her own (Oatley 1994: 57). Having identified these categories, Oatley associates them with particular literary genres and periods, which prototypically evoke one type of emotional response. According to him, a modernist poem, for instance, is mainly based on the text-external emotion of defamiliarization, while a Victorian novel rather aims at text-internal identification.

In the following, we focus on a contemporary genre that draws exactly on this contrast between text-external and text-internal emotions, and which the French author and literary critic Serge Doubrovsky calls "autofiction" (Doubrovsky 2001: 10). While traditional autobiography is normally reserved for famous people, Doubrovsky argues, nowadays everyone is allowed to write down his life – even a professor of literature like himself (2001: 10). An autofictional text relies primarily on the spontaneous and direct communication with a broad audience rather than on refined literary talent or technique. Its author, who is both narrator and principal character of the text, seems to be primarily interested in giving a true and immediate account of verifiably biographical facts, thus aiming to first and foremost evoke text-internal, identificational emotions as a literary response.

At the same time, however, the writer of autofiction tries to escape the narcissistic temptation of traditional autobiography by repeatedly interrupting the relentless confrontation with his own life by critical reflections on the writing process itself, consistently foregrounding the work's fictionality and self-reflexivity. Doubrovsky's *Le Livre Brisé* (1989), for instance, depicts the author's relationship with his young wife Ilse, who suffers from heavy depressions and alcoholism. Ilse's commentaries on their lives and his writing are included in the book, but before the book is finished, Ilse commits suicide and the facts coincide with the moment of their reporting. Autofiction thus approaches in a new and unaccustomed way what constitutes a longstanding issue in literature and literary theory: the dualism of real-life experience and aesthetic distance; or, to speak with John Ireland, "[a]utofiction effects the subtle but crucial metamorphosis by which existential dissatisfaction is converted into textual plenitude" (Ireland 1993:4). It is the dualism of life and literature, reality and reflexivity, identification and distance, emotion and rationality, artist and scholar that forms the core of the two autofictional novels analyzed in the following.

3. Doubrovsky's *Fils*

Serge Doubrovsky's *Fils* – *fils* means 'son' as well as 'threads' – relates a day in the life of the literary scholar Serge Doubrovsky, from his course preparations in the morning via his weekly session with his psychoanalyst up to the evening lecture on the French classical writer Racine. At the core of the novel's plot lies the topic of reading and interpretation. At first, Doubrovsky is presented as a professional reader who puts important critical works on Racine into his briefcase while pondering their main arguments. He then comes across a note about a dream he had the night before: he is at a hotel with a woman, when he suddenly sees a monster creeping out of the sea, wants to shoot it, but cannot do so. He realizes that in this dream, he has mixed up his own memories with the final scene of Racine's drama *Phèdre*, which he is about to discuss in his evening lecture. The scene in question depicts the tragic death of the young hero Hippolyte, who flees into exile, after being accused of incestuous desire for his mother-in-law Phèdre and, consequently, cursed by his father. The curse eventually reaches him, when a sea monster creeps out of the sea and frightens his horses, which subsequently drag him to death.

In the afternoon, Doubrovsky attempts to interpret his dream with the help of his psychoanalyst. He becomes aware that the woman in his dream must be Elisabeth, with whom he had a love affair years ago. At the same time, the woman represents his mother, who died three years ago. The monster, which displays, for the protagonist, androgynous traits both in Racine's text and in his dream, seems to represent the protagonist's repressed bisexuality. The analyst interprets the dream sequence in a fairly traditional way: the presence of the mother represents the protagonist's unresolved Oedipus complex, his incapacity to shoot the monster insinuate his inability to effect closure, which finally would mean to "kill" the mother, while the monster embodies his sexual ambivalence and refusal to become a responsible adult by making his own decisions. After this psychoanalytic session, Doubrovsky is both relieved and frustrated: as much as he feels unable to add anything to the countless interpretations of Racine, he has the impression that his psychoanalysis does not really lead to anything.

His evening lecture starts under unfavorable circumstances. Professor Doubrovsky not only feels unprepared, he is also irritated by the absence of a student with whom he has a love affair and who is an important source of inspiration for his performance. Suddenly unable to read his own handwriting, he begins to improvise. His interpretation is now increasingly affected by the strategies that 'helped' him to partially decipher his dream and the situation that prompted it: he explains the vision of Racine's monster as a traumatic, Freudian ur-scene, including a punishing father and a castrating mother. Through the monster, Doubrovsky argues, Hippolyte is confronted with his own tragic weakness – the weakness of a pseudo-hero always living in the shadow of a dominating father and unable to emancipate himself. By allowing his personal dispositions to penetrate his critical discourse, the protagonist finally succeeds in overcoming traditional interpretations of Racine. At the same time, he transfers his dream-inspired reinterpretation of Hippolyte's tragic insight to his own life: he recognizes that, despite all attempts of his psychoanalyst to cure him, he will remain forever caught in the monstrous ambivalence of his indecisive personality.

In *Fils*, Serge Doubrovsky develops a new, emotional and identificational method of literary interpretation. Indubitably, his protagonist knows the traditional research and explicitly includes it in his interpretation; the intuition of a new way of reading, however, emerges at the moment when the professor relinquishes his control over this rational discourse. We suggest that the novelty of Doubrovsky's depiction of a new way of reading in the novel and his theory lies in the following: first, it prompts a particular cognitive and emotional identification in the reader; second, it advocates the contingency of literary interpretation as well as it offers, third, a new way of construing literary texts, which ultimately consists of actively 'feeding' the text with one's personal experiences.

As concerns the first mentioned novelty – that is, cognitive and emotional identification – the plot of *Fils*, as depicted above, seems to be simple and straightforward. In fact, however, it contains long passages of fragmentary stream-of-consciousness, veiling, at first, who is talking to whom. The splitting of personality, as the core subject of the novel, corresponds to a strong fragmentation of the text, especially in the beginning. According to Hélène Jaccomard, this fragmentation provokes an identificational and "irrational" reading, since the best way to approach the text is to engage with its particular rhythm instead of letting oneself be irritated by comprehension difficulties (Jaccomard 1993: 268, 293). Irritation indeed seems unavoidable at a first reading of *Fils*, since readers are confronted with cognitive gaps that, at best, they can fill imaginatively. The novel, however, goes beyond the evocation of "text-external" emotions, such as irritation – and probably also curiosity – by providing, little by little, the information necessary for the reader to integrate and interpret the text fragments. What is interesting in this process is that, from the reader's perspective, the impression arises that she or he is brought to the same state of knowledge as the protagonist. In the penultimate chapter, it thus emerges that the initial sentence fragments of the novel form the beginning of a book that 'Serge Doubrovsky' is writing at this very moment, and about whose exact meaning he is as puzzled as the reader of *Fils* (2001: 435).

The reader actually becomes a co-interpreter: in the course of the third chapter, in which the protagonist tries to analyze the dream-scene described at the beginning of the novel, both protagonist and reader ultimately 'uncover' its hidden meaning. In the final chapter, then, the reader is able to interpret *Phèdre* together – but also in competition – with the protagonist, since the reader is now equipped with the background information necessary to identify the analogies between Doubrovsky's dream and Racine's tragedy. Like the protagonist, the reader moves from incomprehension to understanding; the reader could even be described as the protagonist's *alter ego* in the ongoing process of cognition and interpretation. *Fils* thus initiates two parallel, but opposite developments: on the emotional level, the text-external responses evoked by the beginning of *Fils* are slowly replaced by text-internal, identificational ones; regarding cognitive processes, however, readers are increasingly able to comprehend – and therefore also to distance themselves from – the protagonist's various attempts at interpreting both text and self.

As concerns the contingency of interpretation, Doubrovsky demonstrates with the final lecture scene that every literary interpretation is influenced by subjective and contingent factors. Certainly, the professor emphasizes that one has to take into account what is actually *in* the text (2001: 471). Nevertheless, his interpretation of Racine is inevitably affected by his psychological disposition, as well as by his completely contingent irritation caused by the fact that his favorite student is absent. Interpretation thus amounts to 'feeding' the text: for the contemporary literary scholar, the method of teaching and interpretation presented in *Fils* may be irritating. Its innovative potential becomes only apparent when taking into account the methodological presumptions that are questioned, primarily those assumptions associated with structuralist criticism.

In his theoretical work *Pourquoi la nouvelle critique* (1966), Doubrovsky already argues against structuralism. Although he defends the (post-)structuralist methods of critics such as Roland Barthes or Charles Mauron against the attacks by proponents of traditional, positivist-biographical approaches, Doubrovsky also distances himself, at least partly, from the methods of the "nouvelle critique". According to him, the old – and, in some points, also the new – modes of literary criticism consider the literary work as something objectively given, and therefore analyzable by purely objective methods, and (in Barthes's case) as an autonomous linguistic entity that can be interpreted, without reference to its existential substratum (Doubrovsky 1966: 90). By contrast, Doubrovsky argues that a literary work is indeed composed of objective parts, but that its totality can only be grasped from a necessarily individual perspective: "Les éléments de composition objectifs (car ils sont bien dans l'objet perçu), ne se livrent qu'à une visée subjective (car toute perception suppose une conscience percevante" [The elements of a composition are objective (since they definitely exist in the perceived object), but they only unfold through a subjective view (since every perception presupposes a perceiving consciousness)] (Doubrovsky 1966: 71, our translation). In order to emphasize that the literary critic cannot avoid the subjective point of view, Doubrovsky cites Sartre to the effect that Blanchot's reading of Mallarmé displays more of Blanchot than of the analyzed poet (1966: 74f.). This already points to Doubrovsky's concept of "autocritique", later practiced in his autofictional novels, which render synonymous the interpretation of the literary text and the interpretation of one's self.

Eleven years later, Doubrovsky takes up his critical position in the final chapter of *Fils*. During his lecture, Professor Doubrovsky repeatedly claims that the methods influenced by structuralism reduce literature to general – that is, mythical, psychoanalytic, or other – structures, reducing the text and depriving it of its individuality. Even though it is not explicitly stated, it emerges that such interpretations ultimately fail in the same way that traditional psychoanalysis fails when it reduces individuals to generalizing schemes. Both structuralism and psychoanalysis have become monotonous insofar as they are unable to say anything new, either about texts or human subjects (Doubrovsky 1966: 116). The strategy to counteract such a simplification, implicitly indicated by Doubrovsky, consists not only of unveiling the text's archetypal skeleton, but fleshing it out, as it were, with the interpreter's subjectivity. In fact, this method of interpretation corresponds to Doubrovsky's conception of literature in general: "La littérature, c'est aussi une œuvre de chair, ça s'écrit avec sa vie, avec son sang, avec son existence – en ce sens-là je suis sartrien jusqu'à la moelle –, c'est le dévoilement d'un être-au-monde, pas uniquement un être de langage" [Literature is also a work of flesh, written with one's life, with one's blood, with one's existence – in this sense, I am Sartrian to the bones –, it is the unveiling of a being-in-the-world, not only a being-within-language] (1992: 133–134, our translation). Consequently, the gap between literature and life constantly referred to in *Fils* can only be bridged when the text is revived by a reading, which, without becoming arbitrary, is based on subjective and emotional involvement.

4. Williams's *Eye of the Father*

David Williams's *Eye of the Father* (1985) is, at first glance, a very different novel. It is the final part of the Canadian literary critic and novelist's Lacjardin trilogy, and is divided into three parts, each of which is introduced by a Nordic rune; the titles (Sigurd, Loki, Baldr) further recall central characters of the *Edda* (Latta 1987). Nordic mythology not only structures *Eye of the Father*, but an interpretation of *Baldr's Dreams* (for which, see Kuester 1999) is the crucial interpretive episode of the novel. *Eye of the Father* centrally engages with how individuals cope with family history. The two chapters that make up the first part relate the stories of Magnus and Hilda. The former flees from his parents (and a pregnant girlfriend) in Norway in order to try his luck in the United States, where he is both abused and abuses people in turn. He eventually settles in Canada, where he terrorizes chiefly his wife Hilda, who intimates the extent of her suffering in the second chapter. The two chapters of the second part are narrated by Magnus's daughter Christine and by Magnus's grandson, Wayne, both of whom attempt to come to terms with their father/grandfather. The last chapter, constituting the third part of the novel, is narrated by Wayne's wife Karen.

As emerges from the foregoing synopsis, *Eye of the Father* is a multi-voiced, somewhat fragmented novel, which, as is the case with Doubrovsky's *Fils*, is frequently 'irritating' for readers; it forces the audience to fill many a cognitive gap – a strategy represented *in extremis* in the previous novel of the trilogy, *The River Horsemen* (1981). Unlike *Fils*, *Eye of the Father* is not an autobiographical or autofictional novel per se. And yet, Williams has emphasized that his novels are partially autobiographical, the penultimate chapter of the novel being a case in point (2000: 58; see also Kuester 1999: 27). Wayne Goodman is not only a college professor of Nordic mythology but also a novelist – with the name possibly being an allusion to Laura Goodman Salverson (Kuester 1999: 33) – who is trying to negotiate his relationship with his grandfather not by communicating with the latter, but by writing a novel about him – that is, "the novel the reader is reading" (Williams 1991: 270). Not least by means of the inclusion of other voices in the novel, Williams evades the narcissistic tendencies of traditional autobiography; Wayne's chapter frequently foregrounds authorial self-reflexivity and fictionality. On our view, the chapter, then, qualifies as autofiction. And like Doubrovsky's novel, the penultimate chapter of *Eye of the Father* involves the reader emotionally and 'identificationally', it advances a contingent mode of literary interpretation, and it specifies that construing literary texts is ultimately a matter of 'feeding' the text with one's own experiences.

The key problem of the various family members, articulating their various thoughts, is how to come to terms with their family history, or, to use Wayne's wife's metaphor, they have to flesh out the "skeletons in the family closet" (Williams 1985: 146). At the same time, Karen and Wayne would like to adopt a child into their Ukrainian-Norwegian-Canadian family. Both have doubts about the latter in the light of the former: are they going to be good parents, given that their own parents were far from ideal? While Karen is willing to confront and, therefore, deal with her overbearing mother, Wayne is doing his best to evade a personal confrontation with his grandfather, who is about to die. He rather wants to write a novel about his grandfather, taking on the problem at one remove – and he is outright angry about Karen's intended visit of her mother:

> "Do you think it's a good idea, though", he said when he had exited safely onto Fourth Avenue, "to mix up a child in our unsolved problems?"
> She looked at him intently.
> "What I mean is, it's hard to consider being father until I do something about my grandfather". "Go face him, then".
> "Is facing your mother going to solve anything this weekend?" (1985: 149)

Karen's mind is set on going through with her plans, and at the ticket counter of the bus depot, she again tries to convince Wayne of seeing his dying relative:

> She turned, head half-bowed, touching him with pity…but the clench of her jaw was her damn mother all over again.
> "Maybe I should ask you first", her eyes didn't waver, "before I pay for a one-way ticket. Will you go to see your grandfather or not?"
> He set her suitcases down smartly and walked off. (1985: 151)

As he leaves the bus depot alone, his thoughts turn to the beginning of their relationship, of his father discovering Karen's underwear in his car, whereupon Wayne decides that it might be good if they did not see each other for a while – only to catch Karen in the arms of another man shortly after, their making up and resolve to get married. These thoughts carry over into his classroom:

> "At first glance", he tried to remember something beyond the rushing bus which carried his wife away, "*Baldrs draumar* looks like a tightly knit poem. But because of one apparent flaw…". Hadn't she ever stopped to consider how dangerous it was for him to look his demon in the eye? "…most scholars think *Baldr's Dreams* is a late imitation of *The Lay of Thrym*". (1985: 155)

Wayne introduces the scene that he wishes to focus on: Baldr intends to learn the meaning of his terrible dreams, which prompts his father Odin to consult the Völva only to learn that his son will perish. Wayne asks his students to identify the "flaw in that little scheme", and Sheila – a dazzling looking female student – singles out Odin's "bad manners" in raising up the sibyl from the dead, "and all he can give her for thanks is scorn". Wayne wonders if the poem accounts for

Odin's bad behavior; the student does not think so, calling Odin a "jerk". Wayne laughs along with the other students, but quickly directs the class away from Sheila's intuitive reading, connecting the insults and riddles in the *Eddic* song to Norwegian mumming practices. Murray, a student Wayne "could always count on" explicates the connection: disguise here functions as a means of articulating one's frustrations, "breaking the shell of convention". Sheila interrupts angrily: "I don't think a god has to take his frustrations out on a woman. I mean, he's the authority. She's the one who could better stand to have things reversed for a while" (1985: 156).

Wayne and Murray try to explain the gender bias of the episode rationally, and point out that Odin could not disclose his true identity, save for a riddle, which also has represented a riddle for Baldr scholars (Kuester 1999: 30). Sheila is increasingly angry, and she rejects such a reading over and over again. Without intending to be "partial", Wayne ultimately sides with Murray. Odin's riddle is not only a way of revealing his true identity, but is implicitly a question "in which the god is looking for maidens who will perform an ecstatic funeral dance for Baldr" (Kuester 1999: 30). Wayne explains:

> The poet has to invent some way for Othin to reveal himself in this unusual chronology. And he has to do it within the usual brief space, so he can only do it by implication. But in effect he asks, "Which girls will be ready to be buried with Baldr?"
> "Isn't that just what I've been saying all along", Sheila sat up indignantly. "The gods of the Vikings are only looking for scapegoats".
> But even as she stared at him, he saw her resentment turn to concern.
> "Professor Goodman? Are you alright?" (1985: 157–58)

Wayne, though, is having a vision of his mother's tears at the occasion of his wedding, which recall – more troublingly – his mother's tears at the funeral for his brother; and Wayne has never forgiven his mother, who has forgiven her father (that is, Wayne's grandfather), for these tears. The sacrificial analogies to the episode just discussed, in combination with Sheila's observation that the gods were just "looking for scapegoats", makes Wayne realize that he, too, had been looking for scapegoats. Coming to himself, "Shame made him dizzy, as if the floor were opening all around him. He could only hope that the old boy was still kicking, that he might get there in time" (1985: 159).

Two emotional readings of the same text converge in the classroom, prompting Wayne not only to face his grandfather (and thus realize his own misbehavior towards Karen), but also – in the light of references to *ragnarök* (see Kuester 1999) that frame the discussed episode – to come to terms with the planned adoption. Neither Wayne nor Sheila make apparent (nor reflect consciously) the emotional processes that lead to their interpretations – at least in Wayne's case, the reader of the novel is keenly aware of the recurrent engagement with his or his near relatives' thoughts about suicide, death, and sacrifice, which have to be seen in context with Wayne's interpretation, as is evident also in the prominent structural position of the recollection of his brother's funeral. The novel thus does make explicit the emotional involvement resulting in an innovative new reading of the difficult riddle in *Baldr's Dreams*. As in Doubrovsky's *Fils*, then, Williams's *Eye of the Father* at once involves the reader emotionally, while portraying two seemingly contingent processes of interpretation of one text, both based on personal emotions, which result, in Wayne's case, in an epiphany that feeds back into Wayne's emotional dispositions.

5. Conclusions

Although there are clearly differences between Doubrovsky's *Fils* and Williams's *Eye of the Father*, we hope to nonetheless have demonstrated several parallels, especially concerning the depictions of interpretative processes. Both texts can be subsumed under Doubrovsky's term "autofiction", insofar as they try to overcome the limits between rationality and emotion, fact and fiction, objective and subjective truth, distancing and identification. This transgression of limits is not only central to the 'reading scenes' analyzed above, but lies at the very core of both novels: Doubrovsky and Williams are literary scholars who cross the line of academic authorship by writing a fictional text about literary scholars who cross the line of traditional academic interpretation. In both texts, it is the adoption of a spontaneous, emotional, 'non-academic' response to classical literature and medieval mythology that transcends limits that generally seem insurmountable to the professional reader – the demarcation separating life and literature.

Challenging the Platonic and Hegelian positions mentioned at the outset of this chapter, which consider emotional reading as an obstacle to either self-understanding or textual interpretation, *Fils* and *Eye of the Father* posit that emotions can prompt readers, to borrow David Miall's phrase, to "go beyond the schema given" (Miall 2006b: 53), an effect that merely rational approaches to literature, if they are possible at all, cannot have – or, at best, to a limited extent. Only by admitting that emotions and identification influence the reading process are the protagonists ultimately able to break with deadlocked structures in their analysis of both prominent texts and personal experiences. Thus, emotions act as catalysts for the process of the mutual illumination of life and art, a function that Oatley equally considers crucial: "One aspect of profound emotions in everyday life [...] is that such emotions can provoke reflections on ourselves, and can change us.

[...] Emotions are potentially transformative for us as readers when we too can respond creatively" (Oatley 1994: 60–61).

In the 1930s, Robin Collingwood employed the catalyst-effect of expressing and producing emotions in the service of a systematic determination of a work's literariness (Collingwood 1938: 114). Today, there appears to be an unbridgeable gap between subjective, emotional creativity and objective, rational 'Literatur*wissenschaft*'. An increasing number of critical schools, such as biographical, historical, sociological, or structuralist criticism, have tried to produce interpretations that are objectively verifiable and, therefore, 'scientific'. From a present point of view, then, structuralist theories and methods must appear somewhat out-dated. We wonder, however, if contemporary cognitive criticism does not likewise fall prey to the problematic presuppositions of a structuralist poetics. Its data-based, statistical methods may become more and more precise; in order to arrive at a meaningful result, however, these data still need to be interpreted. And this act of interpretation is again – inevitably – affected by subjective factors, unless it were left entirely to a machine. The more or less implicit reproach proponents of 'traditional' literary scholarship suffer at the hands of cognitivist critics is that they are only interested in individual views of individual authors, which obliquely expresses that traditional critics indulge in the fantasy of a removed, self-contemplative conversation with the author. This view is mistaken. Every serious, ambitious literature has aimed at transcending the description of an individual state of mind towards general human problems. Hence, every ambitious literary scholarship should aim at elaborating, though by means of various subjective approaches, the elements of this general reflection: the personal identification with the text and its effects enables critics to trace the universal aspects constituting the intersubjective matter of their objects of study. The assumption that subjective, emotional approaches to literature lead to arbitrary readings, it has to be noted in passing, appears all the more baseless since empirical research itself has shown that the reader's emotions have a positive influence on text comprehension and storage (see Miall 1986: 389–397; Miall 2006b: 45, 70, 110).

Since the increasing popularity of texts that one can classify as autofictional indicates that authors – and readers – consider individual-emotional and intersubjective-rational perspectives on literature to be less exclusive than mutually dependent, both traditional literary scholars and cognitive critics should begin to challenge the dichotomy they have established between them. From a methodological point of view, this suggests that cognitivist critics who work empirically should perhaps take into greater account the questions and challenges posed by individual texts themselves instead of exclusively using literature to illustrate and confirm axioms 'established' by cognitive theories. Williams's and Doubrovsky's novels are good examples of literary texts that anticipate issues of cognition that have been explored by empirical research only later, or, as for the differences between professional and non-professional reading, not yet at all (see, e.g. Herman 2007:245–257; Zunshine 2003:270–291). At the same time, traditional literary critics need to relinquish their resistance against the methods used in the cognitive sciences, which, instead of replacing established insights, are primarily supposed to substantiate them (see Hamilton and Schneider 2002:640–658; but cf. Toolan and Weber 2005:107–115). A bi-directional convergence of cognitive studies and literary criticism could thus be a fruitful contribution to existing attempts of either approach to, at least partially, go beyond their respective limits.

References

Adorno, Theodor W. 1998 [1955]. Über das Studium der Philosophie. In *Gesammelte Schriften*, Vol. 20, Rolf Tiedemann et al. (eds.), 318–326. Darmstadt: Wissenschaftliche Buchgesellschaft.

Anz, Thomas. 1998. *Literatur und Lust: Glück und Unglück beim Lesen*. Munich: Beck.

Anz, Thomas and Martin Huber. 2007. Einleitung. *Mitteilungen des Deutschen Germanistenverbandes* 54 (3): 282–285.

Aristotle. 2002. *Nicomachean Ethics*. Sarah Broady and Christopher Rowe (eds. and trans.). Oxford: Oxford University Press.

Behse-Bartels, Grit and Heike Brand (eds.). 2009. *Subjektivität in der qualitativen Forschung: Der Forschungsprozess als Reflexionsgegenstand*. Opladen: Budrich.

Collingwood, Robin G. 1938. *The Principles of Art*. Oxford: Oxford University Press.

Cupchik, Gerald C., Garry Leonard, Elise Axelrad and Judith D. Kalin. 1998. The landscape of emotion in literary encounters. *Cognition and Emotion* 12 (6): 825–847.

Curd, Patricia (ed.). 2008. *The Oxford Handbook of Presocratic Philosophy*. Oxford: Oxford University Press.

Dadlez, Eva M. 1996. Fiction, emotion, and rationality. *British Journal of Aesthetics* 36 (3): 290–304.

Damasio, Antonio. 1994. *Descartes' Error: Emotion, Reason, and the Human Brain*. New York: Putham's Sons.

Doubrovsky, Serge. 1966. *Pourquoi la nouvelle critique: Critique et objectivité*. Paris: Mercure de France.

Doubrovsky, Serge. 1989. *Le livre brisé*. Paris: Grasset.

Doubrovsky, Serge. 1992. Introduction à la lecture. In *Autobiographie und Avantgarde*, Alfred Hornung and Ernstpeter Ruhe (eds.), 133–139. Tübingen: Narr.

Doubrovsky, Serge. 2001 [1977]. *Fils*. Paris: Gallimard.

Edwards, Paul (ed.). 1972. *The Encyclopedia of Philosophy*, 4 Vols. (repr. ed.). New York: Macmillan.

Evans, Dylan. 2004. The search hypothesis of emotions. In *Evolution, Rationality and Emotion*, Dylan Evans and Pierre Cruse (eds.), 179–191. Oxford: Oxford University Press.

Fingarette, Herbert. 1969. *Self-Deception*. London: Routledge.

Goleman, David. 1995. *Emotional Intelligence: Why It Can Matter More Than IQ*. New York: Bantam.

Hamilton, Craig A. and Ralf Schneider. 2002. From Iser to Turner and beyond: Reception theory meets cognitive criticism. *Style* 36: 640–658.

Hegel, Georg Wilhelm Friedrich. 1993. *Introductory Lectures on Aesthetics*. Michael Inwood (ed.) and Bernard Bosanquet (trans.). Harmondsworth: Penguin.

Herman, David. 2007. Cognition, emotion, and consciousness. In *The Cambridge Companion to Narrative*, David Herman (ed.), 245–257. Cambridge: Cambridge University Press.

Ireland, John. 1993. Introduction: Monstrous writing. *Genre* 26 (1): 1–11.

Izard, Carroll E. 1977. *Human Emotions*. New York: Plenum Press.

Jaccomard, Hélène. 1993. *Lecteur et lecture dans l'autobiographie française contemporaine: Violette Leduc, Françoise d'Eaubonne, Serge Doubrovsky, Marguerite Yourcenar*. Genève: Droz.

Keitel, Evelyne. 1996. *Von den Gefühlen beim Lesen*. Munich: Fink.

Kimmich, Dorothee and Schamma Schahadat. 2009. Vorwort: Positionen der Emotionsforschung. *Arcadia* 44 (1): 3–7.

Kneepkens, E. W. E. M. and Rolf A. Zwaan. 1994. Emotions and literary text comprehension. *Poetics* 23: 125–138.

Kuester, Martin. 1999. Norwegians in the New World: David Williams's syncretistic construction of Norsk-Amerikaner mythology. In *Essays on David Williams: A Novelist from the Canadian Prairie*, B. N. Singh (ed.), 26–37. Bareilly: Prakash.

Latta, William. 1987. Rune-writer. Rev. of *Eye of the Father* by David Williams. *Canadian Literature* 115: 191–193.

Livingston, Eric. 2006. The textuality of pleasure. *New Literary History* 37: 655–672.

Matthews, Gerald and Adrian Wells. 1994. *Attention and Emotion: A Clinical Perspective*. Hillsdale, NJ: Lawrence Erlbaum.

Mele, Alfred R. 1987. *Irrationality: An Essay on Akrasia, Self-Deception, and Self-Control*. New York: Oxford University Press.

Miall, David S. 1986. Emotion and the self: The context of remembering. *British Journal of Psychology* 77: 389–397.

Miall, David S. 2006a. Empirical approaches to studying literary readers: The state of the discipline. *Book History* 9: 291–311.

Miall, David S. 2006b. *Literary Reading: Empirical and Theoretical Studies*. New York: Lang.

Nabokov, Vladimir. 1980. *Lectures on Literature*. Fredson Bowers (ed.). London: Weidenfeld & Nicolson.

Nussbaum, Martha. 2001. *Upheavals of Thought: The Intelligence of Emotions*. Cambridge: Cambridge University Press.

Oatley, Keith. 1994. A taxonomy of the emotions of literary response and a theory of identification in fictional narrative. *Poetics* 23: 53–74.

Plato. 1988. *Republic*. S. Halliwell (ed.). Warminster: Aris & Phillips.

Radhakrishnan, Sarvepalli (ed.). 1953. *The Principal Upanishads*. London: Allen & Unwin.

Richardson, Alan and Francis F. Steen. 2002. Literature and the cognitive revolution: Introduction. *Poetics Today* 23 (1): 1–8.

Roche, Mark William. 2004. *Why Literature Matters in the 21st Century*. New Haven, CT: Yale University Press.

Staiger, Emil. 1957. *Die Kunst der Interpretation: Studien zur deutschen Literaturgeschichte* (2nd ed.). Zurich: Atlantis.

Steinke, Ines. 2003. Gütekriterien qualitativer Forschung. In *Qualitative Forschung: Ein Handbuch*, Uwe Flick, Ernst von Kardorff and Ines Steinke (eds.), 319–331. Reinbek: Rowohlt.
Stockwell, Peter. 2002. *Cognitive Poetics: An Introduction*. London: Routledge.
Stockwell, Peter. 2005. Texture and identification. *European Journal of English Studies* 9 (2): 143–153.
Toolan, Michael and Jean Jacques Weber. 2005. Introduction. *European Journal of English Studies* 9 (2): 107–115.
Ulrich, Dieter and Philipp Mayring. 2003. *Psychologie der Emotionen* (2nd ed.). Stuttgart: Kohlhammer.
Wesling, Donald. 2007. Scholarly writing and emotional nowledge. *Papers on Language and Literature* 43: 363–389.
Williams, David. 1981. *The River Horsemen*. Toronto: Anansi.
Williams, David. 1985. *Eye of the Father.* Toronto: Anansi.
Williams, David. 1991. *Confessional Fictions: A Portrait of the Artist in the Canadian Novel.* Toronto: University of Toronto Press.
Williams, David. 2000. Canadian writers and Canadian literature: An interview with David Williams. Interview with Kong Yin. *Prairie Fire* 21 (1): 48–59.
Zunshine, Lisa. 2003. Theory of mind and experimental representations of fictional consciousness. *Narrative* 11 (3): 270–291.

Cognitive poetics and the negotiation of knowledge

Stephan Freissmann
Justus-Liebig Universität, Giessen

Cognitive poetics is a movement in literary theory that researches the cognitive processes involved in understanding literary texts. Mostly, this research results in models of how processes of comprehension work in the minds of human beings. This chapter investigates the relationship of text understanding with the formation of knowledge on the basis of narrative fiction. The central question addressed is which cognitive processes of text understanding are involved in and contribute to the formation of knowledge about an extra-literary field on the basis of narrative fiction. Two tutor texts that enable an especially fruitful discussion of this question in the exemplary field of knowledge of cognition are David Lodge's novel *Thinks…* (2001) and Richard Powers's novel *Galatea 2.2* (1995).

Keywords: David Lodge, influence of literature on cognitive processes, literary and extra-literary contexts, narrative fiction, Richard Powers, text understanding

1. Introduction

The formation and the use of knowledge are almost omnipresent processes that every human being permanently performs in everyday life. For instance, knowledge is involved in navigating the space of experience, in social interaction, or in fulfilling everyday bodily tasks. In how far literary fiction can contribute to the formation of knowledge is, however, a relatively recent research interest in literary studies. In the wake of the New Criticism, fictional literature has for decades been regarded as a collection of autonomous works of art, the connection of which with the real world is, at best, weak. This doctrine reacted to rampant biographism and intentionalism in literary studies. Nevertheless, the discipline of literary studies is in the process of redefining the relationship between fictional literature and the real world surrounding it by focusing on its potential to generate knowledge that is generally valid, also outside the story world. By way of an example, suffice it to mention the edited volume by Gymnich, Neumann, and Nünning (2006). In this volume, the literary work of art is conceived of as being in a permanently ongoing dialogic relationship with the cultural knowledge of its time of origin. That is, the literary work of art is engaged with the real world in which it is embedded, while nevertheless being determined by artistic rules that go beyond a mere mirroring of the real world. A wealth of recent scholarly writing testifies to this research interest (see Wood 2005; Hörisch 2007; Köppe 2008; Klausnitzer 2008; Ette 2004; 2005, 2010 as well as two exchanges in scholarly journals triggered by Ette 2007a, 2007b; and Köppe 2007). A prominent example for the institutionalization of this research interest is the PhD-Net "The Knowledge of Literature", set up by Berlin's Humboldt University in cooperation with Harvard and Princeton Universities as well as the University of California at Berkeley (see http://www2.hu-berlin.de/wissen-literatur/index.php?setlang=en).

The formation of knowledge is, obviously, also a process that depends on cognition. Without advanced abilities in cognition, no human being could form any knowledge, let alone perform the complex interweaving of knowledge into experience and perception that is a property of everyday human life. Consequently, researching the formation of knowledge through fictional literature necessitates a clear account of the cognitive processes involved. An approach focusing on the cognitive processes that enable the understanding of literary texts is cognitive poetics. This chapter intends to sketch first steps towards an integration of cognitive poetics with the mechanisms of knowledge formation at work in literary texts. Consequently, the central question of this chapter is which cognitive processes of text understanding are involved in and contribute to the formation of knowledge about an extra-literary field on the basis of narrative fiction.

This essay provides a twofold contribution to the debate about bi-directionality in the cognitive sciences. For one thing, the cognitive processes of knowledge formation with respect to the knowledge that can be formed on the basis of narrative fiction are scrutinized with the intention to contribute to a notion of knowledge that goes beyond the concentration on so-called hard or scientifically founded facts. Instead, the notion of knowledge advocated here focuses on the hermeneutic side of knowledge formation, which essentially depends on readers' reactions to textual cues. For another thing, this essay intends to contribute to an appreciation of fictional literature's power as an epistemic force for the generation of knowledge about cognition. The decisive point here is to research routes for forming knowledge of cognition other than the standard scientific, i.e. experimental-empirical, method and bring out the contribution to knowledge about cognition that can be made by fictions of cognition.

In order to reach this goal, this essay proceeds in three steps. After a brief sketch of two cognitive elements of text understanding that are important for the present argument, the third part of this essay is devoted to a discussion of the potential of knowledge formation inherent in literary texts. The fourth and most extensive part of this essay consists of a case study that illustrates the different ways in which literary texts can generate knowledge. Two fictions of cognition are discussed, namely David Lodge's *Thinks…* (2001) and Richard Powers's *Galatea 2.2* (1995). The case study concentrates on the knowledge of cognition that can be generated on the basis of fictional texts, as an example of the general potential of knowledge formation inherent in fictional literature.

2. Two elements of text understanding – cognitive schemata and metaphor

The central question of research in text understanding, since its inception by Iser's pioneering study *The Act of Reading* (1978 [1976]), is how the reader forms a notion of the meaning of a text. Cognitive approaches to text understanding conceive of this operation as the formation of a mental model of the story world (cf. Hallet 2008a, 2008b). In his classic study, Johnson-Laird (1983: 4) characterizes mental models as a kind of "small-scale model […] of reality" that has "the same 'relation-structure' as the phenomenon that it represents" (1983: 11). Mere functional equivalence to the represented phenomenon, thus, is not sufficient for a successful mental model. On the contrary, the decisive point concerning mental models is that their "structures […] are identical to the structures of the states of affairs […] that the models represent" (1983: 419). In the case of text understanding, the formation of a mental model is prompted by the textual data, i.e. by the words on the page.

The decisive point in mental model formation on the basis of a written text is how the textual data contribute to conjuring up an imaginative story world. In other words, a theory is needed that conceptualizes the interaction of textual data with what readers bring with them to understand a text. This relationship is regarded as an interaction between textual data and readers' world knowledge, an interaction that can be grasped by resorting to schema theory (see Stockwell 2002: 75–89).

The central tenet of schema theory is that human beings' generic world knowledge is stored in the mind in the form of cognitive schemata. That is, a cognitive schema does not contain the specifics of a certain place or situation in their temporal-spatial context. Rather, it contains the general traits of places or situations, such as the fact that rooms have walls, ceilings, floors, and at least one door. Due to their generic nature, cognitive schemata can be applied to many different situations. Thus, in a certain sense, they make cognition possible, as is reflected in Rumelhart's (1980) assessment that schemata are the "building blocks of cognition". To transfer the notion of a cognitive schema to the understanding of a literary text, the relevant cognitive schema provides the default values for the generation of the mental model of the story world unless it is otherwise specified, that is, unless the literary text says something else. Cognitive schemata have so-called slots that can be filled in with different values. To stick to the example, neither a room with nor one without windows would surprise the experiencing human being; both variants are possible. In the understanding of literary texts, this means that the mental model of the story world is primarily made up of the default values inherent in the relevant cognitive schemata. In cases in which the story world deviates from such default values, the slots of the cognitive schemata have to be filled in with explicitly given textual data that indicate the deviation from the default values (for a concise account of cognitive schemata and for the example used here, see Brewer 1999).

Another decisive trait of cognitive schemata is that they can be dynamic or static, that is, they can refer to processes or situations. This distinction is related to a difference in terminology. Whereas Minsky (1975), who revived the concept of a schema originally developed by Bartlett (1932) for contemporary cognitive science, called it a *frame* – referring to static situations – Schank and Abelson in their classic study *Scripts, Plans, Goals, and Understanding* (1977) use the term *script*, which refers to generic action sequences such as their famous restaurant script, concentrating on dynamic processes. However, both frames and scripts share essential features with the more general term *schema*, which is why this latter term is used in this chapter.

Another, often underestimated, element that contributes to the formation of a mental model on the basis of a literary text is metaphor. This assessment rests on the insight, formulated by Lakoff and Turner (1989: xi), that, "[f]ar from being merely a matter of words, metaphor is a matter of thought", or, as Lakoff and Johnson put it in their pioneering study *Metaphors We Live By* (1980: 5), that "*[t]he essence of metaphor is understanding and experiencing one kind of thing in terms of another*" (emphasis orig.). Thus, Lakoff, Turner, and Johnson, in their theory of conceptual metaphor (CMT), posit a mental principle that is at work in the background of linguistic metaphors. As their research has shown, metaphor is a far more pervasive element of the human conceptual system than would have been predicted by theories that regard metaphor as a merely literary affair. What is most important is that metaphors are at work not only in a conscious way as in their artistic application, for instance in poetry. Rather, according to CMT, metaphor is a principle of mental organization that works for the most part unconsciously and is quasi ubiquitous.

The unconscious nature of most metaphoric thinking entails, conversely, a mostly tacit and difficult to detect influence of linguistic metaphors on the formation of a mental model of the story world, by presenting something in terms of something else or, put differently, by making graspable something in terms of something else. This tacit influence enables metaphors to add what one could call a spin to what is depicted in the text. Nünning (2008) has demonstrated this effect for metaphors of empire and has detailed its importance for the formation of the imperial mindset. For instance, as Nünning points out, the metaphorical presentation of the British Empire as a family had a subtle influence on the way in which empire was regarded by British citizens, influencing individuals' knowledge about the Empire and, consequently, also the formation of an imperial mindset.

In the process of text understanding, the specific textual data activate generic cognitive schemata to form a specific mental model of the text sense, with a tacit influence exerted by the metaphorical presentation of textual data, a tacit influence that should nevertheless not be underestimated. However, if the mental model of the text sense is specific, how is it possible for a text to participate in the generation of knowledge? After all, knowledge is usually associated with being generally valid.

3. Knowledge formation on the basis of fictional literature

The considerations above suggest that the metaphorical presentation of something in terms of something else does not only add a certain spin to the mental model of the story world formed by the reader, but that it also exerts a subtle influence on the knowledge that readers form on the basis of a certain text. In order to account for the apparent conflict between a specific mental model of the text sense and the necessity of general validity in knowledge, it is necessary to augment the notion of knowledge with some elements that take account of the specificity of literary texts. The notion of knowledge is usually associated with a certain type of knowledge, i.e. knowing-that, a type of knowledge that philosophers call propositional knowledge. Propositional knowledge is the privileged object of enquiry in the philosophical discipline of epistemology (see, for instance, Rescher 2003: xiv), which is variously justified by its being more fundamental than other types of knowledge (see Feldman 2003: 9) or by its being an exclusive property of human beings (see Pritchard 2006: 4). However, starting out from propositional knowledge, the argument of this section is that there is more to human knowledge than knowing-that and, consequently, that one must also take into account other types of knowledge to grasp the full potential of knowledge formation inherent in fictional literature. These other types of knowledge, proposed in the following, are tacit knowledge and narrative knowledge.[1]

Knowing-that means knowing something for a fact. The fact that is known is a proposition that is formulated in "a sentence which says that something is the case" (Pritchard 2006: 4; cf. Shope 1992: 396). The sentence, thus, is a verbal container for the proposition. Such propositional knowledge can be integrated into a work of fiction in quite a concrete sense, namely in the form of intertextual relations. It is obvious that a maximally concrete notion of intertextuality, such as the one proposed by Genette in his *Palimpsests* (1997 [1982]: 1–2), is necessary to determine the presence of items of propositional knowledge in a work of fictional literature. Whether the content of a proposition in a work of fiction is representative of an actual item of extra-literary propositional knowledge can only be determined by an intertextual comparison between the work of literature in which it is integrated and the source discourse from which an item of knowledge has been imported. In this way, one can exclude the possibility that a work of fiction presents merely fictional propositional knowledge that holds true only in the story world but not in the real world.

Many philosophers, however, concede that propositional knowledge is not the only type of knowledge and contrast it with knowing-how, or procedural knowledge (see e.g. Blaauw and Pritchard 2005: 95; Feldman 2003: 11). The distinction between propositional and procedural knowledge goes back to Gilbert Ryle's seminal study *The Concept of Mind* (1949: 25–61). Ryle points out that the intelligent execution of an act displays a kind of knowledge that cannot be reduced to propositional knowledge. Rather, in intelligent actions and behavior a different type of knowledge can be observed: "'Intelligent' cannot be defined in terms of 'intellectual' or 'knowing how' in terms of 'knowing that'; 'thinking what I am doing' does not connote 'both thinking what to do and doing it'" (1949: 32). Instead, doing something intelligently exhibits its own genuine type of knowledge, i.e. procedural knowledge.

Procedural knowledge, the knowledge inherent in doing things, is, however, only one part of the "unformulated" (Ryle 1949: 30) type of human knowledge. Michael Polanyi's concept of tacit knowledge encompasses all knowledge that depends on the tacit power to integrate particulars into a meaningful whole (1966: 6; 1969: 138–140). Polanyi derives his definition of tacit knowing from this phenomenon: "This act of integration, which we can identify both in the visual perception of objects and in the discovery of scientific theories is the tacit power we have been looking for. I shall call it *tacit knowing*" (1969: 140, emphasis orig.).

1. For a fuller treatment of the following ideas, see Freissmann (forthcoming).

Tacit knowledge, in other words, is the feeling that the combination of particulars makes sense and produces meaning. Moreover, as Polanyi maintains, "all knowledge is *either tacit* or *rooted in tacit knowledge*. A *wholly* explicit knowledge is unthinkable" (1969: 144, emphasis in original).[2] That is, according to his theory, tacit knowledge does not usually come to consciousness and is difficult to verbalize. However, it is quasi omnipresent and of utmost importance for the formation of knowledge in everyday life.

Both the notions of procedural and tacit knowledge are included in the implicit type of knowledge that is predominantly characterized by the difficulty to formulate it linguistically. The potential of fictional literature to be the basis for forming implicit knowledge is illustrated in the case study in the next section of this chapter.

First, however, the spectrum of types of knowledge that are active in the formation of knowledge on the basis of narrative literature should be augmented by a third type of knowledge, namely narrative knowledge. Narrative knowledge intends to cover the genuinely narrative aspects of knowledge formed on the basis of fictional literature. According to Pablo Valdivia (2007), who originated the concept, the decisive characteristic of narrative knowledge is its position astride propositional and procedural knowledge. Narrative knowledge cannot be reduced to either of the other types of knowledge (2007: 118). Instead, it unfolds in the process of telling a story, thus being akin to procedural knowledge. It depends on the formal arrangement of a narrative. It is also, however, akin to propositional knowledge in that it depends on the facts deployed in the course of the narrative. The decisive surplus value of narrative knowledge over the other two types of knowledge is its aptitude for creating relations (2007: 117), relations that are usually not explicitly spelt out but that remain implicit in the process of telling a story.

The close linking of narrative knowledge to a cultural technique – the cultural technique of narrative – suggests a close connection between the properties of narrative knowledge and the properties of the cultural technique narrative. The latter are better known as factors that determine the degree of narrativity inherent in a given text. As can be deduced from the impressive body of research that is concerned with the issue of narrativity (see, e.g. Prince 1982; Fludernik 1996; Herman 2002; Echterhoff/Straub 2003; Hutto 2007; Herman 2007, 2009), the factors determining narrativity are the following: narrative can function as a force for conceptualizing intentional action and for understanding causal relations between events that usually grow out of intentional action. By concentrating on those events that are out of the ordinary, narrative can serve to isolate important and

2. Polanyi's term *explicit knowledge* is, in the context of this chapter, synonymous with the term *propositional knowledge*.

interesting events from the stream of everyday routine. Through its emplotment, narrative produces temporal-causal relations between the represented events and leaves the ground of mere factuality. In doing so, narrative has the potential to convey the qualia of experience and tends to concentrate on particular processes instead of general truths. Finally, narrative is the cultural technique most appropriate for conceiving of social matters. The more these factors are present in a text the more likely readers will mentally process this text as a narrative.[3] Moreover, the factors of narrativity can, arguably, also be regarded as influencing narrative knowledge. Thus, narrative knowledge should be seen as pertaining to intentionality, to causal-temporal connections between events, to out-of-the-ordinary events, to experientiality, to particulars, and to social matters. All these different potentials of knowledge formation on the basis of literature can be observed in the two novels that are analyzed in the following case study.

4. Two fictions of cognition

The two novels that the case study focuses on, David Lodge's *Thinks…* (2001) and Richard Powers's *Galatea 2.2* (1995), are characterized by remarkably similar plot structures.[4] In both novels, a novelist meets a cognitive scientist, and both are soon involved in a competition of ideas about cognition. Naturally, while the scientists advocate the scientific approach to researching cognition, the novelists hold more old-fashioned humanistic values. Roughly summarizing the two intellectual positions represented by the pairs of protagonists in one word, they can be labeled the empiricist and the phenomenological positions. This character constellation brings out the mechanisms that enable the formation of knowledge on the basis of narrative literature, by showing the exemplary aspects inherent in the specific story world. The pairing of opposing protagonists makes clearly visible the competition among different kinds of knowledge and their negotiation by literary means. After short plot summaries and characterizations of the pairs of protagonists, it is first of all, established that in both novels the two protagonists represent two opposing ways of thinking about consciousness and cognition. In a second step, the discussion turns towards the opposing kinds of knowledge of cognition that are connected to the opposing protagonists. And finally, readers'

3. This list of characteristics and functions of narrative as a cultural technique emphasizes the cohesive function of narrative. This is not to say that this cohesive function of narrative must necessarily dominate. One can argue that, especially in literary contexts, the more interesting phenomena are found on the margins of the narrative genre, not in its prototypical instances.

4. For quotations, the two novels will be abbreviated as *TH* or *G*, respectively.

cognitive operations in building a mental model of the story world that negotiates the two competing kinds of knowledge of cognition are considered.

In Powers's *Galatea 2.2* the novelist Richard Powers returns to his former university as a visitor to the local Center for the Study of Advanced Sciences (*G* 4) where he meets cognitive scientist Philip Lentz.[5] Lentz soon engages Richard in a dubious AI experiment, aiming at the construction of a computer that would simulate the interpretation of a work of English literature sufficiently well to compete against a human being interpreting the same work – an experiment that, in the end of the novel, proves to have been a hoax. In the process, Lentz and Richard nevertheless succeed in designing an intelligent machine that exhibits astonishing cognitive abilities. Instead of responding properly to the chosen literary passage, however, the computer, intermittently named Helen, seems to have felt that it is not human and turns itself off for good, effectively committing suicide. This AI plot takes place against the background of Richard's accounting for the failed relationship with his girlfriend C. and for his previous literary career.

Accounting for previous actions is an equally strong motive in Lodge's *Thinks…*, namely on the part of the novelist-protagonist Helen Reed. Quite like Richard, she comes to the then fictional university of Gloucester on a year's appointment, substituting the resident creative writing teacher. Once there, she stumbles upon cognitive scientist Ralph Messenger with whom she soon engages in a serious argument about the nature of the mind and the soul. Having resisted his romantic advances for a long time, she finally gives in and begins a passionate affair with Messenger after she has discovered her late husband's infidelities.

How, then, can one conclude that the two pairs of protagonists represent opposing ways of thinking about cognition that go beyond the story world represented in the particular pieces of narrative fiction? The impression of the protagonists' being representative for larger intellectual movements stems from their stereotypical characterizations. This does not mean, however, that the protagonists are necessarily wooden or flat characters. Nevertheless, they hold opinions and beliefs that readers conceive of as 'typical' for novelists or cognitive scientists. A case in point is Ralph Messenger's unflinching belief in the all-encompassing range of rational scientific enquiry. While he is sure that science will one day elucidate the workings of the human mind without an unexplainable residue, novelist Helen Reed constantly argues against this omnipotence of the scientific method, asking whether there were no "areas of human experience where scientific method doesn't apply" (*TH* 229). A similar competition of methods can be made out in *Galatea 2.2*. The rather confrontational scientist Lentz challenges novelist

5. To distinguish between the real-world author Richard Powers and the fictional character of the same name, in what follows I refer to the former as Powers and to the latter as Richard.

Richard into the pseudo-experiment by debasing the knowledge cultivated by literary studies: "What passes for knowledge in your so-called discipline?" (*G* 43)

In the face of such stereotypical scientific self-confidence, the novelist-protagonists of both novels soon find themselves in the defensive position. The assessment is, thus, not unreasonable that they represent the set of humanistic values that has come under critique since the emergence of empirical science. Thus, both Helen Reed in *Thinks...* and Richard in *Galatea 2.2* are engaged in more or less painful processes of accounting for their lives in a moral and an emotional sense, whereas their scientific counterparts do not seem to be weighed down by moral or emotional concerns.[6] This element of characterizing the novelists and scientists, respectively, fits the differing temporal orientations that are ascribed to both groups. Whereas C.P. Snow, in his famous *Two Cultures* lecture (1998 [1959]: 10), designated scientists as having the future in their bones – an assessment reiterated in *Thinks...* with respect to Messenger (*TH* 208) – humanists are, conversely, characterized as hearkening back to literary and philosophical, or even religious traditions. This opposition is mirrored in the different relationships with their past lives that the scientist- and the novelist-protagonists exhibit in the two novels. Thus, there are well-defined intertextual relationships that connect the protagonists' characterizations to larger social discourses and stereotypical images of 'the' scientist or 'the' novelist.

Connected to these pairs of opposing characters and characterizations are opposing ways of thinking about consciousness and cognition. Consequently, the reader is confronted with two opposing types of knowledge of cognition. On the one hand, there is the scientific knowledge of cognition that is intended to be intersubjectively testable and valid as well as independent from the specific subjectivity of a certain individual who possesses this kind of knowledge. Scientific knowledge of cognition is often said to regard cognition from the perspective of the third person (see Edelman 1992: 114–115; Dennett 1991: 70–72). On the other hand, there is the phenomenal knowledge of cognition that emphatically depends on the specific subjectivity of an experiencing individual. In this case, one often talks about regarding cognition from the first-person perspective.

These different kinds of knowledge of cognition are produced in both novels through employing the different types of knowledge formation on the basis of narrative fiction charted above. Especially in their conversations, the two pairs of protagonists refer to propositional knowledge, i.e. to statements concerning facts about cognitive science. A case in point is the great amount of propositional

6. This assessment fits Ralph Messenger better than Philip Lentz; however, in his intention to lobotomize Helen the AI, Lentz displays a sort of ruthlessness that does not pay attention to novelist Richard's emotions about the intelligent machine.

knowledge about the design of artificial neural networks that is included in *Galatea 2.2*. For instance, the reader learns that artificial neural networks are manifestations of the connectionist approach in cognitive science, which originates in the assumption that cognition emerges from the complex interlinking of simple computational units, the neurons (*G* 14–15; see Varela/Thompson/Rosch 1999: 85–103). At the same time, the reader learns that most artificial neural networks are only simulated on computers instead of being implemented in actual hardware (*G* 71).

A similar point can be made for the conceptual background of cognitive science that is treated in *Thinks....* Ralph Messenger does not only give Helen Reed – and with her, the reader – a rundown of the basics of cognitive science (*TH* 35–39). They also discuss conceptual fundamentals, such as thought experiments that put into perspective scientific claims to the complete understanding of an individual's cognition (e.g. *TH* 49–56). Thus, they discuss the question whether it is possible to know what it is like to be a bat, a question posed in Nagel's (1974) famous thought experiment, or the question whether a person who knows absolutely everything about color, yet has spent her entire life in a black and white environment, will recognize a color upon seeing it (Jackson 1982, 1986). These and other thought experiments concern the philosophical foundations of cognitive science, with fundamental questions such as whether it is possible to understand someone's personal phenomenology from the third-person standpoint of cognitive science without an unexplainable residue. Thus, they are items of propositional knowledge that can, by intertextual comparison, be verified as being valid not only in the inter-discourse of the literary work, but also in the real-world discourse of cognitive science.

Being largely independent from the verbal medium, isolating implicit knowledge in literary texts is not an easy task. In *Galatea 2.2*, implicit knowledge results from the contrast between the principles of AI design and the determiners of human behavior. Lentz and Richard construct the intelligent machine in accordance with explicit design principles. These design principles give rise to emergent properties, such as the ability to ask questions which Lentz and Richard did not programme into the machine (*G* 217). Nevertheless, the way in which the computer is constructed is determined by rules that are made explicit. The human beings' behavior, however, follows completely implicit and, most of all, predominantly emotional motives. Thus, both protagonists' drive to create an artificial being results from their loneliness. What this plot construction conveys as implicit knowledge is that one factor favoring the failure of the intelligent machine is its lack of emotions, or, rather, its skewed emotions. Under bomb threat, the machine's reaction is a mere "Helen could die? [...] Extraordinary" (*G* 272). Only towards the end of the design process, when Richard feeds Helen the AI the accumulated news of the past decades, does Helen react in a way that human beings might interpret as depressive – she falls silent (*G* 313–314). It is not explicitly mentioned that the imbalance in emotions – which are, after all, an important element of an organism's self-regulation (see Damasio 1999: 53–56) – is one of Helen's problems. This knowledge is present in the narrative discourse in an implicit way and emerges in the process of reading the novel.

A similar point is discernible in *Thinks…*, this time on a gendered note. Whereas the male scientist Messenger knows the third-person perspective on cognition very well and is familiar with the scientific methods for scrutinizing cognitive processes, the intuitive first-person access to others' personal phenomenologies seems to be a privilege of the female characters, once again connecting the protagonists Messenger and Helen in a stereotypical way to the spheres of science or literature, respectively. Both novelist Helen Reed and Messenger's wife Carrie grasp things intuitively that evade Messenger's rational approach to life. Thus, Carrie immediately senses that the doctor whom her husband is seeing about a sinister lump on his liver is no good (*TH* 277), and Helen immediately senses Messenger's preoccupation about the heap of problems that is suddenly piling up for him. This arrangement, stereotypical as it may be, generates the implicit knowledge that there are areas where scientific method does indeed not apply – especially in the realm of human social life, in which empathic access to others' cognition may be worth more than systematically studying cognition. In other words, the implicit knowledge generated by this plot structure is that human social life is a realm of reality in which the specific is more important than the general.

At this point the considerations of implicit knowledge shade into considerations of narrative knowledge. For one thing, it is the generation of implicit knowledge in the course of a narrative in combination with propositional knowledge that provides the basis for the formation of narrative knowledge. Both novels combine propositional and implicit knowledge in a way that makes both types of knowledge inseparable. For another thing, in both novels narrative knowledge is precisely what is opposed to propositional knowledge. Whereas the items of propositional knowledge remain isolated, allowing only glimpses at overarching theoretical concepts that exist in cognitive science, what ties them together is their interweaving with implicit knowledge in the course of the narrative. In other words, the narrative arrangement produces a holistic view of the different types of knowledge that are included in a work of fiction. Such a holistic view is, arguably, much closer to the way in which human beings use knowledge in everyday life than the scientifically rigorous formation of propositional knowledge. The novels discussed here are, thus, closer to the "'thick descriptions' of real people acting in real-life situations" that Cantwell Smith (1999: 770), in tacit reference to Clifford Geertz (1973), proposes as a medium for accounts of situated cognition, or, to use the title of Hutchins (1996), of cognition in the wild, that is more appropriate than the mere network of propositions that one would conventionally expect from scientific texts. In this respect, the formation of knowledge of cognition in fictions of cognition might contribute to bi-directionality in the cognitive sciences.

Finally, it is time to address the final point intended to be discussed in the case study of the two novels, namely the question of which cognitive operations readers perform in order to build a mental model of the story world that includes knowledge of cognition and that reconciles the competing kinds of knowledge. The decisive point here is which processes of text understanding determine the formation of a mental model that negotiates the different kinds of knowledge.

First of all, the basic cognitive schemata that determine the generation of mental models of the text sense are, obviously, the ones that everyone knows from everyday life. Both novels are realistic novels, situating their actions in a story world that could or that does exist, depicting actions that might have taken place. Thus, nothing about the everyday cognitive schemata of causation, orientation in time and space, living beings etc. has to be revised to form a mental model of these novels' story-worlds. Only *Galatea 2.2* requires some revisions of the everyday real-world cognitive schemata that concern computers and their abilities, which does, however, not affect the basic cognitive schemata mentioned above. What is more extraordinary in both novels is that, in order to form a correct mental model of the story world and the knowledge generated by the novels, readers have to produce a system of hierarchically nested mental models.

On the level of mental model formation above the level of mere cognitive schemata there are mental models for the protagonists that are opposed in the two novels discussed here. These mental models result from the interplay between textual cues, such as Messenger's characterization as having the future in his bones and his belief in progress made possible by a rational scientific mindset, with readers' world knowledge, such as the images of the stereotypical scientist as rationalist, believing in progress and forward oriented that most people, arguably, have in their minds. Connected to the different mental models is the differing knowledge that is generated by the fictional text as was sketched out above.

On the next higher level of mental model formation, the opposing mental models for both protagonists have to be integrated into an overarching mental model that contains the whole story world. On this level of mental model formation, the opposing characters and the opposing kinds of knowledge of cognition coexist. The most likely way mentally to conceive of such a coexistence is to regard both sets of knowledge as two ways to approach, grasp, and understand the same phenomenon, in this case cognition. In the process of synthesizing the conflicting elements, a synthesis of the conflicting kinds of knowledge about cognition occurs, resulting in the type of hierarchically higher knowledge – i.e. narrative knowledge – that includes all the other types of knowledge to form a new knowledge that is not explicitly spelt out in the novels. For the concrete examples in this case study, this new knowledge would be that scientific method does indeed not apply to every area of cognition, let alone to human life and experience.

A force that works in the same direction as the processes of mental model formation sketched above are the metaphors, or the incidents of metaphorical thinking, respectively, that are quite prominent in *Galatea 2.2* and *Thinks....* Metaphors are more conspicuous in Powers's novel than in Lodge's, which is why it will be discussed first.

The whole narrative, told by the first-person narrator Richard, is pervaded by a web of metaphors, only some of which can be selected as examples here, that cross the boundary between science and literature or human experience, respectively. The process of designing an AI is frequently depicted in literary terms, while processes of human existence, such as reading or writing literature, as well as feeding a disabled woman, are depicted in AI terms. The latter example occurs during a visit that Richard and Lentz pay to Lentz' wife who, after a domestic accident followed by cardiovascular injury, lives in a care facility. Observing Lentz' efforts at bringing his wife to eat, Richard characterizes his attempts as "[s]upervised training" (*G* 167), that is, as exactly what he himself does with Helen the AI. It is quite clear that Richard attempts to grasp something in terms of something else. The act of literary creation itself is metaphorically described in terms that put it close to cognition, such as in this account that Richard gives of his writing *The Gold Bug Variations*, a novel published by Powers in 1991 (*G* 214–215): "Some reentrant map inside me ran away with itself, insisting it was real" (*G* 215). In this quotation, the act of writing a novel is depicted by using a self-referential map of brain cells as the correlate for the independent life of a novel in the making.

The metaphors also work in the opposite direction, however, depicting the training of the AI in terms of literature. A case in point is Richard's reference to Whitman's famous poem *A Noiseless Patient Spider* from *Leaves of Grass* (1900) in Helen's literary training and in his subsequent characterization of the understanding mechanisms of the computer: "The list of predicates she might spin out was as extensive as the ways of overlap in her neurodal clusters" (*G* 231). This sentence metaphorically connects several domains. The choice of the expression 'spin out' for Helen's understanding process indicates an understanding of the computer's cognitive activity in terms of Whitman's poem, although the expression itself is not used in it. At the same time, the list of predicates is connected to the overlap of clusters, a description that is reminiscent of theories of conceptual metaphor that conceive of metaphor as a blend of two or more mental input spaces.

What this web of metaphors actually does in the novel is to subvert the clean separation of scientific from humanistic knowledge. Instead, the narrating instance's manifest undertaking of understanding humanistic matters in terms of AI and vice versa points towards the holistic unity of knowledge, at the same time invalidating scientific claims for the only valid kind of knowledge. To establish a concise 'moral' on the basis of these considerations, one would have to assert that the use of metaphors in *Galatea 2.2* indicates that human knowledge is not formed in a methodically flawless way, but rather by understanding unknown things in terms of known ones.

A similar point can be observed in incidents of metaphorical thinking in Lodge's *Thinks....* Especially novelist Helen Reed uses metaphoric thinking – well-developed in "one of England's finest contemporary novelists" (*TH* 250) – to understand what she hears. Thus, upon Messenger's explanations about autistic people sometimes having highly specialized unusual abilities, Helen comments that "[c]omputers are a bit like that, aren't they" (*TH* 240).

Messenger himself is not immune to metaphorical thinking, as can be witnessed when he playfully extends this piece of metaphoric understanding in his thought recording, apparently earnestly characterizing computers in terms of several characteristics of autism. Thus, he mentions that computers "have fantastic memories but no common sense, they have a deficit of affect, they can't tell the difference between a true story and a fiction, they don't know how to lie" (*TH* 251). This is a clear incident of understanding something, i.e. computers, in terms of something else, i.e. the properties of autism.[7]

In fact, the whole thought recording experiment that Messenger performs to find out the structure of random, undirected thought relies on a metaphor, namely on the metaphor of the stream of thought, the probably most famous metaphor in the history of psychology and cognitive science. While its originator William James knew full well that his characterization of human thought is, in fact, of metaphorical nature (1950 [1890]: 239), Messenger seems to take the metaphor seriously, remaining blind to the epistemological foundations of his own 'research'.[8]

Thus, the depiction of metaphorical thinking in *Thinks...* can be seen to aim in the same direction as in *Galatea 2.2*. It puts into perspective the separation of different kinds of knowledge, displaying scientific knowledge as depending on the same mechanisms of thought as humanistic knowledge or as human knowledge in

7. Obviously, Messenger's image of autism is an oversimplification.

8. The scare quotes around the word 'research' are used to signal that this experiment is, as Mars-Jones (2001) writes, "an unlikely piece of research for a professor to undertake", representing only one of several inconsistencies in Messenger's characterization as a leading cognitive scientist. The most probable explanation for these inconsistencies is that they serve the real-world author Lodge as a vehicle for debunking the partly grandiose claims made by some cognitive scientists.

general. Thus, it qualifies science's claims to being the only source for 'real', for valid knowledge. At the same time, the use of metaphors and metaphorical thinking in both novels illustrates the implicitness of the power that metaphors have over the formation of knowledge. After all, not only do the characters use metaphors to understand something in terms of something else, they also do so without openly acknowledging it. What is more, readers also do it. Being offered metaphorical explanations, readers form mental models of the knowledge inherent in the novels that include the tacit spin that the metaphors have added to the representation of the story world. These two points indicate that metaphors are indeed a tacit force in the formation of knowledge on the basis of fictional literature.

5. Conclusion

Several points could be demonstrated in the above discussion. First, fictional literature, though concerned with specific and, what is more, invented events, can function as an active force in knowledge formation. As the case study of two fictions of cognition has shown, fictional literature offers many cues for the cognitive operation of knowledge formation. By intertextual comparison one can identify items of propositional knowledge that are woven into the work of fiction and, thus, form elements of the story world. Great amounts of implicit knowledge are equally part of the story world, for instance manifest in characters' actions. In the process of narrating, narrative knowledge is formed that combines the other two types of knowledge into a whole. That is, narrative fiction offers in itself many triggers for the cognitive operation of forming knowledge that goes beyond its mere validity in the story world.

Secondly, by using the concrete case of fictions of cognition as an example it becomes clear that scientific knowledge is not the only kind of valid knowledge of cognition, but that narrative fiction does contribute a kind of knowledge that, arguably, remains foreclosed from scientific enquiry. Thus, fictions of cognition, beyond being mere secondary products of the imagination, can participate in a genuinely bi-directional endeavor to researching cognition, an impact that Daniel Dennett (2008: 161) has acknowledged for Powers's novel: "I know that many people in the field [of cognitive science] have read *Galatea 2.2* and enjoyed it as much as I did, and since that is so, there is bound to be a subtle contributing effect", a contributing effect that can "push[...] the scientific imagination into new places" (2008: 160).

References

Bartlett, Frederic. 1932. *Remembering*. Cambridge: Cambridge University Press.

Blaauw, Martijn and Duncan Pritchard. 2005. *Epistemology A–Z*. Edinburgh: Edinburgh University Press.

Brewer, William F. 1999. Schemata. In *The MIT Encyclopedia of the Cognitive Sciences*, Robert A. Wilson and Frank C. Keil (eds.), 729–730. Cambridge, Mass.: MIT Press.

Cantwell Smith, Brian. 1999. Situatedness/Embeddedness. In *The MIT Encyclopedia of the Cognitive Sciences*, Robert A. Wilson and Frank C. Keil (eds.), 769–771. Cambridge, MA: MIT Press.

Damasio, Antonio. 1999. *The Feeling of What Happens: Body, Emotion and the Making of Consciousness*. London: Vintage.

Dennett, Daniel C. 1991. *Consciousness Explained*. London: Penguin.

Dennett, Daniel C. 2008. Astride the two cultures: A letter to Richard Powers, updated. In *Intersections: Essays on Richard Powers*, Stephen J. Burn and Peter Dempsey (eds.), 151–161. Champaign, IL: Dalkey Archive Press.

Echterhoff, Gerald and Jürgen Straub. 2003. Narrative Psychologie: Facetten eines Forschungsprogramms. *Handlung Kultur Interpretation* 12 (2): 317–342.

Edelman, Gerald M. 1992. *Bright Air, Brilliant Fire: On the Matter of the Mind*. New York: Basic Books.

Ette, Ottmar. 2004. *ÜberLebenswissen: Die Aufgabe der Philologie*. Berlin: Kulturverlag Kadmos.

Ette, Ottmar. 2005. *ZwischenWeltenSchreiben: Literaturen ohne festen Wohnsitz*. Berlin: Kulturverlag Kadmos.

Ette, Ottmar. 2007a. Literaturwissenschaft als Lebenswissenschaft: Eine Programmschrift im Jahr der Geisteswissenschaften. *Lendemains: études comparées sur la France* 32 (125): 7–32.

Ette, Ottmar. 2007b. Lebensfuge oder Eine Philosophie des ÜberLebenSchreibens: Laudatio für Jorge Semprún. *Lendemains: études comparées sur la France* 32 (126/127): 193–207.

Ette, Ottmar. 2010. *ZusammenLebensWissen: List, Last und Lust literarischer Konvivenz im globalen Maßstab*. Berlin: Kulturverlag Kadmos.

Feldman, Richard. 2003. *Epistemology*. Upper Saddle River, NJ: Prentice Hall.

Fludernik, Monika. 1996. *Towards a 'Natural' Narratology*. London: Routledge.

Freissmann, Stephan. Forthcoming. *Fictions of Cognition: Representing (Un)Consciousness and Cognitive Science in Contemporary English and American Fiction*. Trier: Wissenschaftlicher Verlag Trier.

Geertz, Clifford. 1973. Thick description: Toward an interpretive theory of culture. In *The Interpretation of Cultures: Selected Essays*, Clifford Geertz, 3–30. New York: Basic Books.

Genette, Gérard. 1997 [1982]. *Palimpsests: Literature in the Second Degree*. Lincoln: University of Nebraska Press.

Gymnich, Marion, Birgit Neumann and Ansgar Nünning (eds.). 2006. *Kulturelles Wissen und Intertextualität: Theoriekonzeptionen und Fallstudien zur Kontextualisierung von Literatur*. Trier: Wissenschaftlicher Verlag Trier.

Hallet, Wolfgang. 2008a. The multimodality of cultural experience and mental model constructions of textual worlds. In *The Literary Mind* (REAL Yearbook of Research in English and American Literature, Vol. 24), Jürgen Schlaeger and Gesa Stedman (eds.), 233–250. Tübingen: Narr.

Hallet, Wolfgang. 2008b. Literarisches Verstehen und Kognition: Mentale Modelle und Visualisierungsaufgaben im Literaturunterricht. In *Verstehen und Verständigung: Interkulturelles Lehren und Lernen. Festschrift für Jürgen Donnerstag*, Petra Bosenius, Andreas Rohde and Martina Wolff (eds.), 137–170. Trier: Wissenschaftlicher Verlag Trier.

Herman, David. 2002. *Story Logic: Problems and Possibilities of Narrative*. Lincoln, NE: University of Nebraska Press.

Herman, David. 2007. Introduction. In *The Cambridge Companion to Narrative*, David Herman (ed.), 3–21. Cambridge and New York: Cambridge University Press.

Herman, David. 2009. *Basic Elements of Narrative*. Oxford: Wiley-Blackwell.

Hörisch, Jochen. 2007. *Das Wissen der Literatur*. München: Wilhelm Fink.

Hutchins, Edwin. 1996. *Cognition in the Wild*. Cambridge, MA: MIT Press.

Hutto, Daniel D. 2007. The narrative practice hypothesis: Origins and applications of folk psychology. In *Narrative and Understanding Persons*, Daniel D. Hutto (ed.), 43–68. Cambridge: Cambridge University Press.

Iser, Wolfgang. 1978 [1976]. *The Act of Reading: A Theory of Aesthetic Response*. Baltimore: Johns Hopkins University Press / London: Routledge & Kegan Paul.

Jackson, Frank. 1982. Epiphenomenal qualia. *Philosophical Quarterly* 32: 127–136 [quoted from: Peter Ludlow, Yujin Nagasawa and Daniel Stoljar (eds.). 2004. *There's Something about Mary: Essays on Phenomenal Consciousness and Frank Jackson's Knowledge Argument*. Cambridge, MA: MIT Press, pp. 39–50].

Jackson, Frank. 1986. What Mary didn't know. *Journal of Philosophy* 83: 291–295 [quoted from: Peter Ludlow, Yujin Nagasawa and Daniel Stoljar (eds.). 2004. *There's Something about Mary: Essays on Phenomenal Consciousness and Frank Jackson's Knowledge Argument*. Cambridge, MA: MIT Press, pp. 51–56].

James, William. 1950 [1890]. *The Principles of Psychology*, Vol. 1. New York: Dover.

Johnson-Laird, Philip N. 1983. *Mental Models: Towards a Cognitive Science of Language, Inference, and Consciousness*. Cambridge, MA: Harvard University Press.

Klausnitzer, Ralf. 2008. *Literatur und Wissen: Zugänge – Modelle – Analysen*. Berlin: de Gruyter.

Köppe, Tilmann. 2007. Vom Wissen *in* Literatur. *Zeitschrift für Germanistik* 17 (2): 398–410.

Köppe, Tilmann. 2008. *Literatur und Erkenntnis: Studien zur kognitiven Signifikanz fiktionaler literarischer Werke*. Paderborn: mentis.

Lakoff, George and Mark Johnson. 1980. *Metaphors We Live By*. Chicago: The University of Chicago Press.

Lakoff, George and Mark Turner. 1989. *More than Cool Reason: A Field Guide to Poetic Metaphor*. Chicago: The University of Chicago Press.

Lodge, David. 2001. *Thinks…* New York: Viking.

Mars-Jones, Adam. 2001. It was cognition at first sight: Science versus art… David Lodge roasts some old academic chestnuts in *Thinks… The Observer*, February 18, 2008.

Minsky, Marvin. 1975. A framework for representing knowledge. In *The Psychology of Computer Vision*, Patrick Henry Winston (ed.), 211–277. New York: McGraw Hill.

Nagel, Thomas. 1974. What is it like to be a bat? *The Philosophical Review* 83 (4): 435–450.

Nünning, Ansgar. 2008. On the knowledge and functions of metaphors: Interfacing literature, culture and science. In *La conoscenza della letteratura / The Knowledge of Literature*, Vol. 7, Angela Locatelli (ed.), 195–217. Bergamo: Bergamo University Press / Sestante.

Polanyi, Michael. 1966. *The Tacit Dimension*. Garden City, NY: Doubleday.

Polanyi, Michael. 1969. The logic of tacit inference. In *Knowing and Being: Essays by Michael Polanyi,* Marjorie Grene (ed.), 138–158. Chicago: The University of Chicago Press.
Powers, Richard. 1995. *Galatea 2.2.* New York: Picador.
Prince, Gerald. 1982. *Narratology: The Form and Functioning of Narrative.* Berlin: Mouton de Gruyter.
Pritchard, Duncan. 2006. *What Is this Thing Called Knowledge?* London: Routledge.
Rescher, Nicholas. 2003. *Epistemology: An Introduction to the Theory of Knowledge.* Albany: State University of New York.
Rumelhart, David E. 1980. Schemata: The building blocks of cognition. In *Theoretical Issues in Reading Comprehension: Perspectives from Cognitive Psychology, Linguistics, Artificial Intelligence, and Education,* Rand J. Spiro, Bertram C. Bruce and William F. Brewer (eds.), 33–58. Hillsdale, NJ: Erlbaum.
Ryle, Gilbert. 1949. *The Concept of Mind.* London: Hutchinson.
Schank, Roger C. and Robert P. Abelson. 1977. *Scripts, Plans, Goals and Understanding: An Inquiry into Human Knowledge Structures.* Hillsdale, NJ: Erlbaum.
Shope, Robert K. 1992. Propositional knowledge. In *A Companion to Epistemology,* Jonathan Dancy (ed.), 396–401. Malden, MA: Blackwell.
Snow, Charles P. 1998 [1959]. *The Two Cultures.* Cambridge: Cambridge University Press.
Stockwell, Peter. 2002. *Cognitive Poetics: An Introduction.* London: Routledge.
Valdivia Orozco, Pablo. 2007. Lebensform und Narrative Form: Zur Epistemologie des Vollzugs und zum Lebensbegriff der Literaturwissenschaften. *Lendemains: études comparées sur la France* 32 (128): 109–120.
Varela, Francisco J., Evan Thompson and Eleanor Rosch. 1999 [1991]. *The Embodied Mind: Cognitive Science and Human Experience.* Cambridge, MA: MIT Press.
Wood, Michael. 2005. *Literature and the Taste of Knowledge.* Cambridge: Cambridge University Press.

WRITING IS MEDICINE

Blending cognitive and corpus stylistics*

Beatrix Busse
Universität Heidelberg

This chapter investigates Paul Auster's novels *The Brooklyn Follies* (2006) and *The Book of Illusions* (2002) from a cognitive stylistic and corpus stylistic perspective. The first section shows that, following Fauconnier and Turner's (2002) approach to blending, the blend of WRITING IS MEDICINE, and some variations on it, structure the main characters' mental lives as well as the readers' inferencing processes. In the second part of the chapter, I test to what extent the blend of WRITING IS MEDICINE has a textually foregrounded base. Drawing on corpus stylistic methods, I demonstrate that the input spaces of writing and medicine both implicitly and explicitly show up in the lexico-grammatical structure of *The Brooklyn Follies*. Finally, I stress the need for using a cognitive stylistic approach in conjunction with text-based methods to measure and describe in detail the lexico-grammatical character of a text.

Keywords: textual foregrounding, meaning inference processes, lexico-grammatical text structure, Paul Auster

1. Introduction

This chapter examines Paul Auster's *The Brooklyn Follies* (2006) and *The Book of Illusions* (2002) from two different – probably, at first sight, diametrically opposed – perspectives: a cognitive stylistic approach, which applies Fauconnier and Turner's (2002) theory of blending to the two novels,[1] is fused with a text-based corpus stylistic approach, which uses a corpus comparison tool to

* I would like to thank Paul Rayson for allowing me to use Wmatrix and for his useful advice.

1. In cognitive stylistics, Turner (1996, 2003, 2006), Semino (2002, 2006), and Dancygier (2006) apply the theory of blending to literature. Conceptual blending theory has not been applied to Paul Auster's novels, although some studies approach Auster's work with psychoanalytic theory (Bennett, Luksic, and Huidobro 2007).

empirically investigate key words and key semantic fields in *The Brooklyn Follies* (Auster 2006).

To explore a central aspect of the two novels and the main characters' mental lives as well as of the readers' inference processes, I begin my chapter with an investigation of how the blend WRITING IS MEDICINE structures *The Brooklyn Follies* (Auster 2006) and *The Book of Illusions* (Auster 2002) and how it addresses the readers' cognitive processing during the act of reading and interpretation. I examine how the narrators of the novels not only play with the more conventional associations and ideas of writing and medicine, but also allude to transferred meanings, among which are the ideas that writing is a disease (WRITING IS ILLNESS) and that imagining may function as medicine (IMAGINATION IS MEDICINE).

David Zimmer's comparison in *The Book of Illusions* (Auster 2002) between writing and medicine is one of the most obvious examples illustrating the connection between writing and medicine in Paul Auster's work. David, who has lost his entire family in a plane crash, and who writes a book about Hector Mann, a famous comedian and silent film-maker, sees parallels between writing and medicine because writing has helped him to come to terms with the deaths of his wife and children. This "odd form" of medicine is considered to be a more unconventional cure than the customary intake of pills or liquids. He explains retrospectively that writing "about comedy had been no more than a pretext, an odd form of medicine that I had swallowed every day for over a year on the off chance that it would dull the pain inside me" (2002: 5).

The second half of this chapter is experimental and tests whether this systematically and cognitively based analysis can be supported by an "objective", empirical text analysis. As part of this informed (textual) stylistic analysis, I draw on a corpus stylistic approach and exemplary apply the corpus comparison tool *Wmatrix*, developed by Paul Rayson (2007), to Paul Auster's *The Brooklyn Follies*. In a move back to the text, I examine whether what has been outlined as a blend of WRITING IS MEDICINE is lexically and semantically referred to in the novel and, if so, whether these fields are, objectively speaking, deviant, exceptional, or foregrounded (Leech 2008: 163)[2] so that the reader is subconsciously or consciously guided by them (Toolan 2009). A textually-based framework for cognitive stylistics lends itself to this analysis because foregrounding (deviation, parallelism) depends on the text drawing attention to itself. A cognitive analysis in general and the theory of blending in particular often face the charge of being just mere labeling or a subjective visualization of the obvious, not a real analysis

2. In this paper, foregrounding is not understood to be synonymous with emphasizing. It should be related to the theory of foregrounding, which is crucial to stylistics and, among others, has its base in Russian formalism (see Short 1996, Douthwaite 2000, and van Peer and Hakemulder 2006).

of the text.[3] Hence, my aim is to some extent the reverse of Burke's (2005), whose article "How cognition can augment stylistic analysis" expands a stylistic analysis in the direction of cognitive issues. Although it is indispensable to pay attention to psychological and cognitive factors in the reading process, I propose to augment a cognitive approach by an informed stylistic analysis, and I suggest a fruitful cooperation between textual investigation and cognition. From the perspective of cognitive stylistics the question is whether, by focusing on the text, there are ways of proving linguistically, textually, and/or stylistically as well as empirically that the blend WRITING IS MEDICINE actually features in the text. The questions to be answered are whether the text (in its foundational definition as a "weaving"),[4] for example, contains this cognitive metaphor in its material of grammatical phenomena and of lexical words explicitly or implicitly denoting writing, medicine, or mental processes; whether words are statistically prominent when their frequency is compared to that in other Auster novels, for example, or other narrative fiction; what the relationship between quantitative prominence and stylistic meaning-making is in the novels. It is necessary to analyze whether there are possibilities to measure/describe, or to put on a sounder footing, how blending functions *linguistically*.

Since function and form interact in a highly complex way, this approach must not exclude the reader. One may assume that linguistic features that are deviant (e.g. significantly more frequent than in "normal") will come to the attention of the informed and sensitive reader. Of course, a variety of readers will have to be tested. But there is too much ground, which relates to the explicit textual base, to be covered first before these tests can be performed.

2. Writing and medicine – some conventional associations

> The words on the page that we read and make sense of, find the story in, and derive expectations and feelings and understanding from, always come to us integrated within a context of situation and of culture. A variety of kinds of background knowledge informs our reading of a present narrative text […]. (Toolan 2009: 5)

In addition to what Toolan (2009) outlines in the quotation above, reading is also influenced and guided by the readers' own encyclopedic knowledge, personal values, their information about the author of a narrative and so on: "Every literary

3. Toolan and Weber (2005) also mention that, despite the overwhelming attention that cognitive approaches have received, when it comes to the application of a cognitive approach, generally speaking, "the unconvinced, or skeptical, or antipathetic remain numerous" (2005: 107).

4. *OED* "text, *n.*[1]" 1., and Leech (2008).

narrative episode is susceptible to a 'colouring in' by the reader" (Toolan 2009: 7), which allows the reader to draw implicatures. Of the equation of writing with medicine it can conventionally be said that it contains a number of stronger or remoter associations. The equation of writing with medicine is probably a familiar (almost well-worn) trope: writing is seen as therapeutic, healing, and curative. Yet this trope may certainly also have to be understood as culture-specific because, in its narrow sense, it would not be valid for oral cultures or communities. Conventional literal associations for both medicine and writing suggest that they are non-life-threatening. Medicine can be a solid or more liquid consumable whose function is to combat minor illnesses and diseases (see also *OED* "medicine, *n*". 1.a and 1.b.). When referred to as a science, medicine evokes associations with scientific research and medical progress for the good of human beings. In both, doctors or medical authorities are involved as agents. Medicine, however, may also carry more general associations with healing. Cultural history tells us that medicine was used as a metaphor in the Middle Ages for various not exclusively physical activities including love and the enjoyment of art. In a transferred sense, medicine then involves a change in well-being, which can be either positive or negative. It can be sweet or bitter; it can alleviate and heal but also provoke painful purgation.

Writing, in turn, is readily associated with an instrument, i.e. a pen, used for writing, with typing or a computer screen, a sheet of paper or e-mail. But it is also associated with the production and the reception of writing, i.e. with reading documents and books, and also fiction. It is connected with the writer, or concepts of a reporter, a narrator or a storyteller and so on. The process of writing may have the function of reporting, informing, and narrating etc. It may also be a private coming to terms with one's own thoughts, but it also communicates an intentional reaching out to a recipient. In fact, Pennebaker (1997) has shown that writing is a method used in psychology to help patients to overcome their mental and physical deceases. In relation to the production of fiction, writing can be located on several fictional levels. Both characters and extra-fictional narrators write and thus call attention to the process of narration itself. Dietrich von Engelhardt (1991) touches on the multiplicity of interdisciplinary links between literature and medicine, noting three functions: the function of medicine in fiction; the scientific function of literature; and the genuine function of "literalised" medicine (von Engelhardt 1991). What is called bibliotherapy or poetry therapy (Nevanlinna 2005: 129f.) – therapeutic means of healing with the help of literature – is one expression of these interrelationships between writing and medicine to which I return towards the end of my chapter. Paul Auster's novels display a preoccupation both with writing and with characters who have experienced personal crises. As Auster himself points out in an interview with *Neue Zürcher Zeitung* (4 Oct 2008):

> Ich versuche stattdessen, zu der Realität des inneren Lebens vorzudringen, Antwort auf die Frage zu finden, was es an einem beliebigen Ort und zu jeder beliebigen Zeit bedeutet, am Leben zu sein.
>
> [I try to reach the reality of one's inner life, to find answers to the question of what it means, at a given place or a given time, to be alive.] (My translation)

3. Fauconnier and Turner's (2002) theory of blending

According to Fauconnier and Turner's (2002) approach, we construct meaning by creating "mental spaces" (Fauconnier and Turner 2002: 40, 102), which are temporary representations that recruit structure from many conceptual domains during the online process of meaning creation. It may also recruit entrenched mappings or frames. A conceptual integration network, a blend, is a dynamic and flexible construct and is the product of the interaction of different mental spaces (Fauconnier and Turner 2002: 49f.).

A minimal conceptual integration network consists of four spaces: two input spaces, a generic space, which links the input spaces, and the blend. In contrast to conceptual metaphor theory outlined by Lakoff and Johnson (2003), there is no hierarchical distinction between source and target domain, because both input spaces take on the same role or part (democratically). In conceptual metaphor theory, the target is always understood "in terms of" the source. The blend arises from the fusion or blending of the material from the input spaces, but it may also contain additional structures projected from neither of the inputs (Evans and Green 2006: 403). The result is a new blended space that has its own conceptual structure, and it is not uni-directional or one-dimensional. It is in the entire blend that meaning is generated.

Brandt and Brandt (2005) criticize Fauconnier and Turner's model because of its lack of attention to context. Instead they add two spaces to their more semiotically based model: a base space for the immediate or situational context, and a relevance space "that provides relevant thought content for the construction of the blend, and ultimately, for the emergence of reasonable inferences" (Brandt and Brandt 2005: 234f.). Coulson and Oakley (2005) also take over Brandt and Brandt's base space as a "grounding box".

Fauconnier and Turner (2002: 48f.) argue that three processes are involved in the emergent meaning of the blended space. These processes are composition, completion, and elaboration. Composition is always involved in the process of blending, because conceptual content from two or more mental spaces is projected into the blended space (Ungerer and Schmid 2006: 259). In the process of completion, information is always recruited unconsciously, and background frames are also projected unconsciously from the inputs. The process of elaboration, which is often dynamically seen as "running the blend" (Ungerer and Schmid 2006: 260), creates the unique structure of the blend.

Input spaces are also mapped onto each other through various cross spaces, which are based on what Fauconnier and Turner (2002: 89–111) call "vital relations". Fauconnier and Turner (2002: 89–111) distinguish between compression of relations within the input space and compression of relation between mental spaces: "Compression is in fact the ultimate goal of the whole blending process. The crucial effect of compression is that the conceptual complexity of the inputs from several sources is reduced considerably" (Ungerer and Schmid 2006: 260). These additional matching specifications, these "vital relations", can be relations of identity, space, time, representation, change, role value, cause and effect, or part-whole.

To sum up, although blending theory shows links with several other paradigms (see Forceville 2004), which are not always explicitly stated, blending theory helps explain emergent structure and how human beings create concepts by integrating information from different inputs.

4. The blend WRITING IS MEDICINE in Auster's *The Brooklyn Follies* (2006) and *The Book of Illusions* (2002)

4.1 Running the blend

While in Auster's *The Book of Illusions* (2002) the relations between writing and medicine are overtly established, in *The Brooklyn Follies* (Auster 2006), the parallelism is more implicitly construed in a transferred sense of both medicine and writing as causing a positive mental or physical change, and of medicine and writing functioning as a pleasurable drug. Nathan Glass begins to create the integrational network of WRITING IS MEDICINE as a means of thinking about and preserving past events (Auster 2006: 7). He explains the physicality and hypnosis of the parallelism between writing and medicine almost iconically through recurrent parataxis and the repetition of the preposition *by*:

> Every time I sat down to write, I would begin by closing my eyes and letting my thoughts wander in any direction they chose. By forcing myself to relax in this way, I managed to dredge up considerable amounts of material from the distant past, things that until then I had assumed were lost forever. (2006: 7)

Nathan's passion for reading books is also one reason why he starts to write:[5]

> – but I had never lost my interest in books. Reading was my escape and my comfort, my consolation, my stimulant of choice: reading for the pure pleasure of it, for the beautiful stillness that surrounds you when you hear an author's words reverberating in your head. (2006:7)

The enumeration of noun phrases with a similar structural amplification – deictic element and head noun – as, for example, visible in "my escape" or "my consolation", emphasizes the force of the process of reading. Nathan's aims are illustrated at length in the following quotation:

> I called it *The Book of Human Follies* and in it I was planning to set down in the simplest, clearest language possible an account of every blunder, every pitfall, every embarrassment, every idiocy, every foible, and every inane act I had committed during my long and checkered career as a man. [...]
>
> I called the project a book, but in fact it wasn't a book at all. [...] I was compiling what amounted to a collection of random jottings, a hodgepodge of unrelated anecdotes that I would throw into a cardboard box each time another story was finished. There was little method to my madness. (2006:5f.)

The fragmentary character of Nathan's book is also stressed more negatively in *The Book of Illusions*, when David Zimmer describes his: "This is a book of fragments, a compilation of sorrows and half-remembered dreams" (2002:316).

The preceding contexts play an important part in the cognitive processing of each of these quotations and helps develop the blend of WRITING IS MEDICINE. At the beginning of *The Brooklyn Follies*, the reader immediately learns that Nathan Glass, when returning to his home Brooklyn, is a bitter solitary man, who has just been divorced and who seems to await his death, despite a successful cancer operation and good prospects for his survival. By coincidence he meets in a bookshop, owned by Harry Brightman, his equally stranded long-lost but beloved nephew Tom. In *The Book of Illusions*, David Zimmer is a devastated man who lost his family in a plane crash, but because, as a professor, all his work has been "connected to books, language, the written word" (2002:13), he secludes himself into that world, and writing about the allegedly disappeared comedian and film

5. A recent advertising leaflet for one of the famous *Moleskine* notebooks, the City Notebook, reads as follows: "The first guidebook you write yourself", because you get the chance of "go[ing] back over your travels in your memory and your imagination". This advertisement is only one of the many examples that, though rather implicitly, establishes a link between writing and medicine by stressing the mental pleasures of reading one's own travel story and activating one's imagination while reading and writing.

maker Hector Mann is seen as an "odd form of medicine". *The Book of Illusion*, which readers holds in their hands tells his long story about writing this book, his encounter with Hector and his wife and his relationship with Alma, the woman who writes a biography about Hector.

These initial quotations and the contextual information are indicative of the fact that the integrational network WRITING IS MEDICINE captures Nathan's, David's and the other characters' complex developments towards finding new identities. They encourage the reader to go through the various stages of their transformations and of the act of "artistic creation" (Peacock 2006: 56). Auster's work is famous for its multiplicity of narratological levels and its preoccupation with the process of writing fact and fiction. The readers are capable of understanding the blend because of their familiarity with the input spaces, which, in the blend, result in new meaning structures that may be divergent from the input spaces. In Auster's novels, the equation of writing with medicine and vice versa includes the stress on both a physical and a mental process. It emphasizes a holistic religious-spiritual aspect and regards the individual as a reader or a writer, who only subconsciously realizes that he or she interacts with the other (with no expressive mentioning of a physician or a medical authority) through writing and is thereby transformed.

Nathan and David both write their stories retrospectively. In their roles as controlling narrators, the process of healing and recovering – the blend of WRITING IS MEDICINE – is consciously construed in their stories, but also multi-dimensionally and in various nuances. In their roles as characters, they tell their stories subconsciously, and they also undergo this process of change subconsciously.

Figure 1 shows the four spaces and the blend of WRITING IS MEDICINE in Auster's two novels. There is (1) the input space that is writing, (2) the input space that is medicine, (3) a generic space that shares the entities referred to and the time of the process, but is not further pursued here because it does not provide any new information (see also Ungerer and Schmid 2006: 259), and (4) a blend where WRITING IS MEDICINE. The connectors are established by matching (represented by the dotted lines).

It is a double-scope network (Evans and Green 2006: 431), because both inputs contain distinct frames and the blend is organized by a structure taken from each frame. The meanings reside in the entire network of this blend with the spaces contributing to it.

For the interpretation of the two Auster-novels the use of what Brandt and Brandt (2005) call the "base space", which captures contextual information, cannot be underestimated. It is through the way context construes the specific blend of WRITING IS MEDICINE that meaning can be created. This creation includes both pragmatic meanings and effects as deduced from the texts as such

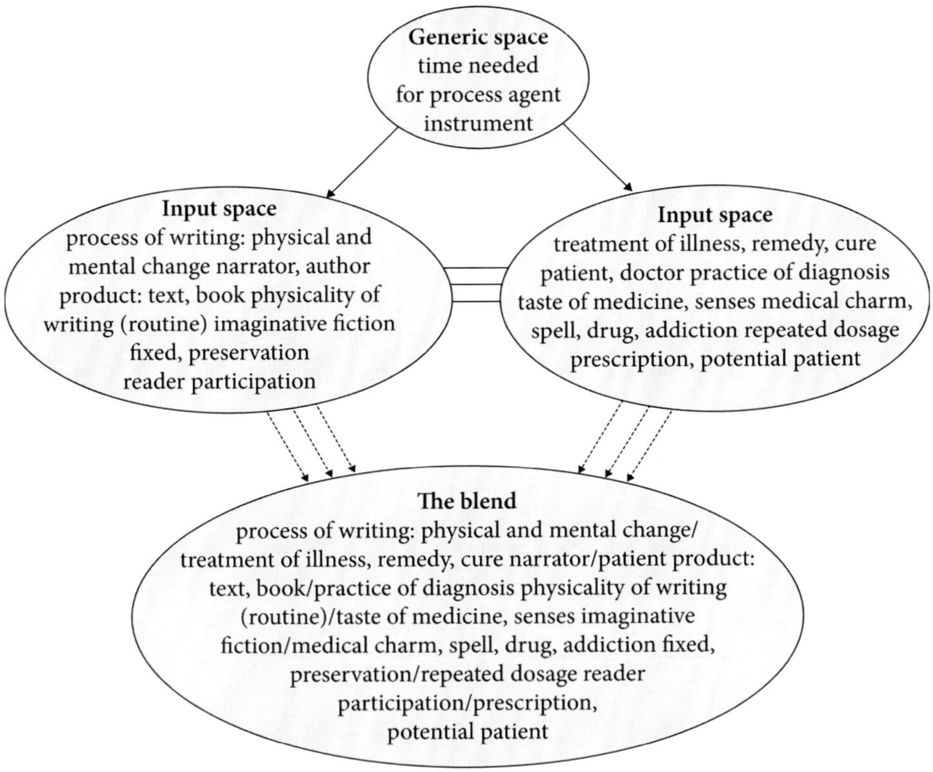

Figure 1. The blend WRITING IS MEDICINE

and encyclopedic knowledge – similar to that outlined in the introduction about writing and medicine. It is in this space where the textual base of each meaning inference process is also stressed.

Having read the novels, the reader creates the mental space "medicine", which may contain the cognitive model that features the treatment of illness, remedy, and cure, a patient, the practice of diagnosis, the taste of medicine, and the consequences for the senses, the medical charm, spell, drug, and addiction, a repeated dosage, a prescription, potential patient, and a doctor or medical authority. Parallel to the medicine space, Nathan's and David's initial references to writing and their high proliferation trigger the second mental space, that of "writing". It is an online conceptual representation, which taps stored information like the process of writing, physical and mental change, the narrator and an author, the product as a text or a book, the physicality of writing (routine), imaginative fiction, something that is fixed and may be preserved, and reader participation.

To evince this line of argumentation further, it should be argued that Nathan's initial subconscious desire to produce and preserve his story results from both his physical and mental struggle: "I'd suffered the long bouts of nausea and dizziness, the loss of hair, the loss of will, the loss of job, the loss of wife" (Auster 2006: 3). So the source of the input space reflects his experience with medicine and his general schematic knowledge of medicine and its effects. Notice how here, as well as in other examples, he describes the negative experience in his life by means of a list of noun phrases, which, in general, according to Halliday (2004: 637–642), construe unchangeable and static facts or commodities. In addition, the enumeration, as an example of diagrammatic iconicity,[6] construes the severity of his situation.

In the novels under discussion, the processes to construe the blend are realized in the following ways. "Composition" (Fauconnier and Turner 2002: 48f.) is involved in the sense that writing, like taking medicine, ideally has the power to improve the health of its writer and his general well-being or generally transforming it. The process of "completion" (Fauconnier and Turner 2002: 48f.) is needed to understand the blend, but not in the sense of underspecification, because the blend WRITING IS MEDICINE runs through the entire novel and demands a continuous process of completion. For example, writing and reading are also seen as routines, with additional specific physical qualities. The dynamic process of "elaboration" (Fauconnier and Turner" 2002: 48f.) is the actual "running of the blend". Initially, Nathan, for example, describes in detail the process of writing and the routines used to prepare for it: closing his eyes, relaxing, and moving to past events. In *The Book of Illusions*, David routinely "swallow[s]" his medicine hoping that "it would dull the pain inside" (Auster 2002: 5). Writing is therefore alluded to as a process of regaining events from the past and preserving them, or as a means of suppressing these memories.

Notice the physical and solitary qualities, but also the sensuality attributed to reading and writing. Thus, in *The Brooklyn Follies*, reading is described as a "stimulant" and as "pure pleasure" (2006: 7), which may be restricted to the individual's experience and may also function as a routine for both reading and writing. The doctor scheme from the medicine space is not overtly mentioned, but Nathan realizes that only in interaction with a "You" he can recover his health. In *The Book of Illusions*, David explicitly denies the help of a doctor, and it is only through the book (but not after the actual completion of writing the book about Hector Mann) that he meets Alma who draws him back to life.

6. Diagrammatic iconicity refers to an "arrangement of signs, none of which necessarily resembles its referent but whose relationships to each other mirror the relationship of their referents" (Nänny and Fischer 2006: 462).

As mentioned, "vital relations" (Fauconnier and Turner 2002: 89–111) can be relations of identity, space, time, representation, change, role value, cause and effect, or part-whole. To begin with identity relations in *The Brooklyn Follies*, there are mappings between the actions of writing a book and taking one's medicine, because these can be seen as having the, admittedly rather optimistic, effect, such as recovery, distraction from pain, and healing. The result of the compression of this identity relation is a new combination of actions, the emergent structure described as "writing as medicine taking" and the goals of the two actions are fused in "writing a book as improving health". But the agent of the input space "writing" turns out to be the patient in need of medical treatment in the blended space, and, at the same time, he is also the physician, "the actor", because no other authority is mentioned. In addition, by writing down their own stories and those of others, Nathan and David (and Alma) are life preserver and healers of others.

Although the narrators in the two novels initially intend to collect their stories for their own personal needs (and to distract themselves, David writes about Hector Mann because he made him laugh), it becomes obvious that there is a further aim – initially subconscious – to it. The reader learns that the blend construes not only Nathan's and David's processes of finding a new identity, but also that of other characters in the novels. For example, in *The Brooklyn Follies*, "hodgepodge" (2006: 6) relates to a magical potion, which at this stage of the novel still alludes to a magical rather than a self-willed change of life. Similarly, the reference to writing as an obsession in Nathan's "There was little method to my madness" (Auster 2006: 6) only subconsciously points to the connection between writing and therapy. In *The Book of Illusions*, Alma, towards the end of the novel, describes her endeavor to write about Hector as "whether it comes out or not, writing the book has been the biggest experience of my life" (2002: 217).

The blend WRITING IS MEDICINE structures the processes of identity construction in which the psychological issues of healing, mental disorders, of transformation, of pleasure but also purely physical experience are involved. It also structures the process of narration in which the narrator/author furthers the process of narration, the activity of writing. Finally, it also structures the product of writing that relates to the reader. The space that results from the blending of the target and the source input is highly creative and idiosyncratic because of the polysemy of meanings of both writing and medicine. Both the production and the reception process of writing and medicine are construed as and also portrayed with complex properties. The blend develops its own structure and creates a rich narrative scenario. It shows that narrative is what Fludernik has called "experientiality" (1996: 10) that is, here, the projection of the consciousness and subjective experience of both Nathan and the reader. WRITING IS MEDICINE projects an intramental construct (Palmer's [2004] term, describing individual cognitive functioning) that is limited to Nathan as narrator, author or patient, but WRITING IS MEDICINE is also an intermental phenomenon that connects interdependent cognitive functioning of groups of characters in the novel (Semino 2006: 57).

The sub-features of the mental spaces contribute to the blend some of the characteristics that Nathan in *The Brooklyn Follies* perceives in writing and medicine or wishes to attribute to it. The blend alludes to the stories as such, to the process of writing and the process of taking medicine as a physical experience, as well as to the product of writing and that of taking medicine as introspection for the writer/narrator and as the healing of the writer/narrator. The blend also alludes to the creation of a different reality and the preservation/immortalization of life, personality and identity, especially in the stories Nathan writes down, and in his wish at the end of the novel to preserve individual people's lives by writing down their biographies and internalizing/preserving the follies of people's lives:

> My idea was this: to form a company that would publish books about the forgotten ones, to rescue the stories and facts and documents before they disappeared – and shape them into a continuous narrative, the narrative of a life. (2006: 303)

In *The Book of Illusions*, David sees one function of Alma's biography about Hector also in preserving Hector's life: "but at the same time Hector gets to live on through you. Not because of his films – which won't even exist anymore – but because of what you have written about him" (2002: 217). Nevertheless, Alma is not able to survive. She commits suicide because Frieda destroys her book. Therewith Frieda destroys Alma's process of finding her identity by means of re-writing Hector's life. Due to David's reverence for Alma, and because he understands the physical and mental strains she must have undergone when writing about Hector's life and reading David's book about Hector, as well as thereby reflecting her own life, he decides that Alma's identity of being a writer should be engraved on her gravestone. As such, writing is seen as medicine beyond life. We find: "ALMA GRUND 1950–1988 WRITER" (2002: 313). David also stresses that

> [e]xcept for the twenty-page suicide note she sent me on the last night of her life, I had never read a word she had written. But Alma had died because of a book, and justice demanded that she be remembered as the author of that book. (2002: 313)

The writing space provides the difference between fact and fiction, and the chance of introspection as well as the spell or addiction to books/medicine as expressed in Nathan's and David's passions for books and the physical and mental effects they attribute to them. The medicine space provides the patient-medicine scheme, which is generally assumed to be successful, as the patient generally gets better and the reader will have that assumption. The blend partly enables Nathan and

David as well as Alma to transform reality into a scheme they can cope with, to keep reality alive and permanent as well as to create another fictional/artificial reality as he is the narrator.[7]

Nevertheless, the subconscious healing or transforming the power of writing also expresses Nathan's and David's subjective experience with the society around them and the people close to them. Hence, several of the other "vital relations" prevail, such as the compression of time and space, and the relation of books and biographies to fact and fiction. The author (or Nathan as both the narrator and author) cannot capture everything, but the reader understands that writing is one way of preserving other people's lives. Also, the vital relations of representation and part-whole are construed here as books may present real lives and may stand for real lives. As such, the blend WRITING IS MEDICINE is further enhanced in *The Brooklyn Follies* by the introduction of more conventional metaphors (Fauconnier and Turner 2002: 279ff.), such as LIFE IS A JOURNEY or UNDERSTANDING IS SEEING (Lakoff and Johnson 2003).

The "vital relation" of analogy is created through various intertextual relations in both novels. For example, in *The Book of Illusions* David's translation of Chateaubriand's *Memoirs* is crucial in its parallelism with David's life. One of Chateaubriand's statements quoted in the novel is "Moments of crisis produce a redoubled vitality in men" (2002: 238). David's vitality only seems to come back towards the very end of the novel, after he completes the book on Hector Mann, which leads to his relationship with Alma and his meeting with Hector. All of this helps him to overcome the loss of his family. In *The Brooklyn Follies*, the collocation of method and madness in "There was little method to my madness" (2006: 5f.) is a WRITING IS MEDICINE reverberation of Shakespeare's *King Lear*, where Edgar finds "Reason in [Lear's] Madness" (4.6.175). In *The Brooklyn Follies*, Lessing's play *Nathan the Wise* is echoed not only by the choice of names for both the narrator Nathan and his daughter Rachel (alluding to Biblical figures in the Old Testament and to Recha, Nathan's daughter in Lessing's play), but also in an overt reference to *Nathan the Wise* and the attributes Nathan transfers to himself after his gradual transformation through writing. In Lessing's play, one

7. But as Toolan (2001: 130) points out: "the appearance or illusion of character control should not be overstated: behind all the fictional individuals, however reported, stands the controlling teller". The sensual attributes and the healing power of writing and reading also allude to the belief of Native Americans in the almost mythical force of stories. Also, the importance of writing and texts alludes to some prominent features of contemporary American Jewish writing, such as the role of the (Jewish) artist, the role of stories within stories, the mixing of fact in fiction (Bus 1997: 434).

example of how Nathan proposes an enlightened humanity is his parable of the three rings that represent the tree major religions. In Auster's novel, Nathan is turned into a counselor of the people he loves as in his writing he reflects his and their situations. Thereby he becomes a medieval knight who protects his women, for example, Rachel, Rory, and Joyce:[8]

> [B]ut she [Aurora] said no [to consulting a psychologist], she'd rather just talked to me. Me. The bitter, solitary man who had crept home to Brooklyn less than a year earlier, the burnout who had convinced himself there was nothing left to live for – knuckleheaded me, **Nathan the Unwise**, who could think of nothing better to do than quietly wait to drop dead, no transformed into a confidant and counselor, a lover of randy widows, and a knight errant who rescued damsels in distress. (Auster 2006: 286, my emphasis)

4.2 Elaborating on the blend: WRITING IS ILLNESS

In *The Brooklyn Follies*, further running of the blend is done by Tom, Nathan's nephew. This running of the blend is also based on contextual and situation information provided by the text. Tom extends the blend WRITING IS MEDICINE in partly negative terms, at first. At the beginning of the novel, Nathan illustrates Tom's attitude to his job as a taxi-driver, and makes it clear that due to his experience with or failures in the world of literature, he detests the written word and the lifeless character of any book or fictional story. To Tom, writing and the consumption of fictional worlds, which he used to like when he started his studies, is not medicine, but a disease. It is "sick-making". This negation of the blend WRITING IS MEDICINE can be incorporated because Fauconnier and Turner (2002) argue that mental spaces are online mental constructions prone to include the individual mental representations of a particular input space. Similar to *The Book of Illusions* where David describes his book/autobiography (which the reader holds in his/her hands) as "a book of sorrows, of half-remembered dreams" (2002: 315) and therefore illusions, the following example of direct speech also illustrates Tom's preference of real life over fictional stories: "No book can duplicate those things. I'm talking about real transcendence, Harry. Leaving your body behind you and entering the fullness and thickness of the world" (2006: 31).

Later in the novel, however, Tom seems to have been healed by the events in his life – e.g. the re-intensified relations to his family through his uncle and the arrival of Lucy, Tom's niece, the daughter of Tom's lost sister and his future

8. Allusions to religion also reverberate the concept of Jewish home writing.

wife Honey – and by working in a bookshop, another reference in life. He still describes writing as a disease, but refers to it as an addiction that always demands introspection and the disclosure of individual and existential deficiencies. Therefore, he praises Nathan for taking the risk: "You are a writer, Nathan. You're becoming a real writer" (2006: 149). Nathan reports about Tom's outburst in indirect speech and free indirect speech:

> There were no rules when it came to writing, he said. Take a close look at the lives of poets and novelists, and what you wound up with was unalloyed chaos, an infinite jumble of exceptions. That was because writing was a disease, Tom continued, what you might call an infection or influenza of the spirit, and therefore it could strike anyone at any time. (2006: 149)

Tom even acknowledges the healing power of writing and the creation of a new, valuable reality when he expresses his fascination with Kafka and Kafka's attempt to save a sad little girl who had lost her doll. It is said of Kafka that he wrote fictional letters from the doll addressed to the little girl he met in a park in Berlin in order to prepare the girl for the separation from her doll. What we get is one of the most moving nuances of the elaborations on WRITING IS MEDICINE and WRITING IS ILLNESS in Tom's direct speech:

> He isn't about to cheat the little girl. This is real literary labor, and he's determined to get it right. If he can come up with a beautiful and persuasive lie, it will supplant the girl's loss with a different reality – a false one, maybe, but something true and believable according to the laws of fiction. (2006: 155)

Hence, Tom not only introduces a further input space – that of WRITING IS ILLNESS – which is an extension to WRITING IS MEDICINE, because disease and the taking of medicine are conventionally interdependent. He also further changes it to the positive by arguing that as long as the production of the written has a serious task behind, everybody is allowed to do it. In addition, Tom seems to suggest that although a writer may be said to be insane, addicted or ill, he may still heal others. This elaboration on the blend leads to another contextual framework in which the blend works and that is IMAGINATION IS MEDICINE.

4.3 Further elaborating on the blend: IMAGINATION IS MEDICINE

In both novels, the act of writing and imagining more or less rapidly leads to the foundation of a non-fictitious re-creation of identity and the coming to terms with past devastating events. The quoted example from Kafka is one of the many cases in point. IMAGINATION IS MEDICINE as an elaboration of the blend WRITING IS MEDICINE is further elaborated on in both novels.

In *The Book of Illusions*, the process of writing and the actual product represent an imaginative retreat for David to cope with the world:

> I was in the book, and the book was in my head, and as long as I stayed inside my head, I could go on writing the book. It was like living in a padded cell, but of all the lives I could have lived at that moment, it was the only one that made sense to me. I wasn't capable of being in the world. (2002: 55)

He also stresses: "All through the writing of the book, I intentionally put off thinking about the future" (2002: 56). As mentioned, it is only much later in time that he is able to write his story.

In *The Brooklyn Follies*, during a meal with Tom and Nathan at a French restaurant, Harry introduces to them his childhood dream of a happy society and of what he then called the "Hotel Existence". This conversation is transferred to the recipient in direct speech, in form of a play text. These modes of speech presentation not only strengthen the immediacy and the authenticity of what is said, but also, for the recipient/reader, broaden the role of the reader to that of the audience.

As a small child passing by a hotel during the war, Harry at first sees life and his existence as a hotel, which provides its guests with a comfortable refuge and shelter to make them happy.

HARRY:	(*closing his eyes; pressing his forefingers against his temples*): It's all coming back to me now. The Hotel Existence. […] The sole purpose of a hotel was to make you happy and comfortable […] A hotel represented the promise of a better world, a place that was more than just a place, but an opportunity, a chance to live inside your dreams.
NATHAN:	That explains the hotel part. Where did you find the word *existence?*
HARRY:	I heard it on the radio that Sunday afternoon. I was only half listening to the program, but someone was talking about *human existence,* and I liked the way it sounded. *The laws of existence,* the voice said, and *the perils we must face in the course of our existence.* Existence was bigger than just life. It was everyone's life all together, […]
TOM:	I still don't follow. You invent a place called the Hotel Existence, but where is it? What was it for?
HARRY:	For? Nothing, really. It was a retreat, a world I could visit in my mind. That's what we're talking about, no? (2006: 102f.)

Due to his war experience, Harry then modifies his idea of the "Hotel Existence" and turns it into a refuge for lost children, and he sees his own role as a savior of the lost and starving boys (2006: 103f.). Harry's somewhat pathetic, altruistic, and noble idea of himself as the hero and what he calls "the product of a rich inner life" is destroyed by his own superficialities, betrayals, eccentricities, and sexual encounters in his early life as a husband and a father. He is as stranded as Nathan and Tom, because of his frantic daughter, his time in prison after having forged the paintings of a dead painter, and the betrayal of his great love. Tom criticizes Harry's dream as "adolescent jerk-off material" (2006: 106) and describes his own vision of his "Hotel Existence" as a community of people he loves, he has lost during his life of crisis, but is related to:

TOM:	[...] What's my Hotel Existence, Harry? [...]
HARRY:	A commune.
TOM:	No, not a commune – a community. There's a difference.
HARRY:	And where would this little utopia of yours be?
TOM:	Somewhere out in the country, I suppose. A place with a lot of land and enough buildings to accommodate all the people who wanted to live there. (2006: 106)

In *The Chowder Inn* in Vermont, the image of the Hotel Existence comes back but even more dream-like. The narrator tells the reader, this time in the present tense that: "The sign at the edge of the driveway says *THE CHOWDER INN*, but a part of me already understands that we have come to the Hotel Existence" (2006: 169).

When Nathan realizes that Harry has given all his money to his lover Rufus and to Tom, Harry turns out to be the real human hero and his "Hotel Existence" becomes alive. Despite Harry's own deficiencies as a human being, he rescues Tom: "For all his wisecracking irreverence, for all his peccadilloes and falsehoods, he had never stopped believing in the principles of the Hotel Existence. Good old Harry Brightman. Funny old Harry Brightman" (2006: 215).

IMAGINING IS MEDICINE then functions as an elaboration of the blend of WRITING IS MEDICINE. Harry does not write anything but he takes refuge in his thoughts, which he later on shares with Tom and Nathan and which Nathan writes down. Although Harry's idea of the "Hotel Existence" is, as he calls it, "a retreat, a world I could visit in my mind" (2006: 103) and therefore not as visible a product as is a book, Harry's idea has strong resemblances with writing because both acts include processes of thinking and imagining.

5. Identification and tracing of lexical and semantic networks in Auster's *The Brooklyn Follies*: A corpus-based approach to WRITING IS MEDICINE

5.1 Preliminaries

This section tests whether and how the blend of WRITING IS MEDICINE can be measured and described in a text-oriented analysis. The assumption is that there are textual signals that establish the relation between writing and medicine and that draw the readers' attention to the blend. Auster's *The Brooklyn Follies* serves as a test-bed. I am inspired by Toolan's study of narrative progression and his classification of textual elements that contribute to narrativity (Toolan 2009: 1). Following a corpus stylistic approach, Toolan shows that the core ingredients of narrativity (suspense, surprise, secrecy or gaps, mystery, tension, obscurity) are created through mainly eight crucial features to which "high narrativity prospection" can be attributed (2009: Chapters 7–8). Among them are, for example, sentences containing high frequency keyword character names or sentences carrying lexical keywords and clusters.[9] My study focuses on lexical patterns, that is, on foregrounded keywords and key semantic fields. The question is whether there are crucial particularities of the lexical structure of *The Brooklyn Follies* that create and confirm the textual signaling towards WRITING IS MEDICINE.

5.2 Method

To identify key semantic fields and key words that are statistically foregrounded, I make use of Paul Rayson's (2007) *Wmatrix* program that is a tool for text comparison and investigates linguistic features, like words, parts of speech and semantic fields for their statistical keyness in the text to be investigated.[10] In order to be able to relate the statistical notion of keyness to deviation, a reference corpus underlies the notion of a norm of comparison. Words are then counted in the two texts and the frequency lists are automatically compiled as well as compared statistically to arrive at words listed in order of keyness (which refers to a "distinctive value"). A word that is statistically prominent in the first text appears at the top of the list. In other words, the words in the keyness list are ordered according to their keyness value, calculated as a positive score of log-likelihood (LL). Log-likelihood is a measure of significance, which can be compared to the chi-square test in the following way: "Any score at or above 3.8 is significant to the level of $p < 0.05$ in chi-square terms; any score at or above 6.6 is significant to the level of $p < 0.01$" (Leech 2008: 165). Leech prospects Toolan's idea of the function of textual elements as primes for narrative progression:

9. The idea of narrative progression, that is, the idea "that earlier sentences in a narrative shape the expectations we carry forward to our reading" (Toolan 2009: 15), alludes to other concepts, such as cohesion, adjacency pairs, Emmot's (1997) contextual frames, or Hoey's (2005) theory of lexical priming, which stresses collocational principles on all levels of language from the word to the genre and the context of usage.

10. For further information about Wmatrix, see http://ucrel.lancs.ac.uk/wmatrix/.

> As a heuristic strategy, it will be possible to run the program automatically over the text, and arrive at a set of linguistic features which are empirically derived 'good bets' to follow up in undertaking a subsequent stylistic analysis in terms of foregrounding. (2008: 164)

Similar to Leech's procedure of a Wmatrix analysis for Virginia Wolf's short story "The Mark on the Wall", my analysis of the blend of WRITING AS MEDICINE preceded the application of Wmatrix to *The Brooklyn Follies*.

As regards the reference corpus, Leech favors what he calls a "scatter-gun approach" (2008: 167) and opts for a multiple comparison with the different reference corpora, including, for example, different genres (see also Toolan 2009: 25–28, for the choice of a "comparator corpus"). *The Brooklyn Follies* was published in 2006. Therefore, a comparison with subcorpora of the *British National Corpus* (BNC) (e.g. BNC sampler imaginative or written) is possible. A reference corpus consisting of other Auster novels (American English) published between 1999 and 2006 has been created as well. The following tripartite base of comparison is suggested:

1. A comparison of *The Brooklyn Follies* with a corpus of three other novels of Paul Auster consisting of *The Book of Illusions* (2002), *Oracle Night* (2003), and *Timbuktu* (1999)
2. A comparison of *The Brooklyn Follies* with the BNC sampler imaginative
3. A comparison of *The Brooklyn Follies* with the BNC sampler written.[11]

As reference corpora, the BNC sampler written and the BNC sampler imaginative allow me to compare Auster's novel with other fictional and non-fictional *writing*, despite the fact that the BNC samplers do not represent the variety of American English. Therefore, *The Brooklyn Follies* is also compared with other novels written by Paul Auster. However, it should be stressed here again that the method chosen – that is, testing with the help of *Wmatrix* whether what has been qualitatively and cognitively described as a blend of WRITING IS MEDICINE can be measured quantitatively via recourse to a text-oriented and corpus linguistic analysis – is an initial attempt at illustrating the interplay (if not bi-directionality of) between textual connections and the effects these may have on the mental processes of the reader.

11. There are a number of flaws related to Wmatrix when using it as a statistical identifier of, for example, key semantic fields (Leech 2008: 167). Wmatrix does not yet analyze such qualitative features as metaphorical usage. But it provides us with a statistical profile of a whole text. Locally situated foregrounded features may not be immediately (statistically) recognized. Furthermore, linguistic features to be investigated in Wmatrix are word-based. The notion of clusters, patterns, constructions, for example, cannot be observed immediately. The qualitative potential of a computer program is limited, and, it is mentioned that there is an 8 per cent error rate for the UCREL Semantic Analysis System (USAS) tagging, which is rather high.

5.3 The results: Keywords and key semantic fields in Auster's *The Brooklyn Follies* (2006)

5.3.1 *Personal names*

There is no avoiding the evidence that both personal names as well as the first-person pronoun *I* most frequently occur by wide margins in Auster's *The Brooklyn Follies*. This becomes obvious in Tables 1, 2, and 3 provided in the Appendix, and regardless of whether the novel is compared with other Auster novels, the BNC sampler imaginative or the BNC sampler written. Also, Table 6 in the Appendix, which shows the most frequently occurring semantic fields in *The Brooklyn Follies* when compared with other Auster novels, lists personal names at the top. Although one would expect the names of protagonists to come high because they appear only in specific texts (that is, here *The Brooklyn Follies*), the fact remains that the novel abounds in references to personal names that are key. *The Brooklyn Follies* is a first-person narration and the first-person pronoun is quantitatively over-represented as well as reference is made to personal names (Toolan 2009: 53) (and also *The Book of Illusions*, which is among the novels in the reference corpus and is also a first-person narration).[12]

The high proliferation of personal names at the top ranks of the lists is one way of naming a frequently focalized character. The narrator does not avoid any direct reference. Towards the end of the novel, Nathan gives a possible reason for this strategy because he is able to express much more clearly what his goals are for the future. He aims at writing biographies to preserve other people's lives, which is expressed in the following quotation already cited above:

> My idea was this: to form a company that would publish books about the forgotton ones, to rescue the stories and facts and documents before they disappeared – and shape them into a continuous narrative, the narrative of a life. (2006: 303)

Naming is one linguistic strategy of identity attribution. *Tom* is the personal name, for example, which carries particular thematic importance. It is through Tom and writing about their reunion that both Nathan and Tom seem to improve their mental and physical health. Similar thematic importance can be attributed to the names *Harry* and *Rory*, which also appear high on the list in Tables 1 and 2.

The frequently occurring personal pronoun *I* foregrounds Nathan's preoccupation with himself and his life, as is visible in the examples quoted in Sections 4.1 and 4.2. Clusters like *I say*, *I mean*, *I suppose*, and *I think* are linguistic markers of stance, which show how Nathan expresses and modifies his statements.

12. In many of the 20th-century short stories analyzed by Toolan (2009) a personal name is also the top key word.

The highly frequent reference to mental states and processes is also illustrated in Table 8, where semantic fields in *The Brooklyn Follies* are compared with the BNC written sampler (position 26, but a LL of 63.66). Table 2 also illustrates the keyness of such words as *grieve*, *silence*, *insane*, *follies*, *thoughts*, and *realize*, which reflect the narrator's constant process of thinking and his writing about himself and the people around him. Nathan links his subjective and personal autodiegetic observations to the observations he makes about people. Nathan's constant reference to his family members is reinforced by an over-usage of words from the semantic field of kinship relations, which is illustrated in Table 6.

The concept of memory is crucial and can be related to the blend of WRITING IS MEDICINE. Naming is a direct and personal mode of reference. By way of implicature (Grice 1975), the foregrounded repetitive structure of naming that is characteristic of *The Brooklyn Follies* adds to Nathan's goal of memorizing other people's lives and their biographies. In a transferred sense, the preservation of Nathan's life and that of the people around him can also be seen as a kind of medical treatment. The function of the highly frequent personal names and the first-person personal pronoun is to transfer to the reader a sense of Nathan's self-conscious discussion with himself and others, on the one hand, but also a structured and coherent narration about report-worthy and emotional events in his life, on the other. These are exactly the alleviating functions that both literary critics like von Engelhardt (1991) and psychologist like Pennebaker (1997) have attributed to writing. Pennebaker (1997: 164f.), for example, gives ample evidence that when people write about their emotional experiences, significant improvements of physical and mental health follow. Experiments have also shown that in the patients' reports linguistic indicators predict improved health. Among them are, for example, coherently written texts and the inclusion of what is called "insider" words. The over-used forms of personal names and the first-person pronoun can be seen to fulfill both of those functions in *The Brooklyn Follies*. They are coherent by constantly reminding readers of the textual frame and functioning as its signals. At the same time, they are subjective linguistic indicators of who is important in Nathan's life.

5.3.2 *Statistically prominent reference to writing, time, place names, and specific grammatical forms*

In *The Brooklyn Follies*, over-used reference is also made to lexemes that denote writing, time, and place names. Although many of those direct references to, for example, *books*, *bookstore*, *manuscript*, *literature*, or *biography* occupy a lower position in Table 1 and 2, their figures of LL show that these words occur in a high frequency. For example, reference to *books* has a LL of 50.91 (Table 2). Table 7, which lists the most frequent semantic fields when *The Brooklyn Follies* is compared with the BNC sampler imaginative, shows that reference to "The Media: Books" is over-used with a LL of 100.64. The same holds true for reference to "Language, Speech and grammar" with a LL of 37.20. Table 2 also illustrates that over-used reference is made to famous writers such as James Joyce or Franz Kafka (LL of 47.40), although, for example, the name *Kafka* only occurs nineteen times in the novel. Yet, it is key when *The Brooklyn Follies* is compared with the BNC sampler written. The occurrence of this name clusters around a special but highly important part of the narrative, which is highlighted in Section 4.3.

Place names are over-used in the novel in the same way as references to time. This can be seen in Table 7, where *The Brooklyn Follies* is compared with the BNC sampler imaginative. The semantic fields of "Time: General" and "Time: Period" are most frequently occurring by a wide margin. A similar result can also be seen in Table 5 where the foregrounded status of temporal nouns, such as *months*, *years*, *days*, *minutes*, or *weeks* interact with the use of cardinal numbers, which help modify the reference to time in such examples as "thirty-three years of living under the same roof" (2006: 3). While the adverbial constructions of place and time reinforce the biographical character of this novel as well as the narrator's attempt to situate his narration within a time frame, the continuous references to the semantic field of writing construe and reflect the novel's preoccupation with writing.

What can I say about the textual base of the mental space of medicine in *The Brooklyn Follies*? Simply because direct textual reference has not emerged at the top of the keyword list and that of semantic fields, I cannot argue that it does not feature in the text. It is more indirectly created. Mention has already been made of the foregrounded reference to mental actions and processes (Tables 7 and 8). Furthermore, Table 7 shows that the semantic field of "alive" is key, which contains reference to *life*, *alive*, or *immortal*. Table 8 also shows that reference to the human body ("Anatomy and physiology", *eyes*, *body*, *hands*, or *tear*) creates a broader field with which medicine is preoccupied. In addition, however, I would like to draw attention to some grammatical features, which are key in the novel and which, though implicitly, contribute to the blend of WRITING IS MEDICINE. Among those are the conjunctions *if* and *but* and the use of negation (see Tables 4 and 5). *If* occurs 309 times in the novel, with a LL of 28.92 when the novel is compared with the BNC sampler imaginative (Table 5). The conjunction conventionally introduces conditional clauses (Carter and McCarthy 2006: §448) that deal with imagined situations: "Some are possible, some are unlikely, some are impossible. The speaker imagines that something can or cannot happen and then compares the situation with possible consequences or outcomes, or offers further logical conclusions about the situation".[13] It is through the choice of tense that differences in the likelihood of the situation, possible or imagined, are expressed. In *The Brooklyn Follies*, many of the *if*-clauses used are real conditionals (and not expressions of politeness, see Carter and McCarthy 2006: 456), and many of them are in the past conditional (in which the past tense occurs in the *if*-clause and the main clause contains the auxiliary verb in the past tense). In *if*-clauses in the past conditional, the speaker or writer responds to a possible or hypothetical situation by indicating a possible outcome. The condition must be fulfilled for the present or future to be different.

The repeated *if* in collocation with conditional clauses reflect the narrator's attempt to evaluate as well as to weigh up what he writes against other possible outcomes. An example is: "If the opportunity came along to add another notch to my belt, I wouldn't have said no, but it was hardly a matter of life and death" (2006: 84). Another example shows the narrator reflecting on what he considers to have done if Lucy had not liked their holiday journey to Vermont: "If she didn't like it there, we gather up her things and drive back to New York" (2006: 144). Both examples are apt ways of conveying the narrator's concern and preoccupation with particular parts of his life and of evaluating possible outcomes.

The frequent usage of *but* as coordinating conjunction has a similar function to *if* because, interpersonally and ideationally, it also construes contrast, while from a textual point of view, it creates coherence. It occurs 824 times in the novel with a LL of 83.40 when the novel is compared with the BNC sampler imaginative (see Table 5). According to Biber et al. (1999: 81), *but* is frequent in conversation and in fiction, and should also be seen in correlation with the high frequency of negatives in conversation, because negation and contrast are related concepts. Both these phenomena appear to be foregrounded in *The Brooklyn Follies*. Nathan uses *if* and *but* to interactively modify his statements or report about the modifying element in other people's discourse:

> God knows what regrettable words would have escaped my lips if I *hadn't* had those ten or twelve seconds to absorb the shock. I'm referring *not only* to the improbable fact that he was there, working as an underling in a secondhand bookstore, *but* also to his radically altered physical appearance. Tom had always been on the chunky side. (Auster 2006: 19, my emphasis)

13. Biber et al. (1999: 776) also treat *if* under the heading of stance adverbials and particularly circumstantial adverbials, which, generally speaking, express the attitude of the speaker.

Reporting to the reader about his feeling when running into his nephew Tom after a very long time, Nathan attempts to transfer his surprise to the reader by linking the phenomenon of recognizing Tom despite his altered physical appearance. It is through this linguistic technique that the reader is informed about his contrastive evaluation of the situation and about the ways Nathan retrospectively attempts to come to terms with his self through writing.

Significantly, negation is foregrounded in the novel. While it is true that negation is much part of spoken conversation, here it is frequently used not only in the stretches where spoken discourse is presented, but also in narrated passages. This strategy also adds to the reflective style of the novel. Especially syntactic negation, as in "[b]ecause Tom is too pure. He's too good, and he *doesn't* have a head for business" (2006: 128), is foregrounded in the novel.

The high proliferation of syntactic negation can be linked with the blend of WRITING IS MEDICINE. Nørgaard (2007) points out that negatives are more difficult to process, as one needs to imagine the positive, and there must be a good pragmatic reason why it is used (Leech 1983). The negated concept is given particular significance, because an entity is deleted, but at the same time evoked. As such negation creates a text-world in Werth's (1999) sense. In the example mentioned above, it is necessary to imagine Tom being a businessman before it can be understood that he is no businessman at all and, morally speaking, "a good person". According to Werth's text-world theory, the negative may here be seen as "an instruction to modify the world-building parameters that have already been set up" and as "a redefinition of the text-world" (1999: 252). In Werth's view, "there are forms of negation […] which not only delete an entity, but also introduce it at the same time" (1999: 253–254), thereby creating a textual sub-world within a given text world. In the following example of indirect thought, Nathan attempts to pretend that he is not aware of the fact that Lucy could not stay with him forever:

> It *hadn't* occurred to me that things would happen so fast. I felt relieved, of course, glad that our problem had been solved in such a quick and efficient way, but a part of me also felt disappointed, perhaps even a little bereft. I was beginning to like Lucy, and all during our shopping trip through the neighborhood, I had gradually warmed to the prospect of having her around for a while – days, I imagined, perhaps even weeks. It *wasn't that* I had changed my mind about the situation (she couldn't live in my apartment forever), but a short period would have been more than bearable for me. I had missed so many opportunities with Rachel when she was young, and now, suddenly, here was a little girl who needed looking after, who needed someone to buy clothes for her and put food in her stomach, who needed an adult with enough time on his hands to pay attention to her and try to draw her out of her baffling silence. (2006: 141, my emphasis)

Here, negation is part of the conversational style; it is also foregrounded because Nathan emphasizes that he is rather shocked by Tom's decision to take Lucy back to her mother. By stressing the opposite through the use of negation, Nathan's attempt at leveling arguments at one another becomes obvious. At the same time, the use of negation also illustrates how the positive always has to be implied in the creation of text worlds.

5.4 Statistical over-use of words and semantic fields as a direct pathway to foregrounding

I would like to ask again what it means if words appear, from a statistical point of view, in an abnormal frequency or, more specifically, if the personal pronoun *I* is one of the most over-used linguistic forms in the novel? It is doubtful whether a reader of Auster's novel consciously recognizes the statistical prominence of the keywords highlighted here. Accordingly, the results achieved in the keyword analysis as well as the general prominence of the blend of WRITING IS MEDICINE have to be tested against other readers' responses. This testing will be a future step, because text-based evidence needs to be established first.

The meaning-making, stylistic impact of quantitatively prominent language has to be correlated to the stylistic notion of foregrounding. In stylistics, foregrounding is seen as one of the most fundamental meaning-making devices that also has aesthetic effects because it is "a form of noticeable excess, markedness, prominence, contrast or departure from one or more of the text's multiple norms" (Toolan 2009: 71, see also Douthwaite 2000, van Peer and Hakemulder 2006). Foregrounding can take place on all levels of language, and it embraces both deviation from and parallelism to one or several norms. The theory of foregrounding, based on the Russian and Prague School of formalists, has most frequently been applied to literature. Foregrounding also has an immediate impact on the reception of a text, its interpretation and evaluation. Most recently, the psychological and cognitive factors that interact with the foregrounding have been further elaborated.

The relations between keywords and foregrounding in a stylistic sense are complex and cannot be immediately be equated on a one-to-one basis. Nonetheless, if a word or a semantic field is quantitatively prominent in a text under investigation this is a major attribute, although I am also not saying that these are the core words in the novel that simply and directly point to a thematic prominence or a specific motif.

Following Toolan (2009), I would like to argue that keywords in Auster's *The Brooklyn Follies* are also both explicit and implicit meaning-making signposts or signals (and, of course, not the only strategies that relate to the blend of WRITING IS MEDICINE, but I have focused on those). The ways readers interpret stretches of discourse are based on meaning inference processes that rely on both the explicit and the implicit. Corpus methodology fosters particular attention to the explicit lexical patterns. It is therefore assumed that keywords of a text and their co-texts are closely (but in a very complex way) interlinked with the meaning inference process they trigger in the reader. Keywords are attended to (Toolan 2009: 75) and, as illustrated, they point to some of the thematic preoccupations in the text. But this link does not always have to be direct, because "given that literary word-choice is often crucial to the particularity of effect and is made with extreme care, it seems reasonable to expect that there may be indirect links with both theme and plot" (Toolan 2009: 87). In addition, the signposting role of keywords and their repetitive re-occurrence may also function as simple reminding cues of what is foregrounded. In the course of a story, the reader has already been primed (Toolan 2009: 74) towards particular textual connections, which, when repeated, serve as means of reassurance. To draw an analogy with music or lyric poetry: these keywords function like a musical motif or a refrain that is recognized by the listener after having listened to it for a few times. However, there may also be variation on a musical, hence, textual theme. Also, some keywords in my analysis indirectly allude to the blend of WRITING IS MEDICINE. This is because language never happens in a vacuum, but in context, and meanings of words and phrases are inferred not only on the basis of a word's or sentence's stated meaning or denotation alone. Gricean implicature and other pragmatic inferences cued by the text are part of our reading. What readers bring to texts – inferences, schema and genre expectations, a variety of backgrounds, and cultural knowledge – plays an equally crucial role, because words on the page come to us integrated within a range of context of situation and culture. In order to meet Forceville's (2004: 87) demand that it "needs to be made clearer what counts as an input space [in Fauconnier and Turner's [2002] terms]" in a linguistic framework it is generated by complex textual foregrounding that interacts with context.

6. Conclusion

The blend WRITING IS MEDICINE in *The Brooklyn Follies* expresses the protagonist's writing as both a physical and a mental experience. It describes a healing project that is not initiated and conducted by any outward medical authority but by the individual in interrelationship with a "You". Auster's preoccupation with healing human beings in crises and re-attributing a sense of identity to them by giving them a voice through biography is holistic and comprises a view of medicine that is spiritual, almost transcendental at times. The process and the product of writing transform and immortalize the stories of the writer. They are therefore both alleviating and healing.

The blend, which functions on the intratextual level of the novel and also addresses the reader, who, by being a potential patient-participant, contributes to the blend, also alludes to psychological tests done in this area (Pennebaker 1997), bibliotherapy and poetry therapy – a method within the fields of bibliotherapy and poetry therapy, which uses literature (both its receptions and its production / or: both text production and text reception) as a process of healing diseases (Nevanlinna 2005). Although Auster in an interview with Céline Curiol in 2006 stresses that "[w]riting can certainly be dangerous. Dangerous for the reader – if something is powerful enough to change his view of the world – and dangerous for the writer" (Curiol 2006: 18), a therapeutic method – as used by bibliotherapy and poetry therapy is rather implicitly construed in the Auster novels analyzed here. In *The Book of Illusion* there is a much stronger critical stress on the illusion of the success of writing a story in fact and fiction than in *The Brooklyn Follies*, and it is difficult to clarify whether the effect that the blend expresses for the narrator in general and Nathan in particular is of equal status to the reader. Yet, in *Man in the Dark* (Auster 2008), the protagonist tries to temporarily create a world in which 9/11 never took place in order to come to terms with the death of his wife. In the German version of the interview with *Neue Zürcher Zeitung* (4 Oct 2008), Auster also stresses:

> Aber mir ist beim Schreiben stets sehr bewusst, dass die Sprache die Wirklichkeit immer verzerrt. Selbst wenn sie die Realität einerseits erschafft, verzerrt sie diese zugleich, und das ist eine der grossen Paradoxien des Schreibens. Sprache schenkt uns die Welt und nimmt sie uns zugleich, weil sie Kategorien schafft, während die Welt der Erfahrung keine festen Kategorien kennt und immer im Fluss der Bewegung bleibt.
>
> [When writing, I am always aware of the fact that language always distorts reality. Although it creates reality, it distorts it at the same time, and that is one of the great paradoxes of writing. Language presents the world to us, but at the same time takes it away, because it creates categories, while the world of experience does not know any fixed categories and always remains in the flux of movement.]
> (My translation)

One reason for the creation of the blend of WRITING IS MEDICINE is the explicit and implicit textual elements that are used and over-used in the novel. Although statistical keyness does not automatically point to foregroundedness, it has been illustrated that particular key words, parts of speech and semantic fields help construe as well as reflect the blend of WRITING IS MEDICINE in *The Brooklyn Follies*. Many of these key features, such as the conjunctions *if* and *but* as well as the high proliferation of negation, can be implicitly linked to the foregrounded features of the novel as they create the particular reflective and healing features of writing. Some of these features, I argue, are brought to the fore only by the Wmatrix analysis. Hence, it makes sense to pay attention to the stated message of the texts. One should never underestimate the reading process and, as Nathan in *The Brooklyn Follies* stresses "[o]ne should never underestimate the power of books" (2006: 304).

References

Auster, Paul. 1999. *Timbuktu*. London: Faber and Faber.

Auster, Paul. 2002. *The Book of Illusions*. London: Faber and Faber.

Auster, Paul. 2006. *The Brooklyn Follies*. New York: Henry Holt.

Auster, Paul. 2007. *The Inner Life of Martin Frost*. New York: Picador.

Auster, Paul. 2008. *Man in the Dark*. New York: Macmillan.

Auster, Paul. 2008. Sprache schenkt uns die Welt. *Neue Zürcher Zeitung*. 4 October 2008.

Bennett, C. Carlos, Yerka Luksic and M. Consuela Huidobro. 2007. The mirror of emptiness: Post traumatic stress disorder and dissociative phenomena in the literature of Paul Auster [El espejo del vacío: Trastorno por stress post traumático y fenómenso disociativos en la narrative de paul auster]. *Revista Chilena de Neuro-Psiquiatria* 45 (1): 51–58.

Biber, Douglas et al. 1999. *Longman Grammar of Spoken and Written English*. London: Longman.

Brandt, Line and Per Aage Brandt. 2005. Making sense of a blend. *Annual Review of Cognitive Linguistics* 3: 216–249.

Burke, Michael. 2005. How cognition can augment stylistic analysis. *European Journal of English Studies* 9 (2): 185–195.

Bus, Heiner. 1997. Jüdisch-amerikanische Literatur. In *Amerikanische Literaturgeschichte*, Hubert Zapf (ed.), 426–437. Stuttgart/Weimar: Metzler.

Carter, Ronald and Michael McCarthy. 2006. *The Cambridge Grammar of the English Language*. Cambridge: Cambridge University Press.

Coulson, Seana and Todd Oakley. 2005. Blended and coded meaning: Literal and figurative meaning in cognitive semantics. *Journal of Pragmatics* 37: 1510–1636.

Curiol, Céline. 2006. The making of *The Inner Life of Martin Frost*. – Céline Curiol in interview with Paul Auster. In *The Inner Life of Martin Frost*, P. Auster, 1–19. New York: Picador.

Dancygier, Barbara. 2006. What can blending do for you? *Language and Literature* 15 (1): 5–15.

Douthwaite, John. 2000. *Towards a Linguistic Theory of Foregrounding*. Torino: Edizioni dell'Orso.

Emmott, Catherine. 1997. *Narrative Comprehension: A Discourse Perspective*. Oxford: Oxford University Press.

Engelhardt, Dietrich von. 1991. *Medizin in der Literatur der Neuzeit I. Darstellung und Deutung*. Hartgenwald: Guido Pressler Verlag.

Evans, Vyvyan and Melanie Green. 2006. *Cognitive Linguistics. An Introduction*. Edinburgh: Edinburgh University Press.

Fauconnier, Gilles and Mark Turner. 2002. *The Way We Think: Conceptual Blending and the Mind's Hidden Complexities*. New York: Basic Books.

Fludernik, Monika. 1996. *Towards a 'Natural' Narratology*. London: Routledge.

Forceville, Charles. 2004. Book review: "The way we think. Conceptual Blending and the Mind's Hidden Complexities". *Metaphor and Symbol* 19 (1): 83–89.

Grice, Herbert P. 1975. Logic and conversation. In *Syntax and Semantics*, Vol. 3: *Speech Acts*, Peter Cole and Jerry L. Morgan (eds.), 41–58. New York: Academic Press.

Halliday, Michael A. K. 2004. *An Introduction to Functional Grammar* (3rd ed. rev. with Christian Matthiessen). London and New York: Arnold.

Hoey, Michael 2005. *Lexical Priming. A New Theory of Words and Language*. London: Routlegde.

Lakoff, George and Mark Johnson. 2003. *Metaphors We Live By* (2nd ed). Chicago: The University of Chicago Press.

Leech, Geoffrey N. 1983. *Principles of Pragmatics*. London: Longman.

Leech, Geoffrey N. 2008. *Language in Literature. Style and Foregrounding*. London: Longman.

Nänny, Max and Olga Fischer. 2006. Iconicity: Literary texts. In *Encyclopedia of Language and Linguistics*, Vol. 5 (2nd ed.), Keith Brown (ed.), 462–472. Amsterdam: Elsevier.

Nevanlinna, Anja. 2005. Was ist Poesie- und Bibliotherapie? Eine Einführung. In *Literatur und Medizin*, Peter Stulz, Frank Nager and Peter Schulz (eds.), 129–137. Zurich: Chroves.

Nørgaard, Nina 2007. Disordered collarettes and uncovered tables. Negative polarity as a stylistic device in Joyce's 'Two Gallants'. *Journal of Literary Semantics* 36: 35–52.

OED = *The Oxford English Dictionary*. 2nd ed. 1989. *OED Online*. Oxford University Press <http://www.oed.com>.

Palmer, Allan. 2004. *Fictional Minds*. Lincoln and London: University of Nebraska Press.

Peacock, Jim. 2006. Carrying the burden of representation: Paul Auster's *The Book of Illusion*. *Journal of American Studies* 40: 53–69.

Pennebaker, James W. 1997. Writing about emotional experiences as a therapeutic process. *Psychological Science* 8 (3): 162–166.

Rayson, Paul. 2007. *Wmatrix: A Web-based Corpus Processing Environment*. Lancaster: Computing Department, Lancaster University <http://ucrel.lancs.ac.uk/wmatrix/>.

Semino, Elena. 2002. A cognitive stylistic approach to mind style in narrative fiction. In *Cognitive Stylistics: Language and Cognition in Text Analysis*, Elena Semino and Jonathan Culpeper (eds.), 95–122. Amsterdam: Benjamins.

Semino, Elena. 2006. Blending and character's mental functioning in Virginia Woolf's 'Lappin and Lapinova'. *Language and Literature* 15 (1): 55–72.

Short, Michael. 1996. *Exploring the Language of Poems, Plays and Prose*. London and New York: Longman.

Toolan, Michael. 2001. *Narrative: A Critical Linguistic Introduction* (2nd ed.). London: Routledge.

Toolan, Michael. 2009. *Narrative Progression in the Short Story. A Corpus Stylistic Approach*. Amsterdam: Benjamins.

Toolan, Michael and Jean-Jacques Weber. 2005. Introduction. *European Journal of English Studies* 9 (2): 107–115.

Turner, Mark. 1996. *The Literary Mind: The Origins of Language and Thought*. New York: Oxford University Press.

Turner, Mark. 2003. Double-scope stories. In *Narrative Theory and the Cognitive Sciences*, David Herman (ed.), 117–142. Stanford: CSLI Publications.
Turner, Mark. 2006. Compression and representation. *Language and Literature* 15 (1): 17–27.
Ungerer, Friedrich and Hans-Jörg Schmid. 2006. *An Introduction to Cognitive Linguistics* (2nd ed.). London: Longman.
Van Peer, Willie and Jemeljian Hakemulder. 2006. Foregrounding. In *Encyclopaedia of Language and Linguistics*, Vol. 5 (2nd ed.), Keith Brown (ed.), 546–550. Amsterdam: Elsevier.
Werth, Paul. 1999. *Text Worlds: Representing Conceptual Space in Discourse*. London: Longman.

Appendix

Table 1. Keywords in *The Brooklyn Follies* compared with three other Auster novels

	keyword	absolute frequency	log-likelihood
1.	*Tom*	480	1161.45
2.	*Harry*	313	773.38
3.	*Lucy*	195	481.82
4.	*Nathan*	104	256.97
5.	*Aurora*	64	158.13
6.	*Gordon*	64	141.58
7.	*honey*	64	141.58
8.	*Stanley*	56	138.37
9.	*I*	2667	132.95
10.	*Rachel*	53	130.96
11.	*Nancy*	48	118.60
12.	*Rufus*	47	116.13
13.	*you*	1027	106.87
14.	*Rory*	41	101.30
15.	*our*	151	95.41
16.	*dryer*	38	93.89
17.	*Joyce*	37	91.42
18.	*David*	72	90.72
19.	*Nat*	35	86.48
20.	*Uncle*	50	79.33
21.	*Reverend*	32	79.07
22.	*Bette*	29	71.65
23.	*we*	448	70.16
24.	*Marina*	28	69.18
25.	*Avenue*	26	64.24
26.	*Pamela*	26	64.24
27.	*minor*	41	63.19
28.	*your*	190	59.61

	keyword	absolute frequency	log-likelihood
29.	*bob*	23	56.83
30.	*existence*	29	53.80
31.	*Brightman*	21	51.89
32.	*Harry Brightman*	21	51.89
33.	*chowder*	21	51.89
34.	*'m*	229	50.87
35.	*B.P.M.*	20	49.42
36.	*Poe*	19	46.95
37.	*Seventh*	18	44.48
38.	*Trumbell*	18	44.48
39.	*Flora*	17	42.00
40.	*diner*	17	42.00
41.	*husband*	58	41.71
42.	*us*	147	41.66
43.	*necklace*	25	41.25
44.	*Smith*	23	40.29
45.	*Kafka*	19	39.69
[…]			
53.	*bookstore*	13	32.12

Table 2. Keywords in *The Brooklyn Follies* compared with the BNC imaginative

	keyword	absolute frequency	log-likelihood
1.	*Tom*	480	974.99
2.	*Harry*	313	711.46
3.	*Lucy*	195	474.64
4.	*Nathan*	104	259.47
5.	*I*	2667	184.90
6.	*Aurora*	64	159.68
7.	*Gordon*	64	143.10
8.	*Stanley*	56	139.72
9.	*David*	72	137.09
10.	*Rachel*	53	132.23
11.	*Honey*	64	131.96
12.	*New York*	56	130.32
13.	*Had*	893	129.57
14.	*Uncle*	50	124.75
15.	*Nancy*	48	119.76
16.	*Rufus*	47	117.26
17.	*Brooklyn*	43	107.28
18.	*About*	402	103.82
19.	*n't*	896	103.32
20.	*To*	2491	101.62

	keyword	absolute frequency	log-likelihood
21.	*Dryer*	38	94.81
22.	*Apartment*	41	93.52
23.	*Joyce*	37	92.31
24.	*Nat*	35	87.32
25.	*Person*	74	83.70
26.	*That*	1279	82.17
27.	*But*	838	81.06
28.	*Anymore*	35	78.86
29.	*Months*	62	75.01
30.	*City*	30	74.85
[...]			
51.	*books*	46	50.91
55.	*Kafka*	19	47.40
56.	*Poe*	19	47.40
78.	*bookstore*	46	32.43
80.	*manuscript*	17	30.98

Table 3. Keywords in *The Brooklyn Follies* compared with the BNC written

	keyword	absolute frequency	log-likelihood
1.	*I*	2667	3060.63
2.	*Tom*	480	2109.25
3.	*Harry*	313	1464.32
4.	*n't*	896	1339.69
5.	*her*	934	1098.79
6.	*Lucy*	195	950.08
7.	*me*	602	750.88
8.	*she*	813	724.71
9.	*had*	893	617.74
10.	*he*	1170	616.81
11.	*you*	1027	613.94
12.	*my*	626	592.09
13.	*Nathan*	104	502.19
14.	*was*	1450	432.14
15.	*'m*	229	395.02
16.	*him*	384	354.02
17.	*but*	838	316.08
18.	*about*	402	312.72
19.	*Aurora*	64	298.31
20.	*what*	422	287.92
21.	*do*	428	282.22
22.	*Gordon*	64	280.89
23.	*Rachel*	53	261.59

	keyword	absolute frequency	log-likelihood
24.	*David*	72	256.97
25.	*Stanley*	56	253.21
26.	*Uncle*	50	246.78
27.	*that*	1279	246.26
28.	*honey*	64	244.95
29.	*'re*	179	232.12
30.	*Rufus*	47	231.98
31.	*Brooklyn*	43	212.23
32.	*'d*	154	197.31
33.	*Nancy*	48	196.22
34.	*did*	282	194.47
35.	*New York*	56	187.49
36.	*Joyce*	37	182.62
37.	*apartment*	41	180.99
38.	*his*	663	178.11
39.	*just*	245	174.07
40.	*dryer*	38	172.03
41.	*myself*	94	171.78
42.	*know*	179	167.48

Table 4. Most frequent parts of speech in *The Brooklyn Follies* in comparison with three other Auster novels

	part of speech	absolute frequency	log-likelihood
1.	pers. pronoun *I*	2685	136.41
2.	pers. pronoun *you*	1061	116.58
3.	base form lexical verb	1274	82.37
4.	sgl. proper noun	2983	74.55
5.	1st-person pl. pers. pronoun *we*	448	70.16
6.	general adjective	5298	63.72
7.	interjection	211	62.30
8.	*am*	269	58.99
9.	*do* base form	327	44.73
10.	*me*	147	41.11
11.	*is*	340	29.16
12.	*have*	257	20.12
13.	poss. pronoun	2300	14.77
14.	degree adv.	454	13.49
15.	determiner	302	12.00
16.	*wh*-determiner	520	10.37
17.	*for*	783	7.33
18.	adverb of time	6	7.20

Table 5. Most frequent parts of speech in *The Brooklyn Follies* in comparison with the BNC imaginative

	keyword	absolute frequency	log-likelihood
1.	pers. pronoun *I*	2685	167.67
2.	*had*	1007	149.98
3.	temporal nouns	375	136.82
4.	inf. marker *to*	1622	122.47
5.	*but* conj.	824	83.40
6.	*it*	810	83.35
7.	cardinal numbers	661	74.23
8.	*that* conj.	713	61.38
9.	*was*	1481	60.41
10.	*me*	603	47.28
11.	past participle of lexical verb	1811	42.65
12.	*for* prep.	783	40.70
13.	*than* conj.	173	40.00
14.	weekday noun	63	34.95
15.	*of* prep.	1832	31.31
16.	sgl. art.	2389	27.19
17.	*have* inf.	288	27.17
18.	*been*	297	27.00
19.	prep.	71	24.36
20.	num. noun	38	23.79
21.	*wh-ever* adv.	22	23.63
22.	ordinal number	295	22.15
23.	*before* clause marker	13	16.59
24.	prep.	5146	16.53
25.	coord. conj.	3081	16.30
26.	sgl. month	50	16.06
27.	gener. superl. adj.	99	15.66
28.	adj.	5298	15.27
29.	subord. conj.	121	14.55
30.	capable of pronominal function	141	13.43

Table 6. Most frequent semantic fields in *The Brooklyn Follies* compared with three other Auster novels

	semantic category	absolute frequency	log-likelihood	semantic field
1.	S4	714	183.42	Kin
2.	Z1	2238	149.32	Personal names
3.	S9	280	47.91	Religion and the supernatural
4.	F1	428	42.26	Food

	semantic category	absolute frequency	log-likelihood	semantic field
5.	S2	328	38.02	People
6.	Z8	13813	33.52	Pronouns
7.	E2+	223	28.26	Like
8.	G1.2	51	27.73	Politics
9.	T3–	192	26.50	Time: New and young and young
10.	I2.2	210	25.09	Business: Selling
11.	Q2.2	1008	23.41	Speech acts

Table 7. Most frequent semantic fields in *The Brooklyn Follies* compared with the BNC imaginative

	semantic category	absolute frequency	log-likelihood	semantic field
1.	L1+	156	223.02	Alive
2.	A13	61	152.19	Degree
3.	T1.1	64	131.96	Time: General
4.	T1.3	896	104.81	Time: Period
5.	Q4.1	220	100.64	The Media: Books
6.	T1.2	236	93.58	Time: momentary
7.	W2	37	92.31	Light
8.	I2.2	210	76.38	Business: Selling
9.	N5–––	43	65.46	Quantities: little
10.	W2–	26	64.87	Darkness
11.	X9.2–	87	61.16	Failure
12.	N1	811	57.22	Numbers
13.	I3.1	244	53.41	Work and employment: Generally
14.	Z7	354	45.93	If
15.	S7.1––	16	39.92	No power
16.	X2.6–	48	38.69	Unexpected
17.	Q3	247	37.20	Language, speech and grammar
18.	A13.7	103	37.07	Degree: Minimizers
19.	T2+++	45	32.81	Time: beginning
20.	T1.1.1	227	27.09	Time: Past
21.	Z8	13813	26.17	Pronouns
22.	T2+	239	25.87	Time: beginning
23.	A2.2	305	24.64	Cause&Effect/Connection
24.	A5.1+++	81	23.55	Evaluation: Good
25.	X2	33	23.20	Mental actions and processes

Table 8. Most frequent semantic fields in *The Brooklyn Follies* compared with the BNC written

	semantic category	absolute frequency	log-likelihood	semantic field
1.	Z8	13813	5194.48	Pronouns
2.	Z6	1577	623.28	Negative
3.	L1+	156	457.37	Alive
4.	A13	61	301.08	Degree
5.	Z1	2238	265.49	Personal names
6.	B1	937	258.69	Anatomy and physiology
7.	S4	714	223.23	Kin
8.	Q2.1	1088	219.69	Speech: Communicative
9.	X2.2+	470	204.08	Knowledgeable
10.	W2	37	182.62	Light
11.	T1.1	64	172.94	Time: General
12.	X2.1	685	172.26	Thought, belief
13.	A3+	2846	137.46	Existing
14.	W2–	26	128.33	Darkness
15.	S1.2.6–	86	119.16	Foolish
16.	S3.2	137	99.09	Relationship: Intimacy and sex
17.	Z7	354	98.11	If
18.	E4.1+	253	86.62	Happy
19.	T1.2	236	85.25	Time: Momentary
20.	N5–––	43	78.54	Quantities: little
21.	X2.5+	128	72.03	Understanding
22.	E4.1–	188	70.92	Sad
23.	M8	136	70.81	Stationary
24.	X2.6-	48	68.12	Unexpected
25.	F1	428	65.10	Food
26.	X2	33	63.66	Mental actions and processes

PART II

Challenges to and limitations on bi-directionality

Collective aesthetics and the Mere Exposure Effect

Alexandra Kleeman
University of California, Berkeley

Gertrude Stein's seminal essay "Composition as Explanation" analyzes the temporal and social factors contributing to the perception of beauty in a work of art. Readers processing Stein's text, composed of many repetitions of highly convoluted grammatical structures, find that they acquire the ability to comprehend the text through the very process of reading it. Thus, Stein's writing is closely linked to the Mere Exposure Effect (MEE), a psychological phenomenon related to implicit learning and conditioning. These two artistic and empirical bodies of knowledge constitute an interdisciplinary perspective from which we may challenge both the assumptions of the cognitive sciences concerning the "non-aesthetic" dimensions of experimental stimuli, and the premise that aesthetic value may be dissociated from the contingencies of production and distribution.

Keywords: artistic and empirical knowledge, conditioning, experimental stimuli, Gertrude Stein, perception of beauty, text comprehension

> Of course it is beautiful but first all beauty in it is denied and then all the beauty of it is accepted [...] Automatically with the acceptance of the time sense comes the recognition of the beauty and once the beauty is accepted the beauty never fails any one. (Stein 1990: 515)

1. Introduction

In the last several decades, the literary canon has increasingly been viewed as both a contested space in which political power is used to exclude particular races, genders, or classes, and a constructed space that reflects collective values and aesthetic preferences. The construction of the canon, and the possibility of its reconstruction, have thus become a focal point for critics who believe that literature is not merely reflective of the social, but is in fact a vital force in its creation. In this view, the literary canon is both the concrete product and the active producer of collective aesthetics. Actors introduce changes to the canon of their time by adding new works, removing works that have come to be considered irrelevant or inappropriate, and reassessing the canon in relation to new systems of value. These alterations reverberate in the reading practices of subsequent generations, shaping their aesthetic criteria and affecting their own modifications to the corpus of literary tradition.

Driven by this recognition of the bi-directional influence of literary texts and social collectives, new approaches to literary historiography have been outlined by theorists such as Franco Moretti, who argues that a field as large as that of literary production "cannot be understood by stitching together separate bits of knowledge about individual cases, because it isn't a sum of individual cases: it's a collective system, that should be grasped as such, as a whole" (Moretti 2005: 4).

In the first chapter of his book *Graphs, Maps, Trees*, Moretti performs an extensive analysis of the annual output of British novels over a two-hundred year span, sorted according to 44 major novelistic genres (e.g. the Utopian novel, the Epistolary novel, the Courtship novel). Moretti finds that new novelistic genres tend to emerge and disappear in clusters, with five or six new types of novels emerging at roughly the same time, remaining static for about twenty-five years, and then being replaced by five or six new forms simultaneously at roughly the end of that time span. Lacking a clear explanation for this phenomenon, Moretti suggests that these destabilizations could correspond, roughly, to periods of generational stability – to shared mental climate and cultural values, perhaps among the middle-aged members of society, who tend to possess economic and artistic power (Moretti 2005: 21).

This unexpected correlation between population demographics and literary genre points to a larger role for collective influences in literary production and reception than has traditionally been supposed. Though the rhetoric of canonization prefers to portray literary merit as a concrete property discovered within particular texts, new approaches argue that literary value is a construct of social and psychological processes that often escape conscious awareness. Aesthetic evaluation is oblivious to its own blind spots, unable to recognize and identify influences that operate on either a larger-scale collective level, or at the level of subconscious, automatic psychological mechanisms. Compensating for these blind spots calls for an interdisciplinary approach combining the methodology of literary studies with that of sociology and psychology, in an attempt to trace social or psychological dynamics that could help to explain phenomena such as the acceptance of avant-garde literary texts into the canon. A bi-directional view of language processing and literature could make each phenomenon visible as more than a localized, naturalized unit of study. Viewed from the dual perspectives of literary and cognitive studies, language becomes a temporally and affectively rich practice, linking individual readers to an extended society of reading subjects.

Gertrude Stein's essay "Composition as Explanation" describes factors that contribute to the popular perception of beauty in a work of art, while reenacting the movement from refusal to acceptance in the formal and stylistic structure of the text. Her essay speaks to the effect of familiarity on the perceived beauty of an object and of the process of canonization, wherein a work long perceived as unappealing suddenly achieves a beauty in the eyes of its audience that, once installed, cannot be easily undone. Stein notes that public approval of a work of art is closely linked to temporal and social factors, stating that "for a very long time almost everybody refuses and then almost without a pause everybody accepts" (Stein 1990: 515). At the same time, Stein demonstrates to the reader how the acquisition of this acceptance occurs, using formal devices that initially obstruct the reader's access to meaning, but become comprehensible and even beautiful as the act of reading progresses. In this respect, Stein's experimental writing converges with empirical investigations of the Mere Exposure Effect (MEE), a psychological phenomenon linked to implicit learning and conditioning, in which simple exposure to symbols and structures enhances their attractiveness on subsequent viewings. A close reading of Stein's text in conjunction with recent experimental work on the relationship between mere exposure and the artistic canon can help to explain why new forms of art take time to be welcomed into the preexisting canon, how cultural factors affect our experience of beauty, and how basic psychological processes involved in ordinary aspects of human cognition influence our aesthetic evaluation of art.

2. Mere exposure and affective judgment

Researchers in the cognitive sciences have often neglected the aesthetic and affective dimensions of language for reasons both theoretical and methodological. Empirical investigations of aesthetics are hindered by the view that aesthetic impulses are peripheral to cognition proper, as well as by the notion that art is created through individual acts of transcendent inspiration unrelated to everyday behavior. Researchers dealing with aesthetic effects also encounter the difficulty of replicating a complex naturalistic activity such as the appreciation of art within a controlled laboratory setting. As a result, psychological research on aesthetics is underdeveloped, centered mostly on the demonstration of empirical differences in the brain activation of individuals looking at different types of artistic stimuli, or comparing the behavior of experts and nonexperts in passive viewing of artistic objects (Zeki and Lamb 1994, Ramachandran and Hirstein 1999). Such studies succeed in linking their efforts to well-established branches of cognitive science, yet leave the phenomenal experience of being positively or negatively affected by aesthetic stimuli more or less unaddressed.

By contrast, the Mere Exposure Effect is a rare instance of an empirically-researched psychological effect that is both central to the issue of aesthetic experience and well substantiated by a long history of experimental investigation. It refers to instances in which multiple exposures to a stimuli increases subject preference for that particular stimuli, and was first discovered during a series of experiments conducted by the psychologist Robert Zajonc in 1968. Using Chinese ideograms and nonsense words as stimuli, Zajonc showed that subjects' ratings for "goodness of meaning" are positively correlated with the frequency with which they have been exposed to the stimuli, demonstrating a strong link between passive exposure and affect. When presented with a series of nonsense words (*dilikli*, for example) at frequencies ranging from 0 to 25 exposures, subjects rated the stimuli more positively the more times they had seen them. In other words, merely viewing an item under conditions that render the given stimulus available to perception increased its positive affect in subsequent rating tasks (Zajonc 1968).

In the two decades that followed Zajonc's initial discovery, over two hundred studies of MEEs were conducted, examining the relationship upon effect size of parameters such as stimulus recognition, duration of exposure, stimulus complexity, number of exposures, and delay between exposures (see Bornstein 1988). Recent experiments have expanded the scope of the project to include neuroimaging research (Elliott and Dolan 1998), and studies relating affect to aspects of processing dynamics such as perceptual fluency (Winkielman and Cacioppo 2001).

Robust effects of mere exposure upon affective evaluation have been found in nearly every study of the relation between exposure and affect, evidencing its widespread influence across species, genders, cultures, and age groups. The effect has been shown at very low exposure rates, appearing after even a single pre-exposure (Monahan, Murphy, and Zajonc 2000), and has been discovered across experiments using a wide variety of stimulus types, from geometric shapes to musical phrases and abstract art. Structural MEEs showing increased preference for previously exposed structures containing novel content have been demonstrated in grammaticality judgments (Luka and Barsalou 2005) and in artificial grammar learning paradigms, under conditions in which the subject is already familiar with the structural elements of the grammar (Reber et al. 2004). The demonstration of MEEs across these diverse stimuli and participant categories argue that they are highly influential in our assessment of the affective value of stimuli.

The effect of mere exposure upon affective judgments of neutral stimuli has startling implications for the nature of the artistic canon: if exposure does, in fact, play a significant role in the evaluation of art within a real-world setting, then artistic canons may form as the result of a type of feedback loop between society and individual. An individual encountering a particular artwork or style of art multiple times will begin to find that art more appealing, and will therefore help to increase its prevalence within society by placing reproductions of it within his home, increasing its appearance in publication, or attending museums that display the same work of art, or works of art similar in style. The efforts of the individual increase the frequency of the artwork in society, and cause an increase in its frequency of exposure for other members of society – which, in turn, increases its appeal to other individuals who begin to encounter the artwork more often. Thus, mere exposure could lead to a rapid cyclical increase in the perceived value of a work of art, establishing widely supported canons that are based in large part upon contextual factors in addition to factors intrinsic to the work of art.

A recent experiment by James Cutting (2003) has tested these claims, asking whether the laboratory phenomenon of mere exposure can truly be applied to a process as complex as the artistic canon. Cutting's experiment explores the effect of mere exposure through an ecologically inspired design that uses actual Impressionist paintings to investigate the effect of publication frequency and mere exposure on aesthetic evaluation. Cutting selected 66 paintings from the collection of Gustave Caillebotte, a collector of Impressionist art who was widely believed to have impeccable taste. 66 non-Caillebotte paintings were selected as controls and matched to the Caillebotte stimuli in terms of colorization, subject matter, and publication frequency – determined through a careful count of each painting's frequency of appearance in Cornell University library collections. Subjects in three different age groups then viewed sets of slides, displayed for eight seconds, featuring one Caillebotte and one control painting, before stating which painting they preferred. Statistical analyses of preference data showed no preference and no increase in recognizability for images from either the Caillebotte or the control group, and instead showed a reliable correlation between publication frequency and preference. Although subjects did show a slightly higher preference for images displayed in the Musée D'Orsay, this preference disappeared once the effects of publication frequency for the D'Orsay's higher-profile artworks were factored out. Cutting's experimental data effectively demonstrates that preference for complex stimuli such as paintings is linked to the prevalence of artworks in society, and to exposure effects arising from their prevalence.

Because this correlation between frequency and preference could also be explained as an effect of the inherent quality of a painting, with individual viewers and publishers alike agreeing upon the objective merit of the artwork, Cutting followed up these experiments with another using the same paintings as experimental stimuli. In this follow-up investigation, students taking a perception course were shown PowerPoint slides of Impressionist paintings throughout the course of a semester. Low-frequency images were shown four times over the semester, while high-frequency images were shown only once. At the end of the course, students were shown pairs of slides corresponding to the stimuli used in preceding experiments, and given a questionnaire asking which painting they preferred. Subject data showed no correlation between preference and frequency, with a mere 48% of subjects preferring the more common image (compared to 57% in the earlier experiments). Elimination of the correlation between high publication frequency and individual preference through passive exposure is evidence of the powerful influence of mere exposure on aesthetic judgment. In addition, this experiment provides evidence of the robustness of the MEE: even at interexposure intervals of 10 to 80 days, the effect of four two-second long exposures was great enough to negate the sum total of subjects' exposure to high-frequency artworks.

Cutting's use of Impressionist art as experimental stimuli is especially interesting in the context of the history of the Impressionist movement and the incorporation of fringe art into the artistic mainstream. The Impressionist movement in painting provides a classic example of an artistic movement's transformation from refusal to acceptance: at its inception, Impressionism was considered aesthetically offensive to realist sensibilities, and it was only after several years of gradual exposure to Impressionist work that the movement was almost unanimously embraced. When Gustave Caillebotte died, his entire collection of Impressionist artworks was offered to the Musée D'Orsay, which reluctantly took possession of approximately 38 of the 70 or so paintings offered (many of which were thought to be "unsellable", and were bought by Caillebotte in order to support painters that he had befriended). Over time, the Caillebotte collection housed at the Musée D'Orsay came to prominence, and is now thought to reflect Caillebotte's superb taste in art. What Cutting's experiments indicate, however, is that their placement in a prominent art museum contributes as much to their prominent placement in the artistic canon as any internal quality of the art. The process of canon formation is thus revealed to be a self-perpetuating, self-organizing dynamic system, heavily influenced by basic cognitive processes linked to feedback and implicit learning.

3. Mere exposure as explanation

Mere exposure provides a possible explanation for why "everybody refuses and then almost without a pause everybody accepts" (Stein 1990: 515): collective preference for art, on a macro scale, is influenced by cultural factors such as prevalence and existing status within the canon. In this way, the psychological tendency to prefer pre-encountered stimuli functions to maintain pre-existing valuations, or even to further their entrenchment in the aesthetic economy. At the same time, mere exposure as a form of implicit learning can subvert existing systems of aesthetic valuation, given a reader engaged with a text that actively works within itself to re-expose the reader to its novel formal qualities. A reader encountering Gertrude Stein's "Composition as Explanation" for the first time is likely to find himself slowed by its difficult syntactic form and recursive, reiterative phrasing. As the reading progresses, however, repeated engagement with Steinian syntax models modes of re-exposure that characterize more canonical literary works, and thus reenacts its own progression from refusal to acceptance. It produces, in effect, its own skilled reader.

Stein begins her essay with the difficult phrase

> [t]here is singularly nothing that makes a difference a difference in beginning and in the middle and in ending except that each generation has something different at which they are all looking". (Stein 1990: 513)

This phrase situates the reader's initial engagement with the text in a vertiginous space, marking a discontinuity between the style of the essay and those formal qualities that characterize the pre-existing conventions of composition. Stein presents a grammatical object foreign to the pre-exposed patterns of conventional language use, an object that does not participate in an existing exposural feedback loop, an object grounded in a language whose comprehensibility is as yet unassimilated. The "brokenness" of her syntax has its own repetitive structure, a broken syntax that becomes a syntax of fracture, and then simply another mode of syntactic construction and processing. Patterns of breakage and recursion generalize to a positive, secondary pattern that enters the economy of exposure and repetition, facilitating fluent cognitive processing.

This language is heterogeneous in the extreme, using terminology that is at times childlike, at others borrowed from academic discourse. Commonplace nouns possessing a generality that makes them at once abstract and highly concrete play key roles in many sentences:

> [s]o then naturally it was natural that one thing an enormously long thing was not everything an enormously short thing was also not everything nor was it all of it a continuous present thing nor was it always and always beginning again. (Stein 1990: 519)

These "things" and "everythings" link recursively back to pronouns given earlier in the sentence or in earlier sentences, pronouns with referents that appear to shift as the process of reading continues and more of the composition is revealed. Even as the reference of pronouns and general terms appears unfixed, relatively precise terms are rendered unstable by their placement within the same highly recursive, repetitious grammatical structures:

> [a]nd so there was the natural phenomena that was war, which had been, before war came, several generations behind the contemporary composition, because it became war and so completely needed to be contemporary became completely contemporary and so created the completed recognition of the contemporary composition. (Stein 1990: 521)

"Contemporary", a word that generally possesses localizable meaning, comes unmoored within the context of a Steinian syntax. Modified in reference by each new instantiation, "contemporary" becomes multiple, necessitating a distinctly different style of reading that defers evaluation indefinitely.

Stein's distinctive syntax is the scaffold for a language that is unfamiliar, and even unpleasant to read. The lengthy sentences found in "Composition as Explanation" may contain multiple restructurings of subject-object relations as well as conversational interjections and the explicit contradiction of statements made earlier in the sentence. Highly repetitive syntax and word use make it possible for words to function as subject, objective, or modifier, while the length of the sentences in which they are embedded increases the difficulty of holding onto any particular semantic orientation. As consistency is visible even in the differences among and between sentences, the reader repeatedly encounters similarly convoluted patterns, a sequence of stimuli united by their distributed rather than localized syntax:

> [a]nd so now one finds oneself interesting oneself in an equilibration, that of course means words as well as things and distribution as well as between themselves between the words and themselves and the things and themselves, a distribution as distribution. This makes what follows what follows and now there is every reason why there should be an arrangement made. Distribution is interesting and equilibration is interesting when a continuous present and a beginning again and again and using everything and everything alike and everything naturally simply different has been done. (Stein 1990: 521)

Paragraphs like this show a pattern of rhythm, recursion, and reassembly that give the sentences consistency, and therefore have the effect of re-exposing the reader to a grammar that was originally unfamiliar and disruptive.

It is through this internal mechanism of exposure and re-exposure that "Composition as Explanation" performs its own progression from refusal to acceptance. Stein's style, juxtaposing highly repetitious syntax and lengthy sentences with a curious semantic opacity, is present throughout the essay and feels increasingly natural as reading progresses. By the essay's conclusion, her style of writing no longer alarms: the reader has reached his or her own acceptance of this new form of art, replicating in miniature the progression of art from refusal to acceptance. Just as experimental subjects begin to find 'ungrammatical' sentences more acceptable and appreciably after multiple exposures, the reader has gained the ability to read what was previously illegible. Barbara Herrnstein Smith observes that aesthetic judgments never engage "raw" objects:

> [n]ot only are the objects we encounter always to some extent pre-interpreted and pre-classified for us by our particular cultures and languages; they are also pre-evaluated, bearing the marks and signs of their prior valuing and evaluations by our fellow creatures. (Smith 1988: 43)

Ultimately, Stein's text demonstrates that no aesthetic judgment occurs prior to conditioning or escapes it: learning to recognize beauty is, tautologically, a pre-condition for its recognition.

Stein's essay thus has multiple affinities with mere exposure experiments, especially those concerned with exposure effects for grammatical structures and patterns. But how should we address the distance between the carefully controlled mere exposure studies of simple stimuli and the analysis of a complex work of art? Exposure alone obviously cannot provide an exhaustive account of aesthetic value, or address all of its diverse dimensions. One answer may be that exposure plays a significant role in facilitating fluency of processing. Reber et al. (2004) have shown that the perceiver's aesthetic response is more positive when the object perceived can be processed fluently (Reber et al. 2004). After reaching an exposural plateau, the text becomes more accessible: the effort of reading is minimal, and other processes may then play a larger role in evaluating pleasurability. Thus, the Mere Exposure Effect may play a large role in rendering the content of a text accessible to its readers, clearing a space in which other modes of aesthetic evaluation may function. By repeatedly exposing the reader to linguistic units that reflect her own style and approach to narration, Stein makes her unconventional syntax more fluent, more appealing. As the reader monitors his own reactions to the essay throughout the course of reading, she or he notices that sentences no longer feel as "difficult", and that she or he is able to spend more resources attending to the content of the sentences, or to other aesthetic factors related to their form, such as rhythm. As a result, the same structures begin to reveal a previously unacknowledged or inaccessible beauty. Whether this effect is related to perceptual fluency or to a conditioned positive response to stimuli that have been shown to have no negative effects, the end result is the same: the inherent beauty of a refused object is exposed, its refusal turns to acceptance.

Stein's model of literary experimentation takes the text as a mechanism for uncovering the psychology of the reader, even as it reconstitutes the reader as a differently configured mind. The performative effects of literary experiments thus parallel those of empirical investigation.

Having identified a certain homology between literary experimentation and psychological experimentation, we turn toward the question of bi-directional translation between literary studies and psychological science. The relevance of mere exposure to literary scholars and sociologists exploring popular reception of literature and collective aesthetics is more readily identifiable than its utility to scientists studying the mechanisms of cognition. What possible benefits are there that a consideration of experimental literary texts such as Stein's offers the empirical study of the mind?

The first of these possible contributions is the introduction of a new and complex object of study to researchers working especially in the domain of language acquisition and processing. The text that configures its own reader receives much consideration by theorists and practitioners of avant-garde literature, but has attracted little attention from researchers in the cognitive sciences, who tend to avoid the use of longer, more complicated texts as experimental stimuli, and who generally treat language as a stable system that undergoes few changes after language skills have been acquired. Yet these texts create unique and identifiable effects in readers, effects distinct from those of the fragmentary and disjointed stimuli that are most often used in language processing experiments, effects that may be able to offer insights into the global processing of holistic texts. Our understanding of how these unusual texts are handled within the cognitive system is at best hazy – further investigation could help to answer questions such as whether reading a particular avant-garde text facilitates the processing of other similarly difficult texts – in other words, are acquired skills generalized and applied to subsequent texts, or confined to that particular text or its particular stylistic qualities? Does the increased grammaticality of the aberrant text affect the reader's subsequent aesthetic judgment of more traditional texts?

Attending to the structure and dynamics of the experimental text could also provide researchers in the cognitive sciences with a new perspective on current experimental practices and methodologies. Examples of experimental language use provide a contrast to the linguistic stimuli found in most cognitive studies, denaturalizing them, foregrounding the artifice underlying their construction, and revealing the disciplinary linguistic norms they embody. In light of exposural effects operating over extended durations, researchers may want to reconsider using stimuli that lack syntactic variation. Though stimuli are often made grammatically homogenous in order to control for unexpected psychological effects, repetition and re-exposure to syntactic forms may have a significant effect on processing fluency, making such experiments inadvertent investigations of language learning.

Gertrude Stein's "Composition as Explanation" thus provides an example of how mere exposure and cultural prevalence affects the psychology of individuals, while interrogating its effects upon the formation of an artistic canon. In its modification of the affective response of individuals during the process of reading, we see one possible mechanism for moving a work from refusal to acceptance. At the same time, these reading practices and formal textual structures are embedded in a larger social context, in a linguistic environment that works to influence and configure our own individual preferences. Many studies have shown that overexposure to stimuli result in an affective plateau in which subsequent exposures cause no change in affective rating, or a reversal of the MEE in which overexposed stimuli become more disliked with each additional exposure: it is possible that new movements rise to prominence when the older movements have begun their affective decline. As a society, we may replicate the affect-exposure curves of individuals, leading to a distributed collective preference for the new that allows novel forms of art to take hold.

In conjunction with empirical research concerning the exposure-affect curve in mere exposure experiments, Stein's essay offers insight into the ways in which individual psychologies and social structures are mutually interpenetrating, into the collectivity that inhabits the individual reader and the subjective affects that embody collective judgments about matters of aesthetic and social value. If the composition of our time is, as Stein says, "the thing seen by every one living in the living they are doing [...] at the time they are living in the composition of the time in which they are living" (Stein 1990: 516), then to see the mutual influence of individual and collective in the literary text is to see a new way in which the present might be composed.

References

Bornstein, Robert. F. 1988. Exposure and affect: Overview and meta-analysis of research, 1968–1987. *Psychological Bulletin* 106: 265–289.

Cutting, James. 2003. Gustave Caillebotte, French impressionism, and mere exposure. *Psychonomic Bulletin and Review* 10: 319–343.

Elliott, Rebecca and Raymond J. Dolan. 1998. Neural response during preference and memory judgments for subliminally presented stimuli: A functional neuroimaging study. *The Journal of Neuroscience* 18: 4697–4704.

Luka, Barbara J. and Lawrence W. Barsalou. 2005. Structural facilitation: Mere exposure effects for grammatical acceptability as evidence for syntactic priming in comprehension. *Journal of Memory and Language* 52: 436–459.

Monahan, Jennifer L., Sheila T. Murphy and Robert B. Zajonc. 2000. Subliminal mere exposure: Specific, general, and diffuse effects. *Psychological Science* 11: 462–466.

Moretti, Franco. 2005. *Graphs, Maps, Trees: Abstract Models for a Literary History*. London: Verso.
Ramachandran, Vilayanur S. and William Hirstein. 1999. The science of art: A neurological theory of aesthetic experience. *Journal of Consciousness Studies* 6: 15–51.
Reber, Arthur S. and Diane M. Zizak. 2004. Implicit preferences: The role(s) of familiarity in the structural mere exposure effect. *Consciousness and Cognition* 13: 336–362.
Smith, Barbara Herrnstein. 1988. *Contingencies of Value: Alternative Perspectives for Critical Theory*. Cambridge: Harvard University Press.
Stein, Gertrude. 1990. Composition as explanation. In *Selected Writings of Gertrude Stein*, Carl Van Vechten (ed.), 511–524. New York: Vintage Books.
Winkielman, Piotr and John T. Cacioppo. 2001. Mind at ease puts a smile on the face: Psychophysiological evidence that processing facilitation elicits positive affect. *Journal of Personality and Social Psychology* 81: 989–1000.
Zajonc, Robert B. 1968. Attitudinal effects of mere exposure. *Journal of Personality and Social Psychology Monographs* 9: 1–27.
Zeki, Semir and Matthew Lamb. 1994. The neurology of kinetic art. *Brain* 117: 607–636.

Embodied mind and cross-cultural narrative patterns

Yehong Zhang
Tsinghua University, Beijing

The fairy tales collected by the Brothers Grimm are important because of their influence on literary socialization in East Asia. Following a brief survey of the translation history of the tales, I analyze their structure in relation to the characteristics of children's thinking, and suggest that the prototype experience of the fictional world corresponds to the conceptual structure of narrative patterns in the reader's mind. I argue that this interaction derives from embodied cognition, which acts as one of the roots of universal understanding of narratives. The results from an empirical study of cross-cultural reading responses demonstrate how plot prediction and, especially, spatial imagination about Grimm's tales display this embodied cognition.

Keywords: children's thinking, empirical approach, Grimm's fairy tales, spatial imagination

1. Introduction

In the age of globalization, cultural barriers are vanishing and the same stories receive more and more interest throughout much of the world. The genre of fairy tales, which has its origins in European culture, is particularly suited to convey a sense of 'otherness' and thus to serve as a bridge between different cultures. In the second half of the nineteenth century, fairy tales became a cultural export from Europe into the culturally different Far East. It is not only in the West that Grimm's fairy tales are considered classics of children's literature; in the Far East as well, their collection, which is the most frequently translated German book, ranks undeniably among the most important works in this field. The question arises as to why Grimm's fairy tales can apparently be translated and transferred into such a different cultural context without major problems.

2. Grimm's fairy tales and the embodied mind

When the Grimm's tales were first introduced into the literary canon in East Asia, the cultures of Europe and East Asia were still largely unknown to each other. Why did the tales immediately become popular in cultures that had no tradition of this genre? The first English translation (*Children's and Household Tales*, in German *Kinder- und Hausmärchen* [*KHM*]) was published in 1887 and is considered to be the earliest complete collection of the Grimm Brothers' fairy tales in the Far East (Noguchi 1977: 1). The earlier Japanese translations of the *KHM* have been analyzed by Yoshiko Noguchi, who has carried out an extensive study on the reception history of the tales, focusing in particular on their translation process. She compares the entire text of the first fairy tale in the Japanese edition of the *KHM* word by word to the original (Noguchi 1977: 47–51). Her study leads to the conclusion that many small details have been changed in the translated version. These changes primarily concern customs, language structure, a moral coloring, and the softening or omission of brutality and sexuality (Noguchi 1977: 93). In my opinion, her word-by-word comparison shows that the main plot of the tales has actually undergone very little change.

The acceptance of foreign literature depends to a large degree on its comprehensibility and on the interest that it is able to generate in the target culture. The successful transfer and the widespread acceptance of Grimm's fairy tales in East Asia demonstrate their general approachability: the main plot in the first Japanese translation, which is also the first translation in East Asia, remained intact. The same holds true for the first Chinese translation, a philologically accurate collection of the Grimm's works, published shortly after the Japanese version (Liang 1986: 59). Why were Japanese and Chinese readers able to understand the tales from such a foreign and unknown culture without any major changes?

The comprehensibility of the Grimm's fairy tales across cultural boundaries is primarily due to their particular structure. Their plots are embedded in a specific progression of events, of 'beginning – initiated and completed actions – conclusion', while the actions are implemented as "tasks, rules and conditions of exact situations, whereby the reward or the punishment is the conclusion" (Lüthi 1989: 30). The easy-to-follow and linear plots keep readers in suspense, while the simple structure and form enables them to concentrate on the creation of visual images stimulated by the tales. As Lüthi has it, "the fairy tale transports processes of the soul and mind, as well as incidents of fate as an experience of mankind into visual and receivable spheres" (Lüthi 1989: 19). This clear imaginary way in which the stories are told and the fact that they take place in unidentified locations allow East Asian readers to generate their imagination and adapt quickly to them.

The formal simplicity of the fairy tales goes back to the oral tradition in which they were originally told and which is based on a schema-oriented model characterized by expectation and fulfillment. When Japanese and Chinese children read the stories one hundred years ago, they formed expectations. Whether their expectations were similar to those of German children at that time remains unknown, but the above-mentioned features – clear plot and language structure, and rich imagery – facilitated their understanding of the tales.

In addition to this, the easy transfer of the tales can be explained, independently from their formal structure, by the fact that they are a form of life experience. Fairy tales encourage readers to imagine themselves in the scenes depicted in the story and to develop mental images of the situations, people, and actions in order to identify with them (Hoppe-Graff and Shell 1989: 90). The literary ability to identify with a story is an important cognitive prerequisite for literary understanding rooted in the human ability for associative imagination and the understanding of other people's feelings and thoughts. This is an anthropological necessity, as humans are social and cultural beings who depend on a constant exchange with their fellows (Meltzoff 1995). In order to interpret and predict the behavior of others, a person has to be able to reconstruct his or her thinking processes through the use of hypotheses. The theoretical modeling and interpretation of this process is called "theory of mind" (Perner 2000). Within this theory, special attention has been paid to literary competence in general (Zunshine 2006) and its development in early childhood (Meltzoff 2007).

According to Kasper Spinner, children's "[...] understanding has the structure of concrete thinking, in the sense of cognitive psychology" (Spinner 1995: 84). Their imaginative world is 'event-oriented'. Children take everything literally, and for them external and inner events are not as clearly differentiated as for adults. In contrast to the latter, children reconstruct the plot of a story in a linear sequence of events.[1] Furthermore, before children are able to grasp abstract concepts, they comprehend the world through their physical experience. Abstract thinking is developed only from concrete ideas that are based on physical experience.

The phenomenon that mental representations of the world basically rely on concrete experiences is called the "embodied mind". This term comprises the interaction of concrete, physical environment with human cognition (Gibbs 2006; Lakoff and Johnson 1998). The embodied mind is the basis of culture-independent imagination. Although life experiences are influenced by cultural

1. When examining how children develop a comprehension of stories that are told or read to them, it becomes clear that the ability to understand stories as a linear sequence of events is an essential step. Children's understanding of narrative texts was analyzed in Kasper Spinner's study (1995), in which children were to retell passages in children's books.

factors, all humans share a biological structure independent of any cultural background. Thus, the concept of the embodied mind provides an explanation for cross-cultural literary competence. Abstract mental processes are inferred from concrete physical experience (Gibbs 2006; Lakoff and Johnson 1998). The procedure by which people produce fictions is the concrete objectification of the world, which lays the foundation for such abstract mental representation. This objectification makes the distinction between narrative content and real-life situation possible (Sperber 2000; Eibl 2003, 2004). 'Objectified' knowledge crosses cultural borders and is received independently from real situations, because objectification makes it possible to process the presented facts without knowing certain context ('off-line' understanding) (Eibl 2003: 571).

Since fairy tales adapt to this culturally independent mode of thinking (i.e. the embodied mind), they can be easily transferred across cultures. They are built on a common knowledge of objectified real-life situations and convey general life experiences that extend over cultural borders because they stimulate the children's embodied mind and their imagination, which is "[...] part of the literary competence" (Spinner 1995: 86). Because the tales are told in a clear and concrete way, they can be understood and internalized by children (Gibbs 1994, 2006). Moreover, they are based on particular narrative schemata inherited from ancestral expectations of what a plot should be like (Labov 1972: 369). It is according to these schemata that conventional stories are usually structured (see Carroll 2004). The expectation fulfillment provides coherence and makes it possible to understand the tales without significant prior knowledge. In the following, I shall provide empirical evidence for my hypothesis that the Grimm's fairy tales prompts the culture-independent embodied mind in the reading responses of children.

3. Empirical examination

To investigate cross-cultural reading responses, I conducted research on the responses to traditional Chinese fairy tales and Grimm's fairy tales with German and Chinese children (see Zhang 2008). The empirical study was conducted using procedures of psychological research and assisted by developmental and social psychologists at every stage of the project. Standard psychological studies are based on consistent, well-established research procedures. This means that properly designed empirical studies should be based on assumptions derived from established theories in the fields in which they are undertaken. As there is no prevailing theory for the cross-cultural comparative study on literary comprehension, a well-established theory of cultural comparative thinking model serves as the leading theory for assumption-building. The established cultural psychological theory demonstrates that the Western focus is more on objects while the East Asian focus is rather on the whole picture and on the context in which the objects exist (Nisbett 2003; Peng, Ames, and Knowles 2001; Markus and Kitayama 1998; Singelis 1994). The psychological theory of cultural thinking systems can be employed to work in conjunction with cultural comprehension models for literary texts through the assumption that East Asians read a story with society-oriented tendencies, while Europeans read with more individual-oriented tendencies. Although literary comprehension is a very complex process, this frequently examined theory provides a solid foundation for the presented empirical investigation.

The questionnaire I developed was the key element in the study. The construction of the questionnaire was based on the assumption mentioned above. As there were no existing reading response questionnaires based on cultural comparative thinking models, I first carried out several pilot tests with open-ended questions to interpret the figures and activities with text samples. As a text sample for the empirical study, I chose the Grimm's tale "Jorinde and Joringel", which tells the story of a witch living in a forest, who fixes everyone who approaches her castle to the ground and turns maidens into birds. One day, a young couple, Jorinde and Joringel, fall under the witch's spell. Jorinde, the maiden, is turned into a bird and caged, but the boy is freed. He is very sad, goes to a distant village and becomes a shepherd. One night, the boy has a dream about a flower that could break the witch's spell, and he consequently searches for this flower. When he finds it, he returns to the castle and touches the witch with the flower, thus rendering her powerless. He also touches all the birds and transforms them back into women. Jorinde and Joringel are reunited in the end. The baseline survey showed that this story was not familiar to the readers in the experimental and control groups and therefore participants' answers would not be influenced by existing bias. The selected text sample also needed to attract the participants' attention. In the pilot study, this story was enjoyed by the majority of the participants.

As the Chinese counterpart I chose the traditional Chinese story "Cowherd and Weaver Girl", which contains topics and structures comparable to those of the Grimm's tale: A poor young cowherd flies up to heaven on the back of his ox. There, he sees nine fairy sisters bathing in a lake. Following the ox's advice, the cowherd steals the clothes of one of them, the "weaver girl", who weaves colorful clouds in the sky. Only if she agrees to marriage will he return her clothes. Under the mediation of an old willow tree, the weaver girl agrees. After seven days, the girl leaves her husband to return to heaven in order to weave clouds again. The cowherd follows her, but with her hairpin, the girl scratches a broad river in the sky that keeps the cowherd away. Once a year, all the magpies in the world fly up into heaven and form a bridge so that the couple can be together for a single night: the seventh night of the seventh month.

Through a string of pilot tests, I developed a design for the questionnaires. The items in the questionnaires had been designed to correspond to the main categories of narratology: figures, plot, and spatial/temporal perception. The items and scales allow the readers' responses to the literary categories to be measured, and they presumably reflect cultural differences in literary interpretation.[2] The questionnaires for both fairy tales were developed with the same categories.

The questionnaires had to be developed in accordance with developmental psychologists' specifications for school-aged children. The questionnaires had to be appropriate for 12-year-old children who can grasp the story and independently evaluate the items.[3] In this study, Chinese children were chosen as a sample representing East Asian cultures and German children to represent Western models of thought. One hundred sixth-grade German children from the city of Göttingen and one hundred Chinese children in the same grade in Beijing were used as participants. These two hundred participants were randomly assigned to two groups of fifty students each who read either the fairy tale from their own culture or the one from the other culture and completed the respective questionnaire. The participants who read the story from their own culture comprised the control group. Each participant read only one story.

4. Results

The data from the study were evaluated by the software Statistical Package for the Social Sciences (SPSS). According to the statistical analysis of the quantitative data of the Grimm's story ($N = 100$), the observed scores in eleven out of twenty-seven items are significantly different for both groups of participants.[4] However, there was no significant difference in responses to spatial semantics. Moreover, both groups' evaluations of the items related to spatial imagination in the Grimm's tale are strikingly similar; the responses of the Chinese and German participants to questions like 'Can you imagine the castle and its surroundings very well?' are

2. Each item was administered using a 6-point Likert scale from 1 (completely agree), 2 (mostly agree), 3 (slightly agree), 4 (slightly disagree), 5 (basically disagree), to 6 (completely disagree). The participants responded to each item using this scale. A forced-choice, six-point Likert-scale was used to avoid the middle choice option.

3. Although fairy tales may no longer be favorite reading for school children, according to Piaget's (1974) theory of children's abstract logical thinking stages, fairy tale interpretation should be a subject at least from the fifth grade.

4. High statistical significance means that the Chinese and German participants interpreted the items significantly differently. Statistical significance is expressed by the p-value, which must be smaller than 0.5 to indicate that the observed measures are truly different from each other.

virtually the same (see Table 1). This result supports my basic hypothesis of the role of the embodied mind in cross-cultural understanding.

Table 1. Results regarding spatial imagination in the Grimm's story

item	**mean (*SD*) Chinese participants**	**mean (*SD*) German participants**	***p*-value**
I can imagine the castle and its surroundings very well.	2.10 (1.54)	2.10 (2.32)	0.995

Note. $N = 100$. *SD* = Standard Deviation
Ratings can be interpreted with scale labels: 1 = completely agree, 2 = mostly agree, 3 = slightly agree, 4 = slightly disagree, 5 = basically disagree, 6 = completely disagree

I also conducted a confirmatory factor analysis with a rotated component matrix. Four factors (with a loading of greater than forty) were derived: "judgment for figures", "spatial/temporal imagination", "plot causality", and "affection on figures". On the basis of the factor analysis, it is useful to determine whether there was any significant difference between the two reading groups in the domain of each factor. I put the data from each item and each factor into one scale to carry out an Analysis of Variance (ANOVA). The factor "spatial/temporal imagination" is the only factor that is not significantly different in the reactions of Chinese and German participants (see Table 2).

Table 2. Results of factor analysis regarding spatial and temporal imagination in the Grimm's story

factor: "spatial / temporal imagination"	$F(1,99) = 2.854$	$p < .094$

That the Chinese children are familiar with the motifs of the Grimm's tale thus provides further support for my hypothesis. Even Chinese participants could predict the ending of the story significantly better after reading the beginning (see Table 3). The prediction of the story ending is closely connected to inherent plot knowledge.

Table 3. Results regarding plot prediction for the Grimm's story

item	**mean (*SD*) Chinese participants**	**mean (*SD*) German participants**	***p*-value**
I predicted correctly at the beginning of the story that the girl and the boy would have problems later.	2.08 (1.63)	3.22 (1.68)	.001*

Note. $N = 100$. Ratings can be interpreted with scale labels: 1 = completely agree, 2 = mostly agree, 3 = slightly agree, 4 = slightly disagree, 5 = basically disagree, 6 = completely disagree
*$p < .05$ after Bonferroni correction

Compared to the Grimm's tale, the reading of the Chinese story resulted in significant differences between the two groups regarding the familiarity with figures, plot, and spatial (time) semantics. In the Chinese story, temporal and spatial descriptions strongly rely on cultural specifics. The Chinese and German readers answered the items concerning spatial and temporal imagination significantly differently (see Table 4).

Table 4. Results regarding spatial and temporal imagination in the Chinese story

items	**mean (*SD*) Chinese participants**	**mean (*SD*) German participants**	***p*-value**
The weaver girl spent a very long time with the cowherd together in heaven before she left him.	2.76 (1.78)	4.92 (1.45)	.000*
The world of heaven and the world of earth are connected to each other in the story.	2.98 (1.79)	3.77 (1.68)	.019

Note. $N = 110$. Ratings can be interpreted with scale labels: 1 = completely agree, 2 = mostly agree, 3 = slightly agree, 4 = slightly disagree, 5 = basically disagree, 6 = completely disagree
$^{*}p < .05$ after Bonferroni correction

An exploratory factor analysis with a rotated component matrix was also conducted for the Chinese story. Four factors (with loading of greater than forty) were derived: "affection on figures", "plot familiarity", "motivation of action", and "judgment for social behavior". The ANOVA-test was carried out for each factor. The two factors given in Table 5 show significant differences for both groups of participants. The result strongly indicates that the plot schemata in the Chinese fairy tale are unfamiliar to the German readers.

Table 5. Results of factor analysis of plot familiarity in the Chinese story

factor "affection on figure"	$F\,(1{,}109) = 9.122$	$p < .001$
factor "plot familiarity"	$F\,(1{,}109) = 78.935$	$p < .001$

Although participants from neither culture had prior knowledge of the story from the other culture, the Chinese recipients showed far more familiarity with the narrative structure of the German story than vice versa. The results from this empirical study provide first evidence in support of the assumption that the embodied mind serves as a foundation for cross-cultural understanding. Our imagination of space and time, and expectations of event-succession are the result of long-term environmental influences, and thus closely related to embodied cognition.

5. Concluding remarks

The aim of this chapter has been to explore how people from different cultures can understand the same stories. The special attributes of the Grimm's fairy tales make them interesting and understandable for readers from foreign cultures, especially children. With the many references to the physical world and their own experiences, children can easily comprehend the one-dimensional, tangible world in the tales. Their clear, predetermined plots and rich imagery make the tales culturally transferable; the stories are understandable 'off-line' without any situational context. In contrast, the setting of the Chinese story – in a heavenly sphere – makes it less suitable to cultural transfer. Such an otherworldly setting is less tangible and experience-based. Likewise, the plots in the Chinese story are more culturally specific, rendering identification more difficult for readers who are unfamiliar with Chinese culture. The fairy tales of the Brothers Grimm are a manifestation of narrative patterns that adapt to embodied cognition and therefore are cross-culturally transferable. Besides the Grimm's works, other forms of media, such as film and comic books, possess these embedded narrative patterns as well, which can make them popular virtually everywhere in our globalized world.

In order to substantiate hypotheses about embodied cognition and culture-independent reading responses, literary studies must be combined with other experimental disciplines to make their results falsifiable and productive for other disciplines (e.g. Bortolussi and Dixon 2003). Empirical investigation provides traditional literary scholars with evidence about concrete reading responses, although some inevitable problems emerge. If the research procedure for reading response is imported from a well-established experimental science like psychology, empirical literary scholars must use clearly defined categories of text comprehension and a well-established questionnaire model for their survey. However, concrete measuring tools for reading research do not yet exist. In order to establish a test mechanism for the responses of actual readers, a string of tests in the various categories of text comprehension should be designed in an interdisciplinary setting following the research procedure in experimental sciences.

References

Bortolussi, Marisa and Peter Dixon. 2003. *Psychonarratology. Foundations for the Empirical Study of Literary Response.* Cambridge: Cambridge University Press.

Carroll, Joseph. 2004. *Literary Darwinism: Evolution, Human Nature and Literature.* New York: Routledge.

Eibl, Karl. 2003. Vergegenständlichung. Über die kulturstiftende Leistung der Menschensprache. In *Regel der Bedeutung. Zur Theorie der Bedeutung literarischer Texte*, Fotis Jannidis, Gerhard Lauer, Matias Martinez and Simone Winko (eds.), 566–590. Berlin and New York: de Gruyter.

Eibl, Karl. 2004. *Animal Poeta. Bausteine der biologischen Kultur- und Literaturtheorie*. Paderborn: Mentis.

Gibbs, Raymond W. Jr. 1994. *The Poetics of Mind: Figurative Thought, Language, and Understanding*. Cambridge: Cambridge University Press.

Gibbs, Raymond W. Jr. 2006. Metaphor interpretation as embodied simulation. *Mind and Language* 21: 434–458.

Hoppe-Graff, Siegfried and Martin Shell. 1989. The comprehension of literary texts: Developmental considerations. In *Comprehension of Literary Discourse. Results and Problems of Interdisciplinary Approaches*, Dietrich Meutsch and Reinhold Viehoff (eds.), 89–110. Berlin and New York: de Gruyter.

Labov, William. 1972. The transformation of experience in narrative syntax. In *Language in the Inner City*, William Labov (ed.), 354–396. Philadelphia: University of Pennsylvania Press.

Lakoff, George and Mark Johnson. 1998. *Philosophy in the Flesh: The Embodied Mind and its Challenge to Western Thought*. Chicago: The University of Chicago Press.

Liang, Yea-Jen. 1986. *Kinder- und Hausmärchen der Brüder Grimm in China. Rezeption und Wirkung*. Wiesbaden: Harrassowitz.

Lüthi, Max. 1989. *So leben sie noch heute: Betrachtungen zum Volksmärchen*. Göttingen: Vandenhoeck & Ruprecht.

Markus, Hazel Rose and Shinobu Kitayama. 1998. The cultural psychology of personality. *Journal of Cross-Cultural Psychology* 29: 63–87.

Meltzoff, Andrew N. 1995. Understanding the intentions of others: Re-enactment of intended acts by 18-month-old children. *Developmental Psychology* 31: 838–850.

Meltzoff, Andrew N. 2007. 'Like me': A foundation for social cognition. *Developmental Science* 10: 126–134.

Nisbett, Richard E. 2003. *The Geography of Thought: How Asians and Westerners Think Differently – and Why*. New York: Free Press.

Noguchi, Yoshiko. 1977. *Rezeption der Kinder- und Hausmärchen der Brüder Grimm in Japan*. PhD dissertation, University of Marburg.

Peng, Kaiping, Daniel R. Ames and Eric D. Knowles. 2001. Culture and human inference. Perspectives from three traditions. In *The Handbook of Culture & Psychology*, David Matsumoto (ed.), 245–284. New York: Oxford University Press.

Perner, Josef. 2000. Memory and theory of mind. In *The Oxford Handbook of Memory*, Endel Tulving und Fergus I. M. Craik (eds.), 297–314. New York: Oxford University Press.

Piaget, Jean and Bärbel Inhelder. 1974. *The Child's Construction of Quantities*. London: Routledge & Kegan Paul.

Singelis, Theodore M. 1994. The measurement of independent and interdependent self- construals. *Personality and Social Psychology Bulletin* 20: 580–591.

Sperber, Dan (ed.). 2000. *Metarepresentations. A Multidisciplinary Perspective*. New York: Oxford University Press.

Spinner, Kaspar H. 1995. Die Entwicklung literarischer Kompetenz beim Kind. In *Lesen im Medienzeitalter. Biographische und historische Aspekte literarischer Sozialisation*, Cornelia Rosebrock (ed.), 81–95. Weinheim and Munich: Juventa.

Zhang, Yehong. 2008. Culture and reading: The influence of western and eastern thought systems on the understanding. In *New Beginnings in Literary Studies*, Jan Auracher and Willie van Peer (eds.), 218–237. Newcastle: Cambridge Scholars Publishing.

Zunshine, Lisa. 2006. *Why We Read Fiction. Theory of Mind and the Novel*. Columbus: The Ohio State University Press.

The mind and the text / the mind in the text

Dirk Vanderbeke
Friedrich-Schiller-Universität, Jena

Reading is famously an exercise in gap filling, and from select and occasionally contradictory information the reader creates coherent characters and believable stories. Our mental faculties are obviously geared to produce a smooth experience and to homogenize disjunctive information and conflicting sensorial data. Over the last decade, several novels have dealt with neuroscientific topics. Novels like Richard Powers's *The Echo Maker* or Ian McEwan's *Enduring Love*, while explaining the relevant concepts and theories, also employ devices that play on the reader's mind in the process of reading. The "willful suspension of disbelief" is disrupted, and the reader is made aware of the internal processes of construction involved in reading. Such meta-narrative devices may contribute to our understanding of cognition and the mental mapping of the world. Literary studies may thus enter a bi-directional exchange with the cognitive sciences.

Keywords: Ian McEwan, meta-narrative, mental mapping of the world, Richard Powers, reading as gap-filling

1. Fill in the gap, or: Let the reader work it all out

In the first paragraph of the first story of the first book that James Joyce ever published, we find the following sentences:

> Every night as I gazed up at the window I said softly to myself the word paralysis. It had always sounded strangely in my ears, like the word gnomon in the Euclid and the word simony in the Catechism. (Joyce 1967: 7)

Of course, we know what *paralysis* and *simony* mean, but the word *gnomon* will probably baffle the reader and, if it is a conscientious reader, send him to the next dictionary (for the following argument, see also Weir 1991: passim). There he will find out that a *gnomon* in geometry is "the part of a parallelogram left after a similar parallelogram is taken away from one of its corners" (SOED, s.v.). It looks like this:

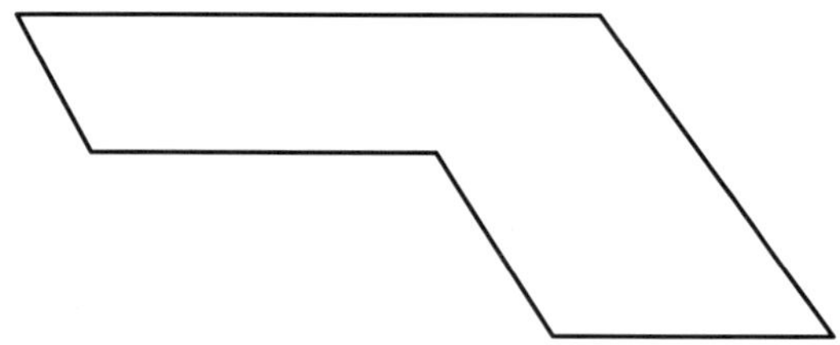

Figure 1. A gnomon

Of course, the question may suggest itself, why this geometrical phenomenon appears so prominently in a story about a dead priest. But as paralysis and simony are of some relevance in this text, it stands to reason that the third item may also have a significant role to play.

We can easily see that the decisive property of the gnomon is the missing corner. The geometrical form is defined by an absence that marks a presence.

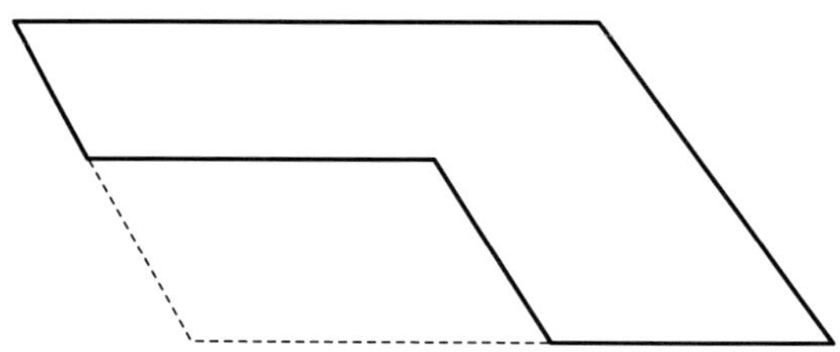

Figure 2. A gnomon with missing corner added

A closer look at the story then shows that it is riddled with omissions and ellipses that require the reader to fill in the gaps and to make sense of the unspoken and possibly unspeakable elements of the text. This then can shed some light on the poetics of Joyce and his concept of the epiphany, a kind of secular mysticism in which a thing or a specific situation may reveal itself to the observer as the pinnacle of aesthetic perception. This phenomenon can be experienced and also possibly be evoked by a work of art, but not put into words. Of course, to realize the analogy is a very complex mental operation, as we have to construct a similarity first between the geometrical figure and the gap-riddled text and then with an artistic concept. It is very much like the act of metaphor – we have to recognize the similar in the difference, the abstract in the concrete.

Further reading reveals that this device does resurface time and again in Joyce's work. His texts are regularly marked by a poetics of absence, narrative gaps that need to be filled by the attentive reader, who may or may not find the necessary information elsewhere.

My favorite example is from the short story "Clay". The story as such is not of relevance here; I just want to quote a short passage in which Maria, an elderly lady, takes a tram.

> She thought she would have to stand in the Drumcondra tram because none of the young men seemed to notice her, but an elderly gentleman made room for her. He was a stout gentleman and he wore a brown hard hat; he had a square red face and a greyish moustache. Maria thought he was a colonel-looking gentleman and she reflected how much more polite he was than the young men who simply stared straight before them. The gentleman began to chat with her about Hallow Eve and the rainy weather. He supposed the bag was full of good things for the little ones and said it was only right that the youngsters should enjoy themselves while they were young. Maria agreed with him and favoured him with demure nods and hems. He was very nice with her, and when she was getting out at the Canal Bridge she thanked him and bowed, and he bowed to her and raised his hat and smiled agreeably; and while she was going up along the terrace, bending her tiny head under the rain, she thought how easy it was to know a gentleman …

All this is quite clear, but then the last sentence ends with: "even when he has a drop taken" (Joyce 1967: 100–01).

The information about the gentleman's slight or not quite so slight intoxication should have introduced the passage, and now it stresses the fact that it was missing at its appropriate place. The reader has to reconsider the whole scene: the upright gentleman with a touch of the military begins to sway, the red color of his face may be due to more than just the cold weather, and the friendly, if one-sided conversation now appears a bit obtrusive. Even more important is the question why the crucial information about the gentleman was previously omitted, and the obvious answer is that it did not really register in Maria's mind. This again says quite a lot about the situation in Ireland and the fact that elderly gentlemen, who were already pickled in the afternoon, must have been so common a sight that their state no longer deemed worthy of notice. The lacuna thus contains quite a lot of information (a slightly different and more extended reading of the passage can be found in Norris 1987: 211).

A more complex gap appears in the Calypso-chapter of *Ulysses*. Leopold Bloom has just carried a tray with breakfast to his wife Molly together with a letter from Blazes Boylan with whom she plans to go on a concert tour. Boylan writes that he will come to see her this afternoon, ostensibly to practice some songs, but actually to go to bed with her as he is also her lover, and Bloom knows, and Molly knows that Bloom knows, and Bloom knows that Molly knows that Bloom knows. It is a touchy situation, the spouses walk around it rather carefully, and the delicate issue of the visit is passed over quickly. Later in the day, the reader with a very good memory realizes that Bloom knows rather more than has been said during this conversation, as, for example, the exact time of Boylan's arrival and that he, Bloom, is not to come home before the evening. Quite obviously, another conversation must have taken place a little later, but it is not part of the text. And it would have been extremely tricky to write. As it is, all we are given are some facts that must have been discussed, the rest is left to the reader's imagination and his or her ability to re-construct a conversation which is consistent with Bloom's and Molly's characters (see Kenner 1987: 48–51).

What I have said here does not yet differ very much from the tenets of reader response theory and Wolfgang Iser's perspective on the act of reading as a perpetual gap filling exercise. However, in Iser's argument this literary phenomenon contributes to the defamiliarization of the text, and while it is also present in pre-modern literature, it is in modernism that it develops into a prominent feature, irritates the reader and thus increases the hesitancy that is evoked by the highly complex texts (cf. for example Iser 1994: 315–27). I certainly do not doubt that in high modernist and even more in postmodernist texts it does, indeed, irritate readers, but I want to argue that this gap filling exercise is far more prevalent than Iser's conceptualization suggests and that it is by no means restricted to the higher echelons of literature.

As a matter of fact, it is one of the defining elements of a literary or semi-literary genre that was still almost unmentionable when reader response theory was hatched. I am talking about the comics, which are built around gaps, i.e. the empty spaces, called the gutter, between panels. These empty spaces constantly have to be filled by the imagination of the reader (cf. McCloud 1994: 68) and, complex as the mental activities involved may be, this process obviously does not require a sophisticated audience. Young kids with hardly any exposure to literature have no problem whatsoever in turning sequential images into a continuous story.

2. The pleasures of reading, or: The mind and its constructs

On the assumption that the typical audience of comics does not really relish aspects of defamiliarization nor to be driven into hesitancy or doubt about the narration, but rather expects a smooth tale with no recognizable impediments to the act of reading, we may have to readjust our concept of gap filling and aesthetic pleasure. Turning to neuropsychological research on aesthetic perception, empirical studies suggest that fluency and smoothness are important aspects in aesthetic perception and favor the evaluation of an object as beautiful or pleasant:

"The more fluently the perceiver can process the object, the more positive is his or her aesthetic response" (Reber et al. 2004: 365; see also Alexandra Kleeman's argument in this volume, that familiarity leads to a favorable evaluation of art works as beautiful; and Cutting, 2003: passim). This suggests that the act of gap filling is not detrimental to smooth perception or, alternatively, that gap filling is itself a smooth mental activity already fully working in children when they first approach literary or semi-literary publications. As such, it seems not only to be performed without us actually noticing the considerable mental work involved, it can quite obviously also be a source of fun and entertainment.

This finds its counterpart in psychological phenomena as presented by Paul Watzlawick. In his book *How Real is Real* (1977), he describes how in an experiment students were treated with incompatible data. In response, they created highly complex and intricate concepts and theories in the increasingly desperate attempt to integrate the conflicting information into a consistent explanation. In fact, the more mental work and energy they had to invest in the course of this process, the more they clung to their frequently arcane conceptualizations even when they were finally informed about the experimental setup and should have dismissed their purely fictional constructions (Watzlawick 1977: 48–53).

Watzlawick's conclusion from this and a host of other experiments is, "that there is no absolute reality but only subjective and often contradictory conceptions of reality" (1977: 53), which ultimately comes down to a restatement of Bishop Berkeley's radical concept that observation is a *sine qua non* of reality with the addition of a multiplicity of dissenting observers. If we agree, for the moment, that the experiment does not confront the participants with the reality they are used to but a manipulated reality that does not adhere to the common rules of causality and coherence, the conclusion may look different.

As the reality we usually deal with is not determined to trick or cheat us with incoherent or contradictory data, we may assume that our minds are generally able to construct a working model of our environment, even if the set of available data is incomplete. Indeed, the more or less permanent construction and adjustment of our working model of reality is a necessity for survival, and thus this ability can in different degrees also be found in some of our biological kin – or, to put it more cautiously, some animal behavior lends itself to the assumption that similar mental activities may be at work.

Research in cognitive functions and psychological *sine qua nons* frequently relies on cases of mental illness or trauma in which the function in question is impaired or destroyed. Roman Jakobson's research into aphasias has been extremely productive for our understanding of metaphors and metonymies not only as rhetorical tropes but also as basic elements in the ways we process information and construct our world (see Jakobson 1956). Similarly, more recent research into the human mind has frequently focused on case studies involving brain injuries or mental diseases. In short, we learn how the mind works when it does not work properly, and it is the complexity of the possible malfunctions that offers a glimpse on the enormous range of mental operations behind the simplest acts of cognition and recognition. The research studies of Antonio Damasio (1998), Oliver Sacks (1985) and V. S. Ramachandran and Sandra Blakeslee (1998) may serve as well-known examples.

3. Cognition in literature, or: The mind observed

These works have also found their echo in recent literature. Ian McEwan, for example, has pointed out that Damasio was one of his sources when he wrote *Enduring Love*, and so was David Enoch and William Trethowan's book on *Uncommon Psychiatric Syndromes* (1991). Here is a vignette, in which the protagonist Joe has the feeling that an intruder has entered his apartment – it is closely modeled on a passage in Damasio's *Descartes' Error* (Damasio 1998: 223–24).

> The primitive so-called sympathetic nervous system is a wondrous thing we share with other species that owe their continued existence to being quick on the turn, fast and hard into battle, or fiery in flight. Evolution has culled us all into that efficiency. Nerve terminals buried deep in the tissue of the heart secrete their nor-adrenalin, and the heart lurches into accelerated pumping. More oxygen, more glucose, more energy, quicker thinking, stronger limbs. It's a system so ancient, developed so far back along the branchings of your mammalian and pre-mammalian past that its operations never penetrate into higher consciousness.
>
> (McEwan 1998: 51)

Rising from his chair and ready to defend himself the protagonist realizes, however, that it is, actually, his wife who approaches him from behind:

> What I saw coming towards me rapidly across the room, with arms outstretched like a cartoon sleepwalker was Clarissa, and who knows by what complex intervention of higher centers I was able to convert plausibly my motions of primitive terror into a tenderly given and received embrace and to feel, as her arms locked round my neck, a pang of love that was in truth inseparable from relief.
>
> (1998: 51)

The instinctive hostile response needs to be reorganized not only into an appropriate response, this also has to be done plausibly, and the resulting emotion merges love with relief, elements that are not easily compatible. The reader is left with conflicting aspects of the protagonist's mind; instead of the usual insights into the rational or emotional faculties the emphasis is on the complex mental attempts to construct a coherent environment and to respond adequately.

In an earlier passage the protagonist has to come to terms with a catastrophic situation – it is the dilemma the book has become famous for. A father and his son land after a flight with a balloon, but when the father leaves the basket a gust of wind once more lifts the balloon with his son and threatens to carry the boy off. Some men watch the scene and try to help pulling down the balloon by some ropes. However the wind begins to lift them up and they are faced with two possibilities: if all of them hold fast they can control the balloon, but if one or two let go the rest will be pulled upwards and probably killed by the inevitable fall. The situation is obviously constructed along the lines of the so called "Prisoner's Dilemma" of game theory, that also plays a role in evolutionary theory and the discussion of the respective advantages of altruism or selfishness (cf. e.g. Wright 1995: 192–93). This is how the process of decision-making is depicted in the moment of despair:

> A child alone and needing help. It was my duty to hang on, and I thought we would all do the same.
>
> Almost simultaneously with the desire to stay on the rope and save the boy, barely a neuronal impulse later, came other thoughts in which fear and instant calculations of logarithmic complexity were fused. We were rising, and the ground was dropping away as the balloon was pushed westward. I knew I had to get my legs and feet locked round the rope. But the end of the line barely reached below my waist and my grip was slipping. My legs flailed in empty air. Every fraction of a second that passed increased the drop and the point must come when to let go would be impossible or fatal. And compared to me Harry was safe curled up in the basket. (1998: 13f.)

The response to the dangerous situation sounds very rational, but the story goes on, and, as it happens, all but one let go – each insisting that he was not the first one to do so – and the balloon is pulled upwards. The boy survives, but the last man to hold fast misses his chance for a safe drop and falls to his death. And now we get a second version of the brief moment of truth in which the fatal decision was made:

> So can we accept that it was right, every man for himself? Were we all happy that this was a reasonable course? We never had that comfort, for there was a deeper covenant, ancient and automatic, written in our nature. Co-operation – the basis of our earliest hunting successes, the glue of our social cohesion. Our misery in the aftermath was proof that we knew we had failed ourselves. But letting go was our nature too. Selfishness is also written in our hearts. This is our mammalian conflict – what to give to the others and what to keep for yourself. Treading that line, keeping the others in check, and being kept in check by them, is what we call morality. Hanging a few feet above the Chilterns escarpment, our crew enacted morality's ancient irresolvable dilemma: us or me. (1998: 14f.)

The reader then is left with the problem of constructing a believable persona from the conflicting accounts: from the rational decision in analogy to the prisoner's dilemma and/or the instinctive response based on our evolutionary heritage. The contradiction cannot be resolved; as in the case of Wittgenstein's 'duckrabbit', two mutually exclusive versions exist side by side. Both accounts are equally valid, but they resist the reader's desire for coherence and their integration into a unified whole.

Of similar interest in this respect is Richard Powers's novel *The Echo Maker*. It is the story of a case of Capgras Syndrome. Capgras Syndrome manifests itself in the delusion that a close friend or relative has been replaced by an impostor who looks exactly the same but nevertheless is not the familiar person. It is, ultimately, a failure of recognition – actually, patients may respond affectionately to the voice of a person on the phone they would reject in a face to face meeting, a fact that Powers also introduces in his novel (see Powers, 204). As the person in question is always closely connected to the patient suffering of Capgras Syndrome, one may suspect that it is rooted in a failure of emotional response to visual contact, i.e. 'I do not have any feeling for the person I see, so it cannot be the person I have always liked or loved'. The result is an inability to construct a coherent image of the person in question, an image that will remain stable in different situations demanding a variety of responses. If recognition requires the ability to realize an underlying sameness of an object or a person in different circumstances and conditions, it also requires the mental work to construct a flexible, but also stable image of the other and to adjust this image to the various conditions in which the object or person is encountered.

In the novel, it is Mark Schluter, a guy as common as common can be, who no longer recognizes his sister after he had a car accident leading to severe brain trauma. For most of the novel he is in the hospital or undergoes treatment in the hope that his condition could be reversed, which on the very last pages it is – and strangely enough, suddenly all the interest invested in him is gone as he turns out to be one of the most boring characters imaginable.

The medical diagnosis and treatment of Mark is aided by one of the leading neuroscientists from New York, a character very much based on Oliver Sacks. It has been criticized that the protagonists of novels dealing with science turn into the embodiments of the respective science and that, for example, in Ian McEwan's novel *Saturday*, the neurosurgeon, even buys his fish neurosurgically, thinking about the "scores of polymodal nociceptor sites just like ours in the head and neck of rainbow trout" (McEwan 2005: 127; cf. Taylor 2005: 32). This phenomenon also runs through *The Echo Maker*. The neurologist Gerald Weber can hardly ever escape, or forget about, his research and the intricacies of the brain. Here is an example in which he quite obviously cannot take a shower without questioning each and every notion of reality as such.

> None of the bizarre neurological insights acquired over the course of his professional life unsettled him more than this simplest one: baseline experience was simply wrong. Our sense of physical embodiment did not come from the body itself. Several layers of brain stood in between, cobbling up from raw signals the reassuring illusion of solidity.
>
> Scalding water streamed over his neck and down his chest. He felt his shoulders relax, but he did not place too much faith in the feeling. The cortex's body maps were fluid at best and easily dismantled. (Powers 2007: 327)

Literature dealing with aspects of science inevitably has to face the most basic problem of representation, i.e., how does it want to present the specific topic to the reader, mimetically or diegetically, by telling or by showing? After all, the text needs to transmit the very principles it tries to incorporate to an audience that will in all probability not be familiar with the latest scientific research. In consequence, the texts occasionally read as if they were copied from James Gleick's *Chaos*, from Stephen Hawking's *A Short History of Time*, or from other works of popular science. In *The Echo Maker* (2007), there are innumerable passages that read like badly disguised paraphrases of Sacks, Damasio, and Ramachandran with some story woven around the case studies. This, of course, can be seen as sloppy writing, with the author trying to show off all his expertise about the state of the art in brain research and in consequence failing in one of the most basic aspects of a fictional text, the construction of believable characters. At one point of the novel, Weber realizes that "his attempt to popularize [neuroscience] had somehow gone wrong" (Powers 2007: 344). And one could ask whether this does not also include some self-referential doubt about the novel itself.

The text, however, also presents the reader with a close analogy between the psychological construction of a coherent environment and the act of narration, when Weber muses that "He was just getting a rough sense of this shattered man, of Mark's attempts to tell himself back into a continuous story" (Powers 2007: 181). And indeed, the strictly diegetic aspects of the novel, i.e. the explanation of the scientific research and case studies, is balanced by mimetic elements, which challenge the reader with some of the psychological phenomena that have already been discussed. In the course of the novel, the characters show increasingly centrifugal tendencies. Weber, the loving husband, successful scientist, and famed author completely deteriorates when his latest book meets with some hostile reviews; at the end of the novel, his previously happy marriage seems to be beyond repair, and the reader may well wonder whether this is, indeed, the same character he encountered 500 pages earlier. The vegetarian environmentalist who always loved Mark Schluter's sister Karen and probably will do so for the rest of his life seems almost too good to be true, but suddenly, at a restaurant, Karen treats him like an unregenerate womanizer, and it is not quite clear whether she has somehow lost it or whether he has suddenly acquired a new set of characteristics. In a later quarrel she yells at him: "Leave me alone. Don't touch me. These aren't real tears. Don't you see, yet? I'm not her. I'm just a simulation. Something you invented in your head" (Powers 2007: 368), and I would argue that this is also a meta-fictional signal for the reader to stay alert in the encounter of the fictional characters.

One more example: the most capable and comforting nurse of the hospital, adored by each and every character for her integrity and loving care, turns out to have previously been a high profile journalist, a "superwoman" (2007: 550) who did on-location stories on Waco, on the Heaven's Gate suicides, and on the orphaned babies after the bombing at Oklahoma city, and she never flinched from sticking a camera in the faces of people in Manhattan after 9/11 before she suddenly suffered from a flameout and ended up bird-watching in Nebraska. Moreover, it turns out that she actually caused the car accident that led to Mark's trauma but never confessed to it for reasons that are anything but clear. Before Mark's eyes, the image he had built of her "disintegrates", she is "erased" (2007: 556). But now it is no longer only Mark, suffering from a mental illness, who cannot maintain a stable concept of the woman he faces. It is also the reader who needs to integrate conflicting aspects of this fictitious character in an increasingly intricate attempt to construct a coherent personality, and by and by one may come to realize that such an endeavor will necessarily require creative work, on the one hand, and a rather selective memory, on the other hand. We constantly have to readjust our perspective not only on this character but also on almost each and everyone if we wish to have a believable set of people. This principle then swaps over to the world at large – the terrorist attacks of 9/11 completely change the country and its inhabitants and once more, we need to create some continuity and coherence rather than experience it. In addition, we are reminded that the concept of terrorism itself is dependent on the perspective of the observer, and that there was a time "when the terrorists were the heroes" (2007: 282), i.e. in the American War of Independence. Ultimately, the novel about a Capgras patient forces the reader into the awareness that coherence may well be a fiction, that our perspective on the world and the people around us is the product of complex mental processes, the presence and functioning of which are quite as bewildering as their absence in a case of delusion.

The basic phenomenon of Capgras, i.e. the sudden disability to recognize familiar and emotionally close persons, can also be detected in various other texts or movies, best known of which are probably the four adaptations of Jack Finney's science fiction novel *The Body Snatchers*. In the text and the movies, the people are indeed replaced by soulless alien invaders. However, behind the story – or rather stories – we can detect the challenge to come to terms with radical changes in the social and cultural environment during the McCarthy era, the post-Watergate era, or in the face of inhuman conformity on a military base. In a short story, "The Father Thing" by Philip K. Dick, exactly the same topic is explored and now is indicative of the emotional tensions in the course of adolescence. In another story by Dick, "The Impostor", the hero we have identified with suddenly realizes that he himself is a living bomb sent to Earth by alien enemies; on the last page, he unwillingly destroys our world with an explosion that can be seen all the way to Alpha Centauri.

All these texts and movies play on the problem that our constructions of stable and coherent characters are highly questionable and that permanence and coherence may well be in the mind of the beholder rather than in our environment or its inhabitants. Ultimately, Capgras Syndrome may appear to be the more adequate response to incompatible experiences or highly variable identities. The fact that we do not readily succumb to notions of alienation and displacement indicates how well our cognitive apparatus is adapted to this kind of mental work.

4. Cognition and literature, or: The reader's mind observed

In his book *Cognitive Poetics*, Peter Stockwell points out that readers need to "track references to characters and events through the process of reading" (Stockwell 2002: 155). Using Coleridge's *The Rime of the Ancient Mariner* as his example, he argues that the mariner may change considerably over time, but nevertheless remains to be the mariner and thus a non-episodic character in the poem (2002: 155). This is, of course, quite true, and Lisa Zunshine pursues a similar point throughout her book *Why We read Fiction: Theory of Mind and the Novel* (2006). It may, however, be worthwhile to add the complementary aspect to the mental endeavor of making sense of literature and also of our experiences as such. In order to achieve a coherent image of characters and our environment, it is occasionally also important, either to lose track of specific traits and events or to engage in a highly developed game of doublethink. Moreover, it is imperative to creatively construct our world from incomplete or incoherent facts and data.

One may also argue that ever since the beginning of the last century, modern literature has entered a kind of contest with its readers. Increasingly, the texts challenge or reject the contract of verisimilitude on which previous fiction was based. Viktor Shklovsky's concept of defamiliarization, Brecht's epic theatre, Joyce's narrative experiments, and a host of other revolutionary literary innovations all contribute to a common meta-fictional objective: to break the illusion, to call into question the suspension of disbelief, and to resist the reader's desire for narrative stability. The inherent tendency of the reader to construct coherent characters in the course of the reading process is persistently countered by the fragmentation of the figures, the resistance against representation, the destabilization of expectations, and the willful evocation of hesitancy and uncertainty in the reader.

Similarly, Reber et al. argue that modern art has broken the contract with the audience and abandoned smooth perception as a basis of aesthetic pleasure:

> Since the emergence of modern art, a piece of art can have aesthetic value without being beautiful (i.e. without producing an experience of aesthetic pleasure). Conversely, a painting that 'pleases the eyes' may be without artistic merit.
> (Reber et al. 2004: 365)

It seems, however, as if the audience responds to these challenges with increased vigor, and even the most fragmented protagonist or story will sooner or later be re-constructed into an integrated whole. Leopold Bloom becomes a "well rounded man", and even Tyrone Slothrop, the protagonist of *Gravity's Rainbow* who suddenly disintegrates after 626 pages and later seems to be "scattered all over the Zone" (Pynchon 1975: 712), feels like a real person to the well-adapted postmodern reader and confirms the notion that our ability to impose order and coherence on the most inconsistent and possibly even chaotic experience can cope with the challenge.

But then, once more we do not have to turn to the most complex novels on our bookshelves for this kind of competition. As a matter of fact, there is at least one literary genre that is defined by the very fact that it turns the automatic internal construction of a coherent fictional reality into a game between the teller and the audience. The tall tale frequently begins as a regular story, but after a while exaggerations and fantastic elements are woven into the narration until, sooner or later, the audience will become restive and realize that it 'has been had'. My favorite example for this kind of game is a spy novel, *Der Gigant* by the Czechoslovakian author Ladislav Mňačko, which opens like many others before with an ex-spy who wants to stage his own suicide in order to escape his previous employer. Spy-novels occasionally tend to be less than fully realistic, so a chase between submarines, one of which is disguised as a whale, does not completely break the suspension of disbelief, but when we later find ourselves on an island in the middle of the Bermuda triangle, where Tarzan, James Bond, and other assorted macho heroes of popular fiction are held prisoners by feminist amazons, we have to ask ourselves, "how the hell did we get here?" The transition is rather smooth, and when I first read the book some time ago, it was not easy to nail down the point when the semi-realistic novel turned into wonderfully absurd nonsense.

Our ability, desire, and obsession to weave the most divergent strands of experience into one tapestry and then ingeniously to detect the 'figure in the carpet' indicates an evolutionary heritage that has attached itself to the relatively new phenomena art and literature. In this respect, the investigation into literature and reading processes may prove to be a fruitful and productive endeavor for cognitive research.

References

Cutting, James. 2003. Gustave Caillebotte, French impressionism, and mere exposure. *Psychonomic Bulletin and Review* 10 (2): 319–343.

Damasio, Antonio. 1998. *Descartes' Error*. New York: Bard.

Dick, Philip K. 1990. Impostor [1953]. In *Second Variety: The Collected Stories of Philip K. Dick*, Vol. 2, 378–393. London: Grafton.

Dick, Philip K. 1994. The father thing [1954]. In *The Father Thing: The Collected Stories of Philip K. Dick*, Vol. 3, 134–146. London: HarperCollins.

Enoch, David and William Trethowan. 1991. *Uncommon Psychiatric Syndromes* [1967]. Oxford: Butterworth-Heinemann.

Finney, Jack. 1998. *Invasion of the Body Snatchers* [1954]. New York: Scribner.

Iser, Wolfgang. 1994. *Der Akt des Lesens* [1976]. München: Wilhelm Fink.

Jakobson, Roman. 1956. Two types of language and two types of aphasic disturbances. In *Fundamentals of Language*, Roman Jakobson and Morris Halle (eds.). The Hague: Mouton, 53–82.

Joyce, James. 1967. *Dubliners* [1914]. Harmondsworth: Penguin.

Kenner, Hugh. 1987. *Ulysses* [1980]. Baltimore: Johns Hopkins University Press.

McCloud, Scott. 1994. *Understanding Comics*. New York: Harper-Perennial.

McEwan, Ian. 1998. *Enduring Love*. London: Vintage.

McEwan, Ian. 2005. *Saturday*. London: Jonathan Cape.

Mňačko, Ladislav. 1978. *Der Gigant*. München: Langen Müller.

Norris, Margot. 1987. Narration under a blindfold: Reading Joyce's 'Clay'. *Publications of the Modern Language Association* 102 (2): 206–215.

Powers, Richard. 2007. *The Echo Maker*. London: Vintage.

Pynchon, Thomas. 1975. *Gravity's Rainbow* [1973]. London: Picador.

Ramachandran, Vilayanur S. and Sandra Blakeslee. 1998. *Phantoms in the Brain: Probing the Mysteries of the Human Mind*. New York: William Morrow.

Reber, Rolf, Norbert Schwarz and Piotr Winkielman. 2004. Processing fluency and aesthetic pleasure: Is beauty in the perceiver's processing experience? *Personality and Social Psychology Review* 8 (4): 364–82.

Sacks, Oliver. 1985. *The Man Who Mistook His Wife For a Hat*. New York: Simon and Schuster.

SOED = Shorter Oxford English Dictionary. 2007. 6th edition on CD-ROM. Oxford: Oxford University Press.

Stockwell, Peter. 2002. *Cognitive Poetics: An Introduction*. London: Routledge.

Taylor, Christopher. 2005. A knife at the throat. *London Review of Books* 27 (5): 1–32.

Watzlawick, Paul. 1977. *How Real is Real?* New York: Vintage.

Weir, David. 1991. Gnomon is an island: Euclid and Bruno in Joyce's narrative practice. *James Joyce Quarterly* 28 (2): 343–60.

Wright, Robert. 1995. *The Moral Animal: Evolutionary Psychology and Everyday Life*. New York: Vintage.

Zunshine, Lisa. 2006. *Why We Read Fiction: Theory of Mind and the Novel*. Columbus: Ohio State University Press.

Verbal irony in Shakespeare's dramatic works

Wolfgang G. Müller
Friedrich-Schiller-Universität, Jena

The present study of Shakespeare's use of irony refers to two theories, first, the echoic mention theory, which assumes that a speech act is ironic when the speaker mentions or echoes an earlier utterance in order to express an attitude of contempt or ridicule towards it, and, second, the theory of irony according to which the ironic speaker pretends to be an injudicious person speaking to an uninitiated audience, intending the addressees of the irony to discover the pretence. Passages I analyze are taken mainly from Shakespeare's early comedy *The Taming of the Shrew* and from his Roman tragedy *Julius Caesar*. Having looked at literary irony from the angle of cognitive theory, two fundamental questions are discussed: (i) whether the recourse to Cognitive Linguistics provides insight into the phenomenon of literary irony and, (ii) what the study of literary irony can contribute to the discussion of irony in Cognitive Linguistics.

Keywords: echoic mention theory, *Julius Caesar*, theory of irony, *The Taming of the Shrew*, relationship between CL and traditional literary criticism

1. Definition

In view of a bi-directional approach in cognitive research, irony is in several ways a highly challenging topic. For one thing, irony is a phenomenon that is as ubiquitous in everyday discourse as it is in literature, and it may be asked whether there are differences in the use of irony in literature and language in general and whether the studies of irony in both fields could mutually benefit from each other and possibly yield some synergistic effect. Yet a reservation as to irony as a communicative phenomenon must be made. Although it would be hazardous to deny the fact that the ability to produce and understand irony belongs to the overall cognitive capacities of humans and is present from around the age of six (Wilson 2006: 1723), we know from experience that there are inveterate ironists and others who avoid or shun irony, in other words, that there are persons who relish producing irony and others who do not understand it or hate it. I believe that cognitive linguistics should also attempt to deal with the problem of irony misfiring or misunderstood. What is interesting in terms of the interdisciplinary study of cognition in language is that failures of ironic speech acts, both on the speaker's and the receiver's side, which occur not infrequently in actual communication, find a counterpart in fictional communication, where these failures tend to be made the object of humor or mockery. This could be another starting-point for bi-directional research. Bi-directionality can also emerge in the inter-relationships of different positions in linguistic approaches to irony, such as Gricean pragmatics and relevance theory. Not all these problems can be dealt with in this article, which concentrates on Shakespeare's use of irony, but a cognitive basis for the analysis of the texts is laid and perspectives for bi-directional work are hoped to be opened. As a whole, the application of cognitive theory to Shakespeare's texts is designed to achieve a more precise and more deep-reaching account of phenomena that have so far been studied only in terms of traditional rhetoric and poetic.

Cognitive research has dealt widely with the phenomenon of irony in the last three decades, as the volume *Irony in Language and Thought* by Raymond Gibbs and Herbert Colston (2007) documents impressively, but other than politeness theory, for instance, which has produced enlightening studies in the field of literature, especially with reference to Shakespeare (Brown and Gilman 1981; Albers 2007), the use of irony in literature remains, apart from occasional references to Swift's *Modest Proposal*, virtually unnoticed in cognitive studies. This holds also true for Cognitive Poetics, which originated about ten years ago and has in the meantime achieved remarkable results (Stockwell 2002; Freeman 2007). While Cognitive Stylistics (Semino and Culpeper 2002; Gavins and Steen 2003), in particular, has given much attention to conceptual metaphor theory or blending theory, irony, which is in the system of tropes related to metaphor, has been practically ignored.

Since this chapter is concerned with predominantly verbal irony, it may be useful to start with a rhetorical definition, i.e. a definition of irony as a trope.[1] As a trope, irony is a use of language that expresses the intended meaning by its opposite, for instance, praise by blame or blame by praise. To put it more precisely, irony substitutes for the proper term (*verbum proprium*) a semantically opposed figurative term (*verbum improprium*) in such a way that the substituted term can be recognized. It is the function of the so-called irony signals to ensure the comprehension of irony. The definition of irony put forward here can analogously be expressed in terms of propositions. Irony depends on the interpretation of a proposition in its literal sense and in a sense contradictory to the literal one.

1. For a wider definition of irony that is particularly relevant to the performance of drama on stage see Greiner (2007).

Since classical times there have been two different definitions of irony: First, Quintilian's narrower definition, according to which irony wants us to understand the opposite of what is stated: "in utroque enim contrarium ei quod dicitur intelligendum est" (*Institutio oratoria*: IX.2.44), and second, Cicero's wider definition, according to which irony says something other than what is meant: "quom alia dicuntur ac sentias" (*De Oratore*: II.67.269). If we deal with irony as part of the system of rhetorical tropes, it is necessary to insist on the criterion of semantic opposition. Otherwise it would not be possible to distinguish irony from allegory. Another trope related to irony is litotes, which affirms something by the negation of its contrary, for instance when we refer to an unshapely man with the words, "He is no Adonis". The ironic equivalent of this proposition would be "He is an Adonis". The semantic contradiction is evoked in litotes only to be deleted by negation, while in irony the contradiction forms the essence of the expression. That is why Linda Hutcheon (1995) appropriately refers to "Irony's Edge" in her study. From the periphrastic nature of litotes the fact derives that this trope can be expanded almost *ad libitum*. I can, for instance, say, "This is by no means what I may be tempted to call an Adonis". Such verbosity would be impossible or inept in an ironic utterance. Verbal irony derives its effect from incisiveness of formulation. Emphatic reductions like in "What an Adonis!" can result in the sharp or bitter tone of sarcasm, which is a subcategory of irony.[2]

2. Communication models

Generally speaking, the receiver's end and cognitive aspects have by and large been left out of consideration in traditional reflections on irony, although a cognitive component is perceptible in Quintilian's above-quoted "contrarium intelligendum est" and in rhetoricians following him. That Renaissance authors were aware of the cognitive aspect of the reception of irony is shown in the following quotation from Gabriel Harvey's polemics against Thomas Nashe: "Now I giue not euery word their litterall sence, and by that you may see how I presume of your good wit, to see if by allusion you can picke the true meaning" (Lichfield 1597: sig. [D.3An$^{v]}$]). The importance of gesture and voice in the comprehension of figurative speech is noted by Abraham Fraunce (1588: sig. II: A): "For without th[e] change of voyce, neither anie Ironia, nor liuely Metaphore can well bee discerned".

2. For sarcasm and its relation to irony see Rockwell (2006).

An important step in the development of the theory and criticism of irony is marked by the advent of communication theory. For communication situations in which irony is directed to one recipient, who is expected to understand the utterance, the simple bipartite model of sender and receiver is sufficient, but for more complex situations in which the ironic utterance is directed to different recipients who receive it in different ways, an expansion of the bipartite to a tripartite model is necessary, which consists of a sender, a recipient, and a third person as the object or victim of irony. An explicit description of this model is given by Fowler:

> Irony is a form of utterance that postulates a double audience, consisting of one party that hearing shall hear & shall not understand, & another party that, when more is meant than meets the ear, is aware of both that more & of the outsiders' incomprehension. [... It] may be defined as the use of words intended to convey one meaning to the uninitiated part of the audience & another to the initiated, the delight of it lying in the secret intimacy set up between the latter and the speaker. (Fowler 1984: 295–296)

Renaissance writers and theorists were acutely aware of such situations. A much-quoted and much-commented example is Terence's comedy *Eunuchus*, in which the parasite Gnatho flatters the braggart soldier Thraso in such a way that all except Thraso understand it as irony. A Renaissance Terence commentator remarks on such a scene, "Ironia, sed quam non intellegit miles" (qtd. in Müller 1989: 193). If ironic utterances are so designed as not to be understood by a party in the communication situation they have to be qualified as lies in relation to this party. In Shakespeare's plays utterances that are poised between irony and lying are quite frequent, as we shall see.

3. Cognitive approaches to irony

Since it is impossible to refer to all of the recent cognitive approaches to irony, I concentrate on two theories that are relevant to my own topic. However, I must make some general remarks beforehand. A tendency of modern irony theory is that the traditional concept of semantic opposition, the assumption "that an ironist uses a figurative opposite to the literal meaning of the utterance" (Jorgensen, Miller, and Sperber 1984: 112), has practically been abandoned. As a consequence, the distinction between irony and understatement (litotes) has been leveled, as is seen for instance in Wilson (2006: 1723). This is a theoretical decision that I cannot follow on account of my historical and rhetorical orientation. The criterion that irony involves a pragmatic violation of the sincerity/

felicity conditions for well-formed speech acts, an aspect that was still acknowledged by Grice,[3] has been played down in recent studies. In my examples from Shakespeare, the sincerity criterion can, however, not be disregarded altogether. Another standard conviction that has been thrown overboard in recent studies is the assumption that irony comprehension passes through two phases, that of the recognition of the literal meaning to that of the understanding of the figurative meaning. The two approaches that are important in relation to my study of irony in Shakespeare are:

1. The echoic mention theory (Sperber and Wilson 1981): The echoic mention theory assumes that a speech act is ironic, when the speaker mentions or echoes an earlier utterance in order to express an attitude of contempt or ridicule towards it. The cognitive performance required from the recipient is to recognize the utterance as familiar and to reinterpret it in terms of the new meaning the speaker intends to convey.

   The status of "echoic mention" requires clarification in relation to relevance theory. Relevance theory elaborates, as an alternative to the classical code model, an inferential model of communication the foundation for which was laid by Grice. According to relevance theory, human cognition tends to be geared to the maximization of relevance. In this context, inferential communication – called "ostensive-inferential communication" by relevance theory – is of paramount importance. Now irony, which conveys the opposite of the literal meaning of an utterance, is a supreme example of the fact that there are forms of communication that cannot be explained by a straightforward code model. Verbal irony conveyed by an echoic utterance involves "the expression of a tacitly dissociative attitude [...] to an attributed utterance of thought" (Sperber/Wilson 2006: 622). The sense of cognition in Grice's understanding of irony differs manifestly from that in relevance theory. For Grice, irony is an overt violation of the maxim of truthfulness, which necessitates a process of first understanding the literal meaning and, in a second step, of recognizing the intended meaning, so that the co-operative principle is ultimately not violated. According to relevance theory, the recipient immediately seizes upon the ironic meaning as the relevant meaning in a cognitive act that is effected by an "ostensive stimulus". This stimulus is what is, in traditional terminology, called irony signal.

3. According to Grice (1975: 53), the ironist flouts the maxim of sincerity, implicating the opposite of what was literally said.

2. The pretence theory of irony (Clerk and Gerrig 1984): According to the pretence theory the ironic speaker pretends to be an injudicious person speaking to an uninitiated audience, intending the addressees of the irony to discover the pretence. The role of pretence was already referred to Grice who said that "[t]o be ironical is, among other things, to pretend [...], and while one wants the pretense to be recognized as such, to announce it as a pretense would spoil the effect" (Grice 1978: 125). This theory is of particular relevance to the use of irony in the Renaissance as a period that was intrigued by all kinds of simulation or pretence. Also it is useful to account for expansions of irony into larger fictionalizing strategies such as Swift's *Modest Proposal*, whose ironic strategy has been called "impersonation irony" (Wilson 2006: 1741). What is sorrily absent from modern theory is the distinction between simulatory and dissimulatory irony, i.e. between the affirmation of the existence of something that does not exist and the denial of the existence of something that exists. In terms of human behavior this opposition could be expressed by the distinction between "behaving as if" and "behaving as if not" (Müller 1989).

4. Shakespeare's use of irony from the point of view of the pretence theory

Let us begin with an example from Shakespeare's *Hamlet*. When king Claudius grows restless during the performance of the play-within-the-play, Hamlet ironically comforts him,

(1) 'Tis a knavish piece of work; but what o' that? Your majesty, and we that have free souls, it touches us not. (III.2.220–221)

Hamlet's ironic utterance can be understood as a kind of pretence. He establishes a temporary identity of the king and himself as being both guiltless. The irony is effected by the use of the plural, which is to drive home to Claudius the singular, namely, that it is he alone who is guilty. In the comprehension of the utterance the king has to perform a cognitive act, he has to understand that Hamlet's use of the plural is a pretence and to realize that he, Hamlet, points at him as the villain.

My main example for Shakespeare's use of irony as pretence is the comedy *The Taming of the Shrew*.[4] The first step in Petruchio's process of "taming" Katherina is an ironic denial of her shrewishness, which turns out to be a carefully prepared

4. For a rhetorical analysis of the use of irony in *The Taming of the Shrew* in the context of Renaissance rhetorical theory and poetics see Plett (2007).

pretence strategy. In a soliloquy that he speaks before first meeting Katherina he refers to his determination to apply the principle of ironic inversion to whatever rudeness or recalcitrance may be in store for him on the occasion:

(2) Say that she rail, why then I'll tell her plain
She sings as sweetly as a nightingale.
Say that she frown, I'll say she looks as clear
As morning roses newly washed with dew.
Say she be mute and will not speak a word,
Then I'll commend her volubility,
And say she uttereth piercing eloquence.
If she do bid me pack, I'll give her thanks,
As though she bid me stay by her a week.
If she deny to wed, I'll crave the day
When I shall ask the banns, and when be married. (II.1.168–178)

Shakespeare here introduces the audience to Petruchio's preparation of his ironic performance. The principle of irony production – particularly the method of semantic inversion – is clearly set out in a series of examples. When Katherina appears on stage he accordingly showers terms of praise on her in contradistinction to her ill reputation as a shrew:

(3) But Kate, the prettiest Kate in Christendom,
Kate of Kate Hall, my super-dainty Kate –
For dainties are all cates, and therefore 'Kate' –
Take this of me, Kate of my consolation:
Hearing thy mildness praised in every town,
Thy virtues spoke of, and thy beauty sounded –
Yet not so deeply as to thee belongs –
Myself am moved to woo thee for my wife. (II.1.185–192)

This is, of course, irony that replaces blame by praise, but in an entirely asymmetrical communication situation praise is rhetorically so strongly intensified that a balance of irony production and irony comprehension cannot establish itself. What is required according to the pretence model of irony – namely, that the speaker's pretence be discovered by the recipient – is not fully realized in this passage. Petruchio's shower of praise is designed to stun the interlocutor. Katherina cannot but seize upon one word, "moved", and attack Petruchio by punning on it: "Moved in good time! Let him that moved you hither / Remove you hence: I knew you at first / You were a movable" (II.1.195–197). The ensuing battle of wits is again interrupted by a longer passage of ironic praise in which Petruchio reacts to Katherina's aggressiveness in their dialogue:

(4) [...] I find you passing gentle.
'Twas told me you were rough, and coy, and sullen,
And now I find report a very liar,
For thou art pleasant, gamesome, passing courteous,
But slow in speech, yet sweet as springtime flowers.
Thou canst not frown. Thou canst not look askance,
Nor bite the lip, as angry wenches will,
Nor hast thou pleasure to be cross in talk,
But thou with mildness entertain'st thy wooers,
With gentle conference, soft and affable.
Why doth the world report that Kate doth limp?
O sland'rous world! Kate like the hazel-twig
Is straight and slender, and as brown in hue
As hazelnuts and sweeter than the kernels.
O let me see thee walk. Thou dost not halt. (II.1.235–249)

The double-faced nature of irony is here more clearly perceptible than in the previous passage. Petruchio ironically inverts the bi-polarity of praise (*laus*) and blame (*vituperatio*). He persistently substitutes for the negative qualities for which Katherina is ill-famed their positive opposites. That this is pretence in the form of a fictionalizing strategy is signaled by the fact that he falsely imputes to Katherina a physical defect, a limp, only to refute it in the context of his eulogy (Plett 2007: 121). Now, if Petruchio's speech creates an elaborate positive counter-image of Katherina, it is not to be understood simply as a sequence of compliments, because in the shrewish qualities that are negated the unpleasant image of a shrew remains perceptible. Although the tone of praise is dominant, the irony of blame by praise is still in the air. Simulation of Katherina's positive qualities goes together with dissimulation of her negative image. Petruchio's irony is more complex than it may seem at first sight.

In the last stage of Petruchio's taming of his wife, to which I am coming now, there seems to be no room for irony, but at the very end of the process of taming, irony comes in significantly. I go directly into Act IV, Scene 6, in which Petruchio's despotism reaches a climax. In contradiction to the sunshine she sees with her eyes, he says, "Good Lord, how bright and goodly shines the moon!" (IV.6.2.). Such a remark could be understood as irony, but in this case, it is an attempt of Petruchio to force his will on his wife. It is actually a violation of cognition. When Katherina contradicts him, "The moon? – the sun. It is not moonlight now" (IV.6.3), Petruchio insists that it is the moon, or star, or whatever he wishes. And he chides her for her crossness and threatens to call off their visit to her father. At this moment, something important happens. Katherina gives in, "And be it moon, or sun, or what you please" (IV.6.13). He insists it is the moon, whereupon she replies, "I know it is the moon" (IV.6.17). He then contradicts his earlier statement: "Nay then you lie, it is the blessed sun" (IV.6.18). And Katherina promptly agrees, "Then God be blessed, it is the blessed sun. / But sun it is not, when you say it is not, / And the moon changes even as your mind" (V.5.18–21). What has happened? Katherina now humors him; she has joined the game. She has obviously realized that her husband is not accessible to the voice of reason, that he just wants to assert his authority. When she says that his mind changes like the moon, she may even indicate that he is mad. It is well known that the words *lunacy/lunatic* are derived from Latin *luna*. If we understand this, we may realize that Katherina actually takes the superior position at this moment in the dialogue. She has obviously found a way of coping with his irrational and tyrannical behavior. What looks like brainwashing or thought control turns out to be a drastic sort of male chauvinism to which the woman is superior. She treats her husband like an idiot who will not listen to reason. She even seems to enjoy the game, as we see a little later in the scene.

Old Vincentio appears onstage and Petruchio greets him, again contrary to the evidence of his eyes, like a beautiful young woman:

(5) Fair lovely maid, once more good day to thee.
Sweet Kate, embrace her for her beauty's sake. (IV.6.34–35)

Now Katherina immediately takes up the cue. She seems to enjoy the game, and even outdoes her husband in the praise of this young woman who in reality is an old man:

(6) Young budding virgin, fair, and fresh, and sweet,
Wither away, or where is thy abode?
Happy the parents of so fair a child,
Happier the man whom favourable stars
Allot thee for his lovely bedfellow. (IV.6.38–42)

This extravagant praise is pure irony on the part of Katherina. Ironic pretence here becomes a game played with gusto. This shows when Petruchio performs yet another cognitive somersault. He contradicts her, pointing out that the person she is praising is not a young maiden, but an old man:

(7) Why, how now, Kate, I hope thou art not mad.
This is a man, old, wrinkled, faded, withered,
And not a maiden as thou sayst he is. (IV.5.43–45)

It is astonishing how flexible Katherina proves to be now. She immediately reverts her earlier judgment:

(8) Pardon, old father, my mistaking eyes
That have been so bedazzled with the sun
That everything I look on seemeth green.
Now I perceive thou art a reverend father.
Pardon, I pray thee, for my mad mistaking. (IV.6.46–50)

What we witness in this scene is the birth of a female ironist. In her referring to her "mad mistaking" there is an ironic allusion to her husband whose "mad" conduct is responsible for her vacillations. In view of the gusto with which Katherina realizes the theatrical potential of ironic pretence, there can be no doubt that her controversially discussed speech on obedience in the play's last act is, with its hyperbolical rhetoric and its pronounced performance quality, also to be understood as ironic.

5. Shakespeare's use of irony from the point of view of the echoic mention theory

While in the analysis of the use of irony in *The Taming of the Shrew* the pretence model of irony could fruitfully be applied, the following discussion of Antony's use of irony in the forum-scene in Shakespeare's *Julius Caesar* refers to the echoic mention theory of irony. This theory explains irony as a result of the echoing repetition or rephrasing of a previous statement. Instances of such irony are, of course, most conspicuous in dialogue. If, for instance, a no longer very young woman says to her husband over breakfast, "I wish I were twenty years younger", and her husband replies, "Yes, I also wish you were twenty years younger", the echoing of his wife's words by the husband has an ironic effect. An example from Shakespeare's *Hamlet* is to be found in the dialogue between Hamlet and his mother in the closet scene:

(9) QUEEN: Hamlet, thou hast thy father much offended.
HAMLET: Mother, you have my father much offended. (III.4.9–10)

This speech exchange shows how effectively the repetition of an interlocutor's words can be used in repartee. It is important, however, that Hamlet's echoing of his mother's utterance is much more than mere repetition. Hamlet sends his mother's words back to her with a crucial change of pronouns ("thou" > "you", "thy" > "my") and meaning. In his utterance the referent of the word "father" is shifted from 'stepfather' to 'real father' and thus the accusation is turned from himself to his mother in what is an acerbic kind of irony that can be called sarcasm. This is an example of the use of the technique of ironic echoing in a tragic context. In his comedies, notably *Love's Labour's Lost*, Shakespeare uses this technique of ironic repartee for comic purposes (Müller 1999).

Let us now turn to Antony's speech in *Julius Caesar.* Among the people of Rome Shakespeare presents in his history play, it is a widespread opinion that Brutus is an honorable man, and Brutus himself attaches great importance to his reputation for honesty, a reputation that Antony desires to destroy. It seems obvious that Antony's repetition of the sentence "And Brutus is an honourable man", in contexts that belie it, is an ironic strategy. It is not only that Antony repeats a general opinion in Rome, but that he keeps repeating the sentence during his speech, actually hammering it into his hearers' ears. Thus the echoing effect that cognitive linguists have attributed to irony is here massively present. The same echo effect applies to Brutus' accusation that Caesar was ambitious, which Antony wants to disprove:

(10) I thrice presented him [Caesar] a kingly crown,
Which he did thrice refuse. Was this ambition?
Yet Brutus says he was ambitious,
And sure he is an honourable man. (III.2.93–96)

This passage echoically refers (1) to Brutus's blaming of Caesar for his ambition and (2) to Brutus's reputation as an honorable man. Almost the whole of Antony's speech can be looked at in terms of an ironic echoing strategy. Antony's repeated reference to Brutus's honorableness has indeed been quoted frequently as a supreme example of irony, but the matter is much more difficult than it may appear at first sight. For the orator does not want his hearers on stage to recognize the irony. He avoids any impression that he is attacking Brutus. He wants them to understand that he really believes Brutus to be an honorable man. What he does is, in fact, to confront his hearers with a cognitive dissonance, which they cannot penetrate and from which they can ultimately only rid themselves by resorting to violence. By placing the word *honourable* in the closest context of murder, he manipulates them into denying it themselves:

(11) ANTONY: I fear I wrong the honourable men
Whose daggers have stabbed Caesar; I do fear it.
5 PLEBEIAN: They were traitors. Honourable men? (III.2.148–150)

It is Antony's rhetorical strategy to make the plebeians – in opposition to his pretended belief in Brutus's honorableness – discover themselvses the falseness of that attribution and act out of their own impulse, as one plebeian's sarcastic remark, "They were traitors. Honourable men!", indicates. This 'recognition' is the result of deception and manipulation on the part of Antony, decidedly not the usual comprehension of irony, which implies an understanding of the ironist's strategy. Where, then, does the much-acclaimed irony in Antony's speech reside?

In order to answer this question, we have to consider the communication situation in the theatre.

It is the audience who penetrates Antony's strategy and recognizes it as irony. What appears as – and actually is – a demagogue's deception and manipulation on the level of the action of the play, may be received as irony by the spectator or reader. It a situation analogous to the one from a Renaissance commentary on Terence quoted above: "Ironia, sed quam non intellegit miles". Possibly the speaker, who enjoys his bravura performance, can also be regarded as his own recipient. This is a hallmark of Shakespeare's villainous deceivers, notably of the self-congratulatory protagonist in *Richard III*, as Wolfgang Clemen has shown in his magisterial commentary on the play (1957).

6. The interaction of the techniques of echoing and pretence

Antony's ironic strategy is similar at the oration's final climax when he denies his rhetorical talent, which is a glaring instance of dissimulatory irony in a speech that shows him all along as a brilliant orator. It is significant that he opens this part of his oration with two references to the conspirators' honorableness, which can be explained in terms of the echoic mention theory:

(12) They that have done this deed are honourable.
What private griefs they have, alas, I know not,
That made them do it. They are wise and honourable,
And will now doubt with reasons answer you. (III.2.203–206)

Having thus once more ironically referred to the conspirators' alleged honorableness, he proceeds to annihilating Brutus's moral character by addressing the latter's use of rhetoric, and in doing so he turns to a strategy that can be interpreted in terms of the pretence theory. He conveys the idea that oratory is an art of deception and that Brutus is an orator (simulation), while he is a plain-dealing man who entirely lacks eloquence (dissimulation):

(13) I come not, friends, to steal away your hearts.
I am no orator, as Brutus is,
But, as you know me all, a plain blunt man
That love my friend; and that they know full well
That gave me public leave to speak of him.
For I have neither wit, nor words, nor worth,
Action, nor utterance, nor the power of speech,
To stir men's blood. I only speak right on. (III.2.207–214)

The irony is supreme. At the very moment when he is about to steal away his hearers' hearts in a grandiose rhetorical effort, he is denying to do so and in powerful rhetoric lists all the qualities of a rhetorician that he pretends to lack: "wit" (*inventio*), "words" (*elocutio*), "worth" (*ethos*), "action" (*actio*), "utterance" (*pronunciatio*) and, most importantly, "the power of speech / To stir men's blood" (*pathos*). In an ingenious twist of his argument he does finally hypothetically invert the roles of himself and his antagonist, suggesting that he would move his hearers to frenzy if he were Brutus:

(14) [...] But were I Brutus,
And Brutus Antony, there were an Antony
Would ruffle up your spirits and and put a tongue
In every wound of Caesar that should move
The stones of Rome to rise and mutiny. (III.2.217–221)

The use of pretence, a constitutive feature of irony according to cognitive linguists, is here developed into the fabrication of false images of the two antagonists, the dissimulatory one of Antony as an anti-rhetorician and the simulatory one of Brutus as an arch-rhetorician. It is the entirely fictional image of Brutus as an expert in the art of stirring the emotions – his speech on the forum is actually devoid of affective rhetoric (*pathos*) – that Antony evokes to whip up the passions of his hearers. Once again: Antony's hearers on the forum are entirely unaware of his manipulative techniques, while certainly none of the speaker's irony is lost on the play's audience.

7. Conclusion

This chapter has looked at difficult and problematic instances of irony in literature. Rather than on simple and straightforward examples, emphasis has been placed on complex cases that are, it is hoped, of a wider significance for the discussion of cognition in ironic utterances. Two fundamental issues remain to be discussed after having looked at literary irony from the angle of cognitive theory. The first is the question whether the recourse to cognitive linguistics helps to provide insight into the phenomenon of irony in literature that is inaccessible to the traditional methods of literary and rhetorical criticism. This question can be answered in the affirmative. The interdisciplinary or bi-directional approach to the study of irony in Shakespeare's works has proved to yield new results. To refer to the cognitive echoic mention theory at first, our analysis leaves no doubt that this theory can be successfully applied to the use of irony in Shakespeare and, for that matter, in other writers. This theory assumes that there is a shared knowledge in the brain of both the sender and receiver of an ironic message, which crystallizes in a judgment or proposition to which the ironist refers. This phenomenon can be observed in everyday as well as in literary dialogues and on a larger fictional level in Antony's strategy of placing the statements "Brutus is an honourable man" and "Caesar was ambitious" in a context that belies them. The irony is here understood by the audience in the theater, while the hearers within the play – who are aware of Brutus's reputation for honorableness and his claim that Caesar was ambitious – do not penetrate Antony's strategy and are thus forced into a cognitive dissonance that explodes in violent action, which is just what the demagogue ultimately aims at. They are the victims of Antony's irony, while the audience in the theater are the recipients of his irony. I believe that the echo theory is a key to a deeper understanding of Antony's refutation of Brutus's claim to be an honest man whose actions are strictly based on honorable intentions. As for the second cognitive theory applied in this article, the pretence theory of irony, the result has been similarly successful. This theory of irony may contribute to a comprehension of ironical strategies insofar as it helps us to recognize fictionalizing strategies that involve processes of positive self-fashioning of the ironist and misrepresentation of his antagonists. The pretence theory actually offers the literary scholar perspectives that can lead him or her beyond mere verbal irony to the realm of larger fictionalizing strategies of a simulatory and a dissimulatory kind. Although linguists may be of the opinion that "typical cases of verbal irony [...] are best analyzed as cases of echoic allusion and not of pretence" (Wilson 2006: 1725), literary critics may appreciate the pretence theory for its wider scope that opens the view into the spheres of impersonation and simulation.

The second issue – the question whether the study of the use of irony in literature can contribute to the discussion of irony in cognitive linguistics – cannot be so easily answered in the affirmative. It may be argued, however, that for a linguist it may indeed be of interest to realize the enormous possibilities that emerge in the representation of the production and comprehension of irony in literary texts. And it may alert him or her to the complexities of the phenomenon of irony that may also emerge in normal speech situations, which, as we know, are frequently much more intricate than they may be expected to be. Also, literary examples of irony may shed light on borderline cases, in which irony imperceptibly merges with strategies of lying and deception. Be this as it may, there is no doubt that the phenomenon of irony opens a field of investigation in which linguists and literary critics may fruitfully cooperate.

References

Albers, Bettina. 2007. *Normabweichungen im Gesprächsverhalten in Shakespeares Komödien*. Trier: Wissenschaftlicher Verlag Trier.

Brown, Roger and Albert Gilman. 1989. Politeness theory and Shakespeare's four major tragedies. *Society and Language* 18: 159–211.

Clemen, Wolfgang. 1957. *Kommentar zu Shakespeares Richard III – Interpretation eines Dramas* (2nd ed.). Göttingen: Vandenhoeck & Ruprecht.

Clerk, Herbert H. and Richard J. Gerrig. 1984. On the pretense theory of irony. *Journal of Experimental Psychology: General* 113: 121–126.

Fowler, Henry W. 1984. *A Dictionary of Modern English Usage*. Ware: Omega Books.

Fraunce, Abraham. 1588. *The Arcadian Rhetorike*. London.

Freeman, Margaret H. 2007. Cognitive linguistic approaches to literary studies: State of the art in Cognitive Poetics. In *The Oxford Handbook of Cognitive Linguistics*, Dirk Geeraerts and Hubert Cuyckens (eds.), 1175–1202. Oxford: Oxford University Press.

Gavins, Joanna and Gerard J. Steen (eds.). 2003. *Cognitive Poetics in Practice*. London: Routledge.

Greiner, Norbert. 2007. Szenische Ironie in Shakespeares Historien. In *Irony Revisited: Spurensuche in der englischsprachigen Literatur*, Thomas Honegger, Eva-Maria Orth and Sandra Schwabe (eds.), 135–150. Würzburg: Königshausen & Neumann.

Gibbs, Raymond W. and Herbert L. Colston (eds.). 2007. *Irony in Language and Thought: A Cognitive Science Reader*. New York: Lawrence Erlbaum.

Grice, Herbert P. 1978. Further notes on logic and conversation. In *Syntax and Semantics*, Vol. 9: *Pragmatics*, Peter Cole (ed.), 113–128. New York: Academic Press.

Grice, Herbert P. 1975. Logic and conversation. In *Syntax and Semantics*, Vol 3: *Speech Acts*, Peter Cole and Jerry L. Morgan (eds.), 41–58. New York: Academic Press.

Hutcheon, Linda. 1995. *Irony's Edge: The Theory and Politics of Irony*. London: Routledge.

Jorgensen, Julia, George A. Miller and Dan Sperber. 1984. Test of the mention theory of irony. *Journal of Experimental Psychology: General* 113: 112–120.

Lichfield, Richard [Gabriel Harvey]. 1597. *A Trimming of Thomas Nashe*. London.

Müller, Wolfgang G. 1989. Ironie, Lüge, Simulation, Dissimulation und verwandte rhetorische Termini. In *Zur Terminologie der Literaturwissenschaft. Akten des IX. Germanistischen Symposiums der Deutschen Forschungsgemeinschaft. Würzburg 1986*, Christian Wagenknecht (ed.), 189–208. Stuttgart: Metzler.

Müller, Wolfgang G. 1999. Dialogue and dialogicity in Renaissance drama. In *Anglistentag 1998 Erfurt: Proceedings*, Fritz-Wilhelm Neumann and Sabine Schülting (eds.), 211–224. Trier: Wissenschaftlicher Verlag Trier.

Plett, Heinrich F. 2007. Rhetorik der Ironie in Shakespeares Komödien. In *Irony Revisited: Spurensuche in der englischsprachigen Literatur*, Thomas Honegger, Eva-Maria Orth and Sandra Schwabe (eds.), 115–133. Würzburg: Königshausen & Neumann.

Rockwell, Patricia Ann. 2006. *Sarcasm and Other Mixed Messages: The Ambiguous Ways People Use Language*. Lewiston: Edwin Mellen.

Semino, Elena and Jonathan Culpeper (eds.). 2002. *Cognitive Stylistics: Language and Cognition in Text Analysis*. Amsterdam: Benjamins.

Shakespeare, William. 1997. *The Norton Shakespeare*. Stephen Greenblatt (ed.). New York: Norton.

Sperber, Dan and Deirdre Wilson. 1981. Irony and the use-mention distinction. In *Radical Pragmatics*, Peter Cole (ed.), 295–318. New York: Academic Press.
Sperber, Dan and Deirdre Wilson. 2006. Relevance theory. In *Handbook of Pragmatics*, Lawrence Horn and Gregory Ward (eds.), 607–631. Oxford: Blackwell.
Stockwell, Peter. 2002. *Cognitive Poetics: An Introduction*. London: Routledge.
Wilson, Deirdre. 2006. The pragmatics of verbal irony: Echo or pretence? *Lingua* 116: 1722–1743.

Invisible, visible, grammaticalization

Liane Ströbel
Heinrich-Heine-Universität, Düsseldorf

The present chapter underlines the claim that grammatical forms are neither "meaningless", nor can be regarded as a "paradoxon" (Pustet 2005: 186). On the contrary, grammar is meaningful and reflects the physical, psychological, and social experience of speakers. Furthermore it is illustrated that the motivation and mental images behind the rise of grammaticalization products can be reduced to the following formula: an abstract action or concept that is invisible in the discourse world is made "visible" with the help of a universal strategy. This strategy is not language-specific, but similar developments can be observed across typologically distinct languages, such as English, French, Spanish, Italian, German, Chinese, Japanese, Turkish, and Ewe (Ströbel 2010). Taking this strategy into consideration can make an important contribution not only to the general cognitive science endeavor, but also to neighboring disciplines such as foreign language acquisition, pedagogy, and teaching.

Keywords: abstract concepts, cross-typological approach, influence of experience on grammar, language universals

1. Introduction

The grammar of a language can be seen as a cognitive achievement and as the result of "problem solving strategies" that generations of speakers of a speech community have used in order to structure their thoughts and to communicate with each other. "Equivalence in difference" (Jakobson 1959: 262) still remains the cardinal problem of linguistics. A cognitive usage-based approach to grammar focuses therefore on the motivation underlying grammatical structures and tries to detect the mechanism(s) at work in a language and, comparatively, across different languages (Bybee, Perkins, and Pagliuca 1994: 1).

In the following I illustrate the transition from "invisible" to (linguistically) "visible" with the help of the verb that denotes the meaning of 'have' in French, Spanish, Italian, English, German, and its near-equivalents in Chinese, Japanese,

Turkish, and Ewe. The study of this verb is an attempt to bring together various guiding principles of cognitive linguistics into a unified case study of the concept 'closeness to the body of the speaker' and its linguistic realizations.

The following study of factors that led to a rise of figurative (complex and abstract) meanings of 'have', may serve to illustrate some of the richness and freshness that a cognitive approach brings to the study of language. It is also an attempt to document and motivate the various conceptual mappings that exist between the source domain, the literal meanings of 'have' and the target domain, the figurative meanings of 'have' (Lakoff 1987: 267).

The phenomenon I refer to as "visible, invisible, grammaticalization" is presented following Langacker's "three-level-pyramid" (2008: 8), starting with a characterization of empty verb + abstract noun constructions in general, followed by the presentation of the prototype 'have' + abstract noun. The discussion ends with a functional explanation and the presentation of a hypothesis concerning the motivation behind the rise of this universal phenomenon of language change.

The present study is closely linked to the bi-directional approach, not only because the human language faculty itself is already a bidirectional system (i.e., it can be used in both directions, to understand and to generate utterances in a language), but because the topic of this study combines not only an existing form with a new meaning, but even more specific, pragmatics with the sensory motor system. By decoding an universal strategy, based on sensory motor concepts, it is a step towards a better understanding of the mechanisms at work of language change. The fact that grammar is not arbitrary, but the result of problem-solving strategies could have an impact on the way it is taught (see the contribution by Alexander Ziem in this volume). This may lead to new perspectives in language acquisition and teaching, due to a better understanding of the processes involved and the results of grammaticalization. Furthermore, the study of the mechanisms and pathways of grammaticalization or language change in general, also represent a fertile research base for further neighboring disciplines, focusing on "embodiment" or "grounded cognition" theories and analyzing the relationship between stimuli and neural spiking, such as philosophy of the mind or neurolinguistics (see Conclusions, below).

2. Characterizing empty verb + abstract noun constructions

According to Bybee, Perkins, and Pagliuca (1994), the goal of linguistics is

> not just to study pathways of change and make predictions about them; it is rather to uncover the actual mechanisms of change that operate in everyday language use that eventually give rise *to grammatical categories*.
>
> (Bybee, Perkins, and Pagliuca 1994: 24, emphasis in original)

In the following, I want to uncover the actual mechanisms that have led to a rise of empty verb constructions in many languages.

Language is a genetically given human ability to combine mental units in a way no other being is able to do. With the help of the linguistic code it is not only possible to speak about something visible in the discourse situation, but also to talk about actions that have either taken place in the past or will happen in the future, or even about abstract entities such as emotions. In the following, I show that it is rather basic vocabulary that takes its source in the human body that is used in order to make an "invisible" action or concept linguistically "visible" in the concrete discourse situation. It is illustrated at the end of this chapter that this strategy ultimately leads to the rise of new grammatical categories.

The grammar of a language "reflects an essential feature of human cognition" (Langacker 2008: 540). The fact that grammar has generally a "bad reputation", is due to the fact that it is often perceived as a "system of arbitrary forms based on abstract principles unrelated to other aspects of cognition or human endeavor" (Langacker 2008: 3). As complex and opaque as the grammatical inventory may appear, the word "system" already evokes that grammar is not arbitrary but a conglomeration of functional elements, which are all linked and interact with each other. Its opacity is due to the fact that grammatical changes take place on the diachronic axis, while the results of grammaticalization surface only on the synchronic axis.

Grammatical morphemes are meaningful, too (Langacker 2008: 3), but in comparison to the lexicon, the original motivation for a certain grammatical morpheme is often lost with its rise in frequency. Still, behind every grammatical structure, behind every form that turned from a lexical and semantically free and autonomous form into a grammatically functional and often bound element, there is a hidden strategy that needs to be discovered. Since figurative language has always been a prominent source and feature of the cognitive approach to language study, I concentrate in the following on the arm/hand-string and illustrate how it enriches the lexicon and mainly the grammar of a language. In Lakoff's experientialist approach (Lakoff 1987: 266) the human body and body parts (such as *head/face, back/belly, leg/foot, arm/hand* etc.) play an important role in our conceptualization of the world:

(1) a. to head in a direction
    b. to look back
    c. to let an opportunity slip through your fingers
    d. to handle a problem
    e. to know by heart

The speakers of a language are constantly extending the existing resources of their language to new situations. The wish for more expressivity is most of the time the cause for the creation of new metaphorical or metonymical extensions. With time and a rise in frequency, these new forms can become conventionalized.

In the following, I present the rise of a new category of empty verb constructions, with the prototype 'have' and possible extensions with verbs that denote 'make/do' and 'give/take', etc. The following examples all have in common that they belong to the arm/hand-string:

(2) a. have
to have… (dis)advantages, a choice, concerns consequences, courage, a discussion, fear, feelings for, a fight with, goal, hope, an idea, influence, the intension to, the impression, the possibility of, the opportunity to, patience, power, privileges, problems, respect for, success, suspicions, sympathies for, talent, time, worries, etc.

b. make
to make… a comment, an effort, an excuse, an experience, an observation, a wish, etc.

c. give
to give… advice, my opinion, orders, information, my word to someone or somebody; a concert, a performance, a lesson, a lecture, an address, a smile, a call, etc.

d. take
to take… a walk, your time, care, etc.

I use the term 'empty' verb constructions instead of 'functional' verb or 'light' verb constructions because most of the empty verb constructions, if not all, can also include functional verb or light verb constructions, but not vice versa. Empty verb constructions share one criterion in that they pass the possessive pronoun test, which is presented in the next section.

3. Finding prototypes

Empty verb-constructions take their source in the human body, with verbs that denote 'be' as a departure point for the arm/hand- and the leg/foot-string. These two strings do not only represent a great departure point for empty verb constructions, but, in general, many of the verbs that tend to go down the grammaticalization path appear in one of these two strings.

In the following I concentrate on the arm/hand-string and use grammaticalization processes taking place on the leg/foot-string just for illustration. Verbs that denote 'have', 'hold', 'make', 'do', 'take', and 'give' etc. belong to the arm/hand-string (see Figure 1). These verbs can be divided into three groups: static verbs ('have'/'hold'), verbs implying a more or less visible movement with the hands ('make'/'do') and directed movement verbs ('give'/'take'). 'Have' can be regarded as a prototype for the arm/hand-string on which I focus in the following. 'Have' can best be characterized by its two predominant features: conceptual simplicity and semantic complexity. Its conceptual simplicity is due to the fact that it belongs to the verbal group of basic level concepts. Basic level concepts are very frequent morphemes of elementary human action such as the English verbs *have, take, give, go, come* etc. (Heine, Claudi, and Hünnemeyer 1991: 32; Di Meola 2000: 20). These morphemes show signs of semantic ambiguity, due to their high frequency of use in everyday communication. Semantic ambiguity represents a fertile ground for different interpretations and can be accounted for by the rise of more complex and metaphorically or metonymically motivated structures. In other words, it is the conceptual simplicity of basic level concepts that can be regarded as the source for the semantic complexity of most (if not all) basic level verbs.

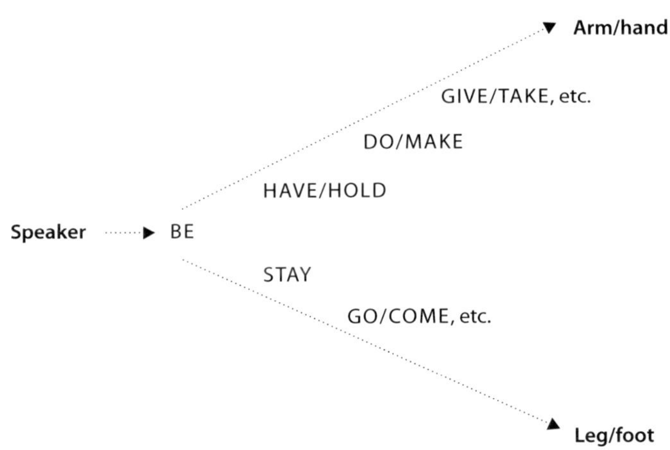

Figure 1. The cardinal points of verbal grammaticalization processes

The prototypical meaning of 'have' is 'possession'. 'Possession' is an action-orientated complex act that contains a POSSESSOR and a POSSESSEE (as in *Paul has a dog, money, a house*) and involves other concepts such as 'hold', 'take', 'seize' (Meillet 1926: 10). Possession is a common feature of all basic level concepts of the arm/hand-string. Newman (1996) for example describes the complexity of 'give' as a telic action, which involves a volitional act of transfer of an object (POSSESSEE) from a 'former' POSSESSOR to a 'new' POSSESSOR. In sum, the abundance and diversity of the "figurative" uses of to 'have' (without any implication that such meanings are considered as such by native speakers), or any other basic level concept can be reduced to two predominant properties of its literal use: conceptual simplicity and semantic complexity.

The diversity of the figurative 'have'-extensions correlates with the internal complexity of this basic morpheme. The semantic complexity inherent in the literal meaning of 'have' is a rich source for potential "figurative" extensions and turns it (like many other basic level concepts) into an attractive candidate and a productive source for a diversity of metaphorical and metonymical understandings of more complex notions.

The figurative meanings of 'have' include meanings that involve something other than the possession of a concrete object, e.g. 'have' in its function as a temporal (3a) or modal (3b) auxiliary:

(3) a. temporal auxiliary

English: The girl has lost the money, which her father doesn't know.
French: La fille a perdu l'argent, ce que son père ne sait pas.
German: Das Mädchen hat das Geld verloren, wovon ihr Vater nichts weiß.
Spanish: La muchacha ha perdido el dinero, lo cual su padre no sabe.
Italian: La ragazza (...) ha perduto il denaro, ciò che suo padre non sa.
(Heatwole and Pei 1977: 109–113)

b. modal auxiliary

English: Paul has to go for a walk with the dog.
French: Paul a un exposé à faire.
German: Paul hat zu tun.
Spanish: Pablo tiene que trabajar hasta las siete.
Italian: Paolo ha da finire i compiti per domani.

Besides the fact that on a diachronic level, in many languages 'have' has been derived via semantic bleaching of active possession verbs such as 'take', 'hold', 'get', 'grasp', and 'seize' etc. (Givón 1984: 103; Heine 1997: 48), many semantic and functional extensions of 'have' in which it is functioning more as a grammatical morpheme than a lexical item can be regarded as the result of one or more grammaticalization processes.

The successive stages of the transition or grammaticalization of a full verb that denotes a concrete act of possession into a verb with an abstract meaning have been described in detail by Heine (1997). All grammaticalized forms of 'have' share the fact that to the degree to which they are deprived of their semantic content or meaning, they gain grammatical functionality.

In these constructions, 'have' represents a grammaticalized form (similar to 'have' + verb constructions, e.g. in English *I have finished* or in French *J'ai fini*), or other grammaticalized verbs of the leg/foot-string, e.g. static ('stay') or (directed) movement verbs ('go'/'come').[1]

(4) a. French *aller* 'go'
Je *vais* *faire* *mes* *courses.*
1PSG go-1PSG do POSS PRON-1PSG shopping
'I am going (to go) shopping'

b. French *venir* 'come'
Je *viens* *de* *faire* *mes* *courses.*
1PSG come-1PSG PREP do POSS PRON-1PSG shopping
'I am (just) coming back from doing my shopping'

Therefore, the main function of the semantically 'empty' verb 'have' in combinations of 'have' and an abstract noun (e.g. *to have doubts*) is to translate a non-predicate (the noun *doubts*) into a predicate. In other words, in the construction 'have' + abstract noun, the verb does not make an independent contribution to the meaning of the sentence, but serves instead to transform a non-predicate into a predicate, similar to the copula (Hengeveld 1992: 32), by strengthening the relation between the subject (X) and the utterance (Y):

(5) a. English: Paul (X) is tired (Y).
French: Paul (X) est fatigué (Y).
German: Paul (X) ist müde (Y).

b. English: The boss (X) is in a good mood (Y).
French: Le patron (X) est de bonne humeur (Y).
(Busse and Dubost 1983: 137f.)
German: Der Chef (X) ist in / hat gute(r) Stimmung (Y).

c. English: Paul (X) is angry (Y).
French: Paul (X) est en colère (Y).
German: Paul (X) ist wütend (Y).

The strengthening of the relation of X to Y and the semantic emptiness can be illustrated with the help of a possessive pronoun as in (6).

1. Abbreviations used in glosses: 1P = first person, 2P = second person, 3P = third person, ACC = accusative, AP = aspect, ART = article, DAT = dative GEN = genitive, MASC = masculine, NOM = nominative, OBJM = object marker, PL = plural, POSS PRON = possessive pronoun, PREP = preposition, QM = question marker, SG = singular.

(6) a. English: Yesterday, we (X) had a discussion (Y).
~ our (X) discussion (Y)
French: Hier matin, nous (X) avons eu une discussion (Y).
~ notre (X) discussion (Y)
b. English: I (X) had a feeling (Y).
~ my feeling
French: Il (X) avait un pressentiment (Y).
~ son (X) pressentiment (Y) (ne le trompait pas)

The advantages of the test with the possessive pronoun are that an equivalent control verb (e.g. *discussion* vs. *discuss*) is not necessary. This becomes especially important in those cases, in which the high frequency of empty verb constructions such as in Example (7) has blocked the rise of a synthetic form:

(7) a. French:

J'	*ai*	*faim.*	[~ ma faim]
1PSG	have-1PSG	hunger	[~ my hunger]

Ø (vs. Latin *es[s]urire*, ~ 'to be hungry')
'I am hungry'

b. French:

Elle	*a*	*peur / crainte.*	[~ sa peur]
3PSG	have-3PSG	fear	[~ her fear]

Ø / *craindre*, ~ 'to be afraid'
'She is afraid'

With the help of the possessive pronoun it is also possible to separate the group of figurative 'have'-constructions (e.g. the English example *to have courage* or *avoir du courage* in French) from metaphorical constructions:

(8) a. I have a foot in the door.
≠ my foot
b. I have/take a hand in something.
≠ my hand
c. Paul has an eye to business.
≠ his eye

In (7), 'have' is nothing more than "a semantically empty supportive device, functioning as a carrier for tense, mode, aspect, and possibly other distinctions" (Hengeveld 1992: 33). Similarly, van Peteghem (1991: 5f.) describes an empty verb as an "élément de surface" ('surface element'), whose presence is mainly due to the grammatical category of the predicate. In Section 7 I present a hypothesis in which empty verb constructions are represented as a possibility for the speaker to emphasize that the utterance has an impact on the discourse situation.

4. Functional explanation A: Advantages

An advantage of empty verb + abstract noun constructions, which might have also favored the spread of empty verbs, is that due to their flexibility they can be used in order to embed foreign material. In Japanese, the high amount of originally Chinese vocabulary, and in Turkish, the high percentage of words coming from Persian led to a rise of empty verb constructions with 'make/do'. The examples in (9) illustrate the ongoing productivity of this functional domain.

(9) a. French: *Je fais du shopping.*
I do shopping.
~ mon shopping
~ my shopping

b. French: *Je fais la sieste.*
I do/make siesta
'I have a siesta'
~ ma sieste
~ my siesta

c. Turkish: *telefon etmek*
phone make
'phone'

d. Turkish: *kontrol etmek*
control make
'control'

e. Turkish: *dans etmek*
dance make
'dance'

Empty verb constructions are neither restricted to a certain language nor do the predominant constructions have to correspond in different languages (e.g. the equivalents of *have* in English, French, Italian, German, the equivalents of *hold* in Spanish and Portuguese, and the equivalents of *make/do* in many Asian and African languages).

(10) Japanese

a. *doraibu suru*
drive make
'to drive'

b. *fukushû suru*
revenge make
'take revenge'

c. *eikyô* *suru*
influence make
'have influence'

d. *gekido* *suru*
anger/rage make
'be enraged/angry'

Nevertheless, 'be' can be regarded as the source or point of departure for all further developments. It makes a difference concerning the choice of the prototype whether the functional domains of 'be' and 'have' are still clearly distinct (as in English, German, French, or Italian) or overlap (as in e.g. Japanese).

Starting with 'be' (which only states the existence of the body), there are mainly two strings that tend to be exploited for grammaticalization: the arm/hand-string and the leg/foot-string. Even though I concentrate on the first string, it is still interesting to show that they operate on the same level going from static verbs such as 'have', 'hold', 'stay' (e.g. Spanish *estar* and Italian *stare*), and 'rest' (e.g. Spanish *quedar*) to verbs implying an action with a more or less visible movement, such as 'make', 'do', 'go', 'come' etc., with (potential) further empty verbs belonging to the group of directed movements verbs such as 'give', 'take', 'return' (e.g. Spanish *volver*) etc., see Table 1.

Table 1. Development tendencies of verbal grammaticalization processes

body	strings	static verbs	verbs implying a visible movement	directed movement verbs
BE	*ARM/HAND*	HAVE HOLD	MAKE DO	GIVE TAKE
	LEG/FOOT	STAY	GO COME	

Heine and Kuteva (2006: 10) already pointed out that 'have' as an auxiliary is not an *Euroversal*, but an "universal phenomenon" (Benveniste 1966: 207; Bybee and Dahl 1989: 98). A comparison of empty verb constructions in inflecting, agglutinating, isolating and serial-verb languages revealed that empty word constructions are very common in many different languages (Ströbel 2010). The difference is that the near-equivalent of 'make' in Japanese (*suru*する), Turkish (*etmek/yapmak*), Chinese (*zuò*做), and Ewe (*wɔ̀*) are more frequent than constructions with 'have'. In Chinese, empty verb constructions are expressed either by *yǒu*有 'have', such as in 有 *yǒu* 'have' 意 *yì* 'intention' ('to have the intention'), or without an empty verb by combining two nouns (besides the possibility of *zuò*做 as in *zuò juédìng* 'make a decision'):

(11) Chinese

a. 荣 róng 幸 xìng
honor luck
'have (the) honor/ have the pleasure'

b. 悔 huĭ 恨 hèn
regrets hate
'have regrets'

c. 趋 qū 向 xiàng
tendency direction
'have the tendency'

Interestingly, typical empty verb + abstract noun constructions in French, Spanish, Italian, or German (such as the combination of the respective words for 'fear' or 'hunger' with 'have' as in French *J'ai peur/faim,* Spanish *tengo miedo/hambre,* Italian *ho paura/fame,* or German *Ich habe Angst/Hunger*) are not expressed with the help of empty verbs (e.g. Chinese *yǒu).* In these particular cases a simple morpheme, e.g. Chinese 惧 *jù (*'fear' / 'to fear'), 饿 *è (*'hunger' / 'to be hungry'*),* or 渴 *kě* ('thirst' / 'to be thirsty') seems sufficient (due to syntactic reasons) to emphasize the relationship between the subject (X) and the object (Y) of the utterance.

In Ewe, a serial-verb-language, the etymology of *le asi* ('have'; *le* 'be' + *asi* 'in hand') seems still fairly transparent and favors the rise of constructions with *dzè* (see Examples (14g) and (14h) further below).

(12) Ewe

Ga le asinye (Westermann 1907: 75)
money be hand-1PSG
'I have money'

In Japanese and Turkish the distinction between 'be' and 'have' is still fluid, e.g. *iru*いる 'be' vs. *aru*ある 'have'; *imek (-dir)* 'be/stay' vs. *var* 'have'. The functional domains of 'be' and 'have' overlap especially in these cases in which the aim of the utterance is to state the existence of something or somebody. A similar development can also be seen in Chinese:

(13) a. Chinese: Yǒu 有 guĭ ma.
have ghosts QM
'Do ghosts exist?'

b. Chinese: Yǒu 有 ren lai le.
have human being come AP
'Somebody came over' (Dezhang 1996: 80)

c. Turkish: *Bende köpek var.*
I dog have
vs.
Köpek benim
dog I-be-DAT
'I have a dog' (Cimilli and Liebe-Harkort 1976: 42)

d. Japanese: *Boku wa kodomo ga aru.*
I-MASC SUBJM children OBJM have
vs.
Boku wa kodomo ga iru.
I-MASC SUBJM children OBJM be
'I have children' (Kuno 1988: 87f.)

e. Japanese: *Kare wa watakushi no hon o motte iru.*
he SUBJM I GEN book ACC hand be/exist
Kare wa watakushi no hon ga aru
he SUBJM I GEN book OBJM have/exist
'He has my book' (Kuno 1988: 87f.)

These favored the rise of empty verb constructions with 'make/do' in these languages.

(14) a. Chinese: *zuò zongjié*
make conclusion
'conclude'

b. Chinese: *zuò juédìng*
make decision
'make a decision'

c. Japanese: *keikén suru*
experience make
'have experience'

d. Japanese: *shínbô suru*
patience make
'have patience'

e. Turkish: *arzu etmek*
wish make
'make a wish'

f. Turkish: *dikkat etmek*
care make
'take care'

g. Ewe: *wò* *ŋkú*
make look
'take a look'

h. Ewe: *dzè* *dzí*
make success
'have success'

5. Functional explanation B: Motivation

In the following, I concentrate on the motivation and hidden strategy that gave rise to grammatical changes in many languages. As Langacker (1991:508) points out:

> [o]nly sound waves actually travel from the speaker to the hearer. Words are not containers for meaning but serve instead to elicit knowledge systems of indefinite expanse, in a flexible and open ended manner.

Grammatical change never occurs out of context, it rather happens in real discourse situations (Kuteva 2001:3, 113, 180). A rather high tolerance scale in a face-to-face communication is favoring linguistic variation. The impact whether a linguistic innovation will lead to a change in language relies, however, upon the transparency of a pattern or universal strategy behind it.

Linguistic utterances (at least in the mother tongue) are produced automatically. In general, the speaker is convinced that the hearer will understand the new form/structure because he/she is aware that the hearer is constantly interpreting during their conversation. The coding of the cognitive pattern that the speaker is using is most of the time easily decoded by the hearer, given the fact that there are certain factors or communication rules at work that influence the expectations of the hearer. In general, the hearer is aware that the submitted content is far more complex than the transmitted linguistic material. The fact that information has to be added to the content is therefore a basic communication tool (Hopper and Traugott 2003:79).

In order to decode the message successfully, the hearer has to be able to understand – consciously or intuitively – the reason why the speaker has chosen a certain structure or form to code the message. In other words: the motivation behind the choice must be obvious, at least partially. Still, the "problem is that 'one swallow doth not a summer make', and one change in the grammar of an individual does not constitute what we think of as a change in 'a language'" (Hopper and Traugott 2003:47; Croft 1995:53). In other words: three factors must be fulfilled in order that one utterance will lead to linguistic change: First, the pattern or the strategy employed by the producer of the utterance should not be specific but of universal nature. Second, the utterance must not only be transparent but at the same time should not lead to more than one interpretation. Third, the new form should be flexible enough to be used in a many different situations.

The motivation behind the rise of periphrastic constructions might originally have been the wish for "une expression intense" (Meillet 1926: 140f.) or expressivity (Thielmann 1885: 519–521). By using a metaphor or metaphorical expression the producer does not only avoid using a simpler form but also adds semantic information to the utterance (e.g. *I have a foot in the door*). Metaphorical processes take place on the paradigmatic level, while metonymic processes take place on the syntagmatic level (e.g. *I have doubts*). The main difference between metaphor- and metonymy-based constructions lies in their flexibility:

(15) I have a *hand, *arm, *leg, etc. in the door.
vs.
I have doubts, concerns, ideas, a dream, faith etc.

Metonymy-based constructions are more flexible than metaphor-based ones, because in this case a new expressive form is not simply chosen, but the relation between the core (*doubts*) and the subject of the utterance is strengthened by using an empty verb (*have*).

As already stated earlier, I concentrate on the metonymy-based constructions in which meaning changes arise out of contextual contiguity, such as 'have' + abstract nouns, in which 'have' is used in order to help "a non-verbal predicate to act as a main predicate in those languages and under those circumstances in which this non-verbal predicate could not fulfill this function on its own" (Hengeveld 1992: 32).

An "invisible" action or concept (in the discourse world) is made linguistically "visible", by presenting it (e.g. an abstract action or concept such as doubts) as it would be in the hand of, near, close to or even related to the speakers body. By using the auxiliary 'have', the speaker does not add a new semantic component to the utterance, but instead strengthens the relationship between the subject and the action by presenting it as immediately related to his/her body. This principle works with one or even more arguments as in *I have doubts* and *I had a discussion with Paul.* In both cases the speaker is using the same strategy: he/she pulls the action closer to him/her by making a connection to his/her body.

When first used these constructions are expressive, but with time, due to the spread of analogous processes and their high frequency of use, the novelty effect and with it the original expressivity will disappear (Diewald 1997: 106). The increase in expressivity goes along with a stabilization and conventionalization of the new construction (Koch and Oesterreicher 1990: 78).

A similar grammaticalization process is the rise of the *going to*-future. In this case a spatial movement is used to express movement in time:

(16) be going [to visit Bill] >
[be going to] visit Bill >
[be going to] like Bill. (Hopper and Traugott 2003: 93)

The reanalysis that goes along with the grammaticalization process does not affect immediately the *ordre linéaire* of the construction, but changes its hierarchical structure (Lang and Neumann-Holzschuh 1999: 7):

(17) a. Latin: Et haec est fiducia, quam habemus ad eum (1. Joh 5, 14)

est	fiducia	→	habemus	fiduciam
be-3PSG	confidence	→	have-1PPL	confidence

'This is the confidence we have in approaching God'

b. French:

Elle	*a*	*peur.*
[Elle]	[a	peur]
3PSG	have-3PSG	fear

'She is afraid'

c. French:

Paul	*fait*	*un*	*faux pas.*
[Paul]	[fait	un	faux pas]
Paul	make-3PSG	ART	faux pas

'Paul makes a faux pas'

Similarly, Langacker (1977: 58) refers to this as a "change in the structure of an expression or class of expression that does not involve any immediate or intrinsic modification of its surface manifestation". As a result of this reanalysis the hierarchy of the sentence has changed and with it its semantic function. In the example of *going to*, the sense of physical motion is lost, but a new meaning (in this case of future prediction or intention) is gained (Sweetser 1988: 392). The strategy of taking a spatial movement in order to express a temporal movement can be traced back to the body of the speaker. By fulfilling a movement towards a future action, the speaker anchors the future action in the present discourse situation. He/she has built a cognitive bridge between the now and then (Heringer 1968: 34; Heine and Kuteva 2006: 141).

In the case of empty verb-constructions the choice of the prototype and its extension is not random, but grounded in the perception that the speaker uses something always present in any given discourse situation in order to refer to something not "visible" in a concrete face-to-face communication: his/her arm/hand. It is the speaker who uses parts of his/her body (in this case the arm/hand-string) in order to turn an invisible action or concept such as doubts 'visible' by presenting it as if it was in his/her possession or hand at the moment of speaking (*I have* [possess, hold in my hand] *doubts*).

The advantages of this strategy are obvious: the body of the speaker is not only always visible and present in the discourse situation, but also already established as a communicative support (through facial expression and gesture). The processes whereby the speaker uses something always present and therefore visible in any given discourse situation can be called processes *in praesentia* in comparison to processes *in absentia*, whereby the speaker uses material from outside the discourse situation (e.g. *door* in the example *I have a foot in the door*).

Processes *in praesentia* are used in order to either background a temporal discrepancy (e.g. 'have' + verb constructions) or, as is shown in the next sections, to focus on the here and now (e.g. 'have' + abstract noun construction).

6. Functional explanation C: Transparency

It is interesting that one morpheme out of the basic level group can be the source for many different grammaticalization processes in only one language ('have' as a perfect and future marker in French, e.g. *j' ai chanté* vs. *chanterai*) and in different languages, e.g. 'go' as a future marker in English (*I am going to sing*), French (*Je vais chanter*), Spanish (*voy a cantar*), and past marker in Catalan (*vaig cantar*), or 'come' as a future marker in Swedish (*komma att*), Hausa (*zà*), and near past in French (*venir de*) or Sesotho (*-tsoa*). These examples all have in common that no matter what basic level concept is chosen in order to fulfill a certain function in a given language, the figurative concept behind the choice of this basic level concept is fairly transparent, e.g. 'hold' or 'stay' in order to express continuity in Imonda (*ula*) or Italian (*stare a*). Nevertheless, the development of basic level concepts rests unpredictable given the fact that not all languages use for example 'go' to express near future, but make different choices:

(18) Paul takes a walk.

~ 'make':	German:	Paul macht einen Spaziergang.
	French:	Paul fait une promenade.
~ 'give':	Italian:	Paul da un passaggio.
	Portuguese:	Paul da um passeio.
	Spanish:	Paul da un paseo.

Therefore, it is important to note that the two-string-model is an idealistic model and that not all steps have to be taken in a certain language. In Spanish and Portuguese, for example, the verbs *tener*/*ter* ('hold') are very strong and have already replaced 'have' in many areas in Spanish (*tengo miedo, hambre* vs. *la chica ha perdido el dinero*), and even completely as an auxiliary in Portuguese (*ter*). In French and Italian, on the contrary, *faire/fare* ('make/do') appear in more empty verb constructions than *tenir/tenere* ('hold'). This has probably to do with the fact that in these two languages the auxiliary section is divided between 'be' and 'have' and that therefore the meaning of 'have' is still strong and does not have to be emphasized with 'hold'. This favors the rise of verbs further away from 'have'. In the case of Spanish, 'have' has taken over the function of 'be' as an auxiliary. By doing so, the verb 'have' has lost some of its semantic content. This semantic bleaching opened the door for a more concrete form 'hold'. Similar developments can be seen in different languages, see Table 2.

From a diachronic perspective it is also interesting to note that the developments of empty verbs + abstract nouns take place in one string (Latin *fames mihi est* [BE] > *famem habeo* [HAVE] > Spanish *tengo hambre* [HOLD]). Constructions such as empty verb/auxiliary + verb can also take place in between the two strings with a slight semantic change involved and a synchronically double presence of the two forms (Latin *cantare habemus* [HAVE] > *chanterons* vs. *allons* [GO] *chanter*). The reason for the overlapping of the two strings might be that the older (synthesized) form turned opaque and therefore the speaker does not associate the form with a particular string any more.

Table 2. Development tendencies in European and non-European languages

	BE/STAY	HAVE	HOLD	MAKE/DO	GIVE
Chinese	shì 是	yǒu 有	→	zuò 做	jǐyǔ 給予
Japanese	iru いる	aru ある	→	suru する	→
Turkish	imek (-dir)	var	→	etmek/yapmak	vermek
Ewe	le, nyé, du/di	le asi →		wɔ̀ (dzè)	→
French Italian English	être essere/stare to be	avoir avere to have	→ → →	faire fare to do	donner dare to give
Spanish	ser/estar	(haber) tener →		hacer	dar
Portuguese	ser/estar	ter →		fazer	dar

In case of the empty verb + abstract noun constructions the cognitive concept is still transparent, similar to other auxiliary + verb constructions such as the English auxiliaries (*have, do*, or the emphatic *have got*). Theoretically, this morphosynthetic transparency could disappear with the synthetization of the empty verb and the noun, similar to the English suffix *–(i)fy* (Latin *ficare* < *facĕre* 'make') as in *pacify* (similarly Flobert 1996: 197), but in many languages, due to syntactic reasons, this potential development seems blocked.

7. Functional explanation D: Strategy

In the following I present three examples in order to illustrate one strategy that can be reduced to the formula "invisible, visible, grammaticalization".

1. An action that is 'invisible' in the moment of speaking, for example an action having taken place in the past, is made linguistically 'visible' either by underlining it with the help of 'have/hold' so that the result of the action is presented as if it was still in the possession of the speaker, e.g. *I have written*, *I have sung*; French *J'ai écrit*, *J'ai chanté*; Spanish *he cantado*, *he escrito*, *tengo escrito*; Italian *ho cantato*, *ho scritto*; by emphasizing it with 'do' to show that an action has been fulfilled, e.g. *I did sing*, or by using 'be' (the source of all strings) in order to underline that the relationship between the action and the speaker is very close, if not equal (X = Y) as in French *je suis allé* and Italian *sono venuto*.

2. An action that is 'invisible' in the moment of speaking, for example an action that takes place in the future, is made linguistically 'visible' by presenting it as if it was already in the possession of the speaker, e.g. French *je chanterai*, Spanish *yo cantaré*, Italian *io canterò*. In these cases, we are dealing with already opaque forms, because the etymology of the suffixes *-ai/-ré/-rò* (< Latin *habeo* 'have') is not transparent any more for most of the speakers of these languages. As we have already seen in the previous section, a decline in transparency is most of the time combined with or the result of a synthetization of two former autonomous forms (e.g. Latin *habeo cantatum*). This can lead to the rise of new empty verbs, even from different strings. In this particular case it led to 'go', a member of the leg/foot-string, e.g. *I am going to sing*, French *je vais chanter*, Spanish *voy a cantar*. In other languages it favored the rise of 'come' (e.g. Swedish *komma att*).

3. A concept that is due to its very nature 'invisible', for example an emotion, is made linguistically 'visible', by presenting it with the help of 'have' or another empty verb as a real object that is in the possession of the speaker, e.g. *I have doubts*, French *j'ai peur/faim/des doutes*, Spanish *tengo miedo/hambre/dudas*, Italian *ho paura/fame/dubbi*, or again by equating the subject (X) with the action (Y), e.g. *I am afraid/hungry*.

These three examples all share one strategy: The speaker tries to turn an action or concept that is 'invisible' in the discourse situation linguistically 'visible' by presenting a past or future action or a concept as in his/her possession or at least close to him/her at a given moment in time or discourse situation. Interestingly, in these examples the speaker may choose a complex predicate instead of a simple one (e.g. *I have doubts* vs. *I doubt*). The advantage of analytic forms is that these can not only be used more flexibly, but the emphasis of an utterance can be put on a certain aspect. In this case, the analytic form does not only strengthen the relationship between the speaker and his/her utterance, but also underlines the importance of the utterance for the discourse situation.

These empty verb constructions or new analytic verbs serve as an explicit present because they focus on the relevance of the utterance for the here and now:

(19) a. English: I have faith in you (~ in this particular situation)
vs.
I have <u>always</u> faith in you.

b. French: J'ai peur, faim, etc. (~ maintenant)
'I am afraid, hungry, etc.' (~ now)
vs.
J'ai toujours peur, faim, etc.
'I am always afraid, hungry, etc.'

It is an implication of the literal use of 'have/hold', 'make/do', 'give/take' that the speaker and the object of the utterance are most of the time present at the same place (e.g. Latin *mihi est* > *habeo*). The speaker is using this feature of the literal use by presenting a concept as if it would be in his/her possession (*I have a problem, doubts* etc.). This can also be regarded as an attempt to elicit a reaction, for example to make the hearer ask him/her: What kind of problem do you have? Why do you have doubts?

In sum, the rise of perfect markers can be traced back to a strategy whereby the speaker is underlining that he/she has already done something by presenting the result of that action as still in his/her possession. The existence of future markers is due to the fact that the speaker is emphasizing that he/she will fulfill a future action by either presenting it as if the action would already be in his/her possession or that he/she is already on the way to the action. Finally, the rise of a new category of predicate and explicit present markers is motivated by a strategy whereby a concept is represented as an object in the possession of the speaker at the moment of speech. As has been illustrated, the fact that, in general, the notion of present implications is not regarded as a prominent feature of empty verb constructions is due to their high flexibility and frequency of use in any given discourse situation.

8. Conclusion

In this chapter, empty verb constructions have been used in order to illustrate that grammatical forms are meaningful, even if they have been deprived of their original semantic meaning. Empty verbs are far more than "morphosyntactic ballast", they fulfill important linguistic functions depending on the linguistic material they are combined with. They do not only strengthen the relationship between the subject and the utterance, but also anchor an action in the discourse situation. The flexibility of these constructions favors even functional extensions.

The example of empty verb construction has illustrated that grammar emerges out of strategies speakers of a language use in order to emphasize a certain part of an utterance (e.g. the relationship between X, Y and the discourse situation). As a comparison of present and future development tendencies in European and non-European languages (see Table 2) has shown, even very different languages make similar choices because one universal strategy can be regarded as the source of empty verb constructions. This strategy is not restricted to empty verb constructions, but can also function as an explanation for the rise and the choice of the linguistic material of synthetic and analytic perfect and future markers, and can therefore make an important contribution to a better understanding of the similarities of even unrelated languages.

Besides the direct benefits that a cognitive approach to grammar represents, there are important indirect benefits to be drawn from the stimulation that it provides to neighboring disciplines, such as foreign language acquisition, pedagogy, and teaching (as illustrated by Ziem's contribution in this volume). For these, the interface between language and thought, and with it intention-reading and pattern-finding, is becoming increasingly important. The existence of a pattern (e.g. from static to directed movement, see Table 1) and the fact that even slight variation in the choice of empty verb constructions (such as illustrated in Example (18)) can be explained by the same strategy, can change the way language and especially grammar is perceived and most importantly the way it is treated in these three disciplines. Realizing that grammar is not meaningless or arbitrarily based on abstract principles can not only lead towards new ways of teaching, but also to the reframing of certain pedagogical issues.

Furthermore, the topic of this chapter can also be regarded as a fertile research ground for further neighboring disciplines of the cognitive endeavor, such as philosophy of mind and neurolinguistics. In both these disciplines, the focus is recently on the phenomenon that many concepts or products of linguistic change are grounded on sensory motor processes. Many philosophical approaches, based on "embodiment" and "grounded cognition" theories, refer to the hypothesis, that primary sensory and motor areas are involved in language comprehension (Barsalou 2008; Martin 2007; Ito 2005; Vosgerau and Synofzik 2010; Vosgerau and Newen 2007; Wilson 2002). The raise of auxiliaries and empty verb constructions out of the arm/hand (and/or leg/foot) string described in this chapter can be used in order to better illustrate this tight link between high-level cognitive abilities and low-level, basal sensory-motor processes.

Recent neurological studies (using fMRI and EEG) have furthermore shown that premotor and motor areas are activated somatotopically when subjects read verbs referring to hand or foot actions (Boulenger, Hauk, and Pulvermüller 2009; Hauk, Johnsrude, and Pulvermüller 2004; Hauk and Pulvermüller 2004). Studies focusing on particular verbs or contrast pairs, such as e.g. German *greifen* 'grasp' vs. *begreifen* 'comprehend' [a prefixed form of *greifen* 'grasp'], have detected that, in contrast to the non-prefixed (and non-figurative) use, the cortex activation in the prefixed (and figurative) use of these verbs diverges considerably or is even absent (Biermann-Ruben et al. 2008; Rüschemeyer et al. 2007). A similar neurolinguistic analysis of the figurative and non-figurative use of the (in both cases non-prefixed) verbs, described in this chapter (e.g. direction inherent motion verbs, such as *to take a book from the shelf, to go to town* vs. *to take a break, I am going to sing*), could help to detect and clarify whether the fade of motor cortex activation is due to prefixation or simply due to the figurative use of these verbs. A transdisciplinary approach to the mechanism at work of language change (combining linguistic, neurolinguistic, and philosophic theories) could therefore lead to a better understanding about the way language is conceptualized, structured, and cognitively stored.

References

Barsalou, Lawrence. 2008. Grounded cognition. *Annual Review of Psychology* 59: 617–645.

Benveniste, Émile. 1966. *Problèmes de linguistique générale*. Paris: Gallimard.

Biermann-Ruben, Katja, Klaus Kessler, Melanie Jonas, Hartwig Siebner, Tobias Bäumer, Alexander Münchau and Alfons Schnitzler. 2008. Right hemisphere contributions to imitation tasks. *European Journal of Neuroscience* 27 (7): 1843–1855.

Boulenger, Veronique, Olaf Hauk and Friedemann Pulvermüller. 2009. Grasping ideas with the motor system: semantic somatotopy in idiom comprehension. *Cerebral Cortex* 19 (8): 1905–1914.

Busse, Winfried and Jean-Pierre Dubost. 1983. *Französisches Verblexikon: Die Konstruktion der Verben im Französischen*. Stuttgart: Klett.

Bybee, Joan and Östen Dahl. 1989. The creation of tense and aspect systems in the languages of the world. *Studies in Language* 13: 51–103.

Bybee, Joan, Revere Perkins and Willian Pagliuca. 1994. *The Evolution of Grammar, Tense, Aspect, and Modality in the Languages of the World*. Chicago: The University of Chicago Press.

Cimilli, Nükhet and Klaus Liebe-Harkort. 1976. *Sprachvergleich Türkisch-Deutsch*. Düsseldorf: Schwann.

Croft, William. 1995. *Typology and Universals*. Cambridge: Cambridge University Press.

Dezhang, Liu. 1996 *Verbergänzungen und Satzbaupläne des Deutschen und des Chinesischen, Eine kontrastive Untersuchung im Rahmen der Verbvalenzgrammatik*. Hamburg: Kovac.

Di Meola, Claudio. 2000. *Die Grammatikalisierung deutscher Präpositionen*. Tübingen: Stauffenburg/Narr.

Diewald, Gabriele. 1997. *Grammatikalisierung: Eine Einführung in Sein und Werden grammatischer Formen*. Tübingen: Max Niemeyer.

Flobert, Pierre. 1996. Les verbes supports en latin. In *Akten des VIII. Internationalen Kolloquiums zur lateinischen Linguistik*, Alfred Bammesberger and Friedrich Heberlein (eds.), 193–199. Heidelberg: Universitätsverlag Winter.

Givón, Thomas. 1984. *On Understanding Grammar*. New York: Academic Press.

Hauk, Olaf and Friedemann Pulvermüller. 2004. Neurophysiological distinction of action words in the fronto-central cortex. *Human Brain Mapping* 21 (3): 191–201.

Hauk, Olaf, Ingrid Johnsrude and Friedemann Pulvermüller. 2004. Somatotopic representation of action words in human motor and premotor cortex. *Neuron* 41 (2): 301–307.

Heatwole, Oliver and Mario Pei. 1977. *A Comparative Practical Grammar of French, Spanish and Italian*. New York: S. F. Vanni.

Heine, Bernd and Tania Kuteva. 2006. *The Changing Languages of Europe*. Oxford: Oxford University Press.

Heine, Bernd, Ulrike Claudi and Friederike Hünnemeyer. 1991. *Grammaticalization: A Conceptual Framework*. Chicago: The University of Chicago Press.

Heine, Bernd. 1997. *Grammaticalization Chains as Linguistic Categories*. Duisburg: L.A.U.D.

Hengeveld, Kees. 1992. *Non-verbal Predication, Theory, Typology, Diachrony*. Berlin: Mouton de Gruyter.

Heringer, Hans-Jürgen. 1968. *Die Opposition von „kommen" und „bringen" als Funktionsverben*. Düsseldorf: Schwann.

Hopper, Paul and Elizabeth Traugott. 2003. *Grammaticalization*. Cambridge: Cambridge University Press.

Ito, Motoo. 2005. Bases and implications of learning in the cerebellum: adaptive control and the internal model mechanism. *Progress in Brain Research* 148: 95–109.

Jakobson, Roman. 1959. On linguistic aspects of translation. In *On Translation*, R. Brower (ed.), 232–239. Cambridge, MA: Harvard University Press.

Koch, Peter and Wulf Oesterreicher. 1990. *Gesprochene Sprache in der Romania*. Tübingen: Niemeyer.

Kuno, Susumu. 1988. *The Structure of the Japanese Language*. Cambridge, MA: MIT Press.

Kuteva, Tania. 2001. *Auxiliation: An Enquiry into the Nature of Grammaticalization*. Oxford: Oxford University Press.

Lakoff, George. 1987. *Women, Fire and Dangerous Things: What Categories Reveal about the Mind*. Chicago: The University of Chicago Press.

Lang, Jürgen and Ingrid Neumann-Holzschuh. 1999. Reanalyse und Grammatikalisierung: Zur Einführung in diesen Band. In *Reanalyse und Grammatikalisierung in den romanischen Sprachen*, Jürgen Lang and Ingrid Neumann-Holzschuh (eds.), 1–17. Tübingen: Niemeyer.

Langacker, Ronald. 1977. Syntactic reanalysis. In *Mechanisms of Syntactic Change*, Charles Li (ed.), 57–139. Austin, TX: University of Texas Press.

Langacker, Ronald. 1991. *Foundations of Cognitive Grammar*, Vol. 2: *Descriptive Application*. Stanford, CA: Stanford University Press.

Langacker, Ronald. 2008. *Cognitive Grammar: A Basic Introduction*. Oxford: Oxford University Press.

Martin, Alex. 2007. The representation of object concepts in the brain. *Annual Review of Psychology* 58: 25–45.

Meillet, Antoine. 1926. *Linguistique historique et linguistique générale*. Paris: Librairie ancienne Honoré Champion.

Newman, John. 1996. *Give: A Cognitive Linguistic Study*. Berlin: Mouton de Gruyter.
Peteghem, Marleen van. 1991. *Les phrases copulatives dans les langues romanes*. Wilhelmsfeld: Gottfried Egert.
Pustet, Regina. 2005. *Copulas: Universals in the Categorization of the Lexicon*. New York: Oxford University Press.
Rüschemeyer, Shirley Ann, Marcel Brass and Angela Friederici. 2007. Comprehending prehending: Neural correlates of processing verbs with motor stems. *Journal of Cognitive Neuroscience* 19 (5): 855–865.
Ströbel, Liane. 2010. *Auf den Spuren der Entstehung einer neuen Kategorie: Leerverben als paralleler Kopulastrang*. Frankfurt/Main: Lang.
Sweetser, Eve. 1988. Grammaticalization and semantic bleaching. In *General session and parasession on grammaticalization*, Shelley Axmaker et al. (eds.), 389–405. Berkeley, CA: Berkeley Linguistic Society.
Thielmann, Philipp. 1885. *Habere* mit dem Partizip Perfekt Passiv. *Archiv für lateinische Lexikographie und Grammatik* 2: 372–432; 509–549.
Vosgerau, Gottfried and Albert Newen. 2007. Thoughts, motor actions, and the self. *Mind & Language* 22: 22–43.
Vosgerau, Gottfried and Matthis Synofzik. 2010. A cognitive theory of thoughts. *American Philosophical Quarterly* 47: 205–222.
Westermann, Diedrich. 1907. *Grammatik der Ewe-Sprache*. Berlin: Reimer.
Wilson, Margaret. 2002. Six views of embodied cognition. *Psychonomic Bulletin and Review* 9: 625–636.

How does the mind do literary work?

Gary Thoms and Stefano Versace
Strathclyde University, Glasgow

In this chapter we present a number of examples regarding literary forms and the way the mind processes them, ranging from meter to deviant syntactic structures in avant-garde literature. In particular, we look into the mismatches between these forms and language. Arguing against traditional positions in literary and linguistic criticism, we show the advantages of divorcing functional and structural relations between language and literary structures. Instead we suggest that the complex relations between literary form and language are best accounted for by building on a modular approach to cognition (Fodor 1983), thus developing a more theoretically and empirically coherent theory of literary forms (e.g. Fabb 2009).

Keywords: avant-garde literature, empirical approach, mental processing, relationship of function and structure, theory of literary forms

1. Introduction

A core aim of bidirectional work in literary linguistics is to draw generalizations about how language works from the forms observed in the language of literary texts. An immediate problem for this approach is that many of these forms might not actually be products of the cognitive domain of language. If it can be shown that a given formal property of a text does not have the basic characteristics of linguistic form, then we cannot draw reliable conclusions about how language works from this property; rather, the formal properties we are observing in texts are telling us about some other domain of cognition, and any relation between the literary form and linguistic form may be indirect at best. The logic of this argument is based on the assumption that there are separate domains of cognition, and such a position is known as a modular theory of mind (Fodor 1983);[1]

1. Recent defenses of variations of this theory include Fodor (2000), Sperber (2002), and Carruthers (2006).

this approach to literature can be called a modular literary linguistics, and it can be opposed to the non-modular work done in the domain-general framework of Cognitive Linguistics and Cognitive Poetics.

In this paper we analyze two unrelated phenomena in literary language and argue that they both present evidence for the modular view of literary linguistics. We examine metrical forms in Italian and German, and deviant syntax in avant-garde English poetry, and we argue that these formal properties constitute two cases of a clear divorce between literary and linguistic form. We show that these phenomena cannot be adequately explained by a domain-general approach, but rather they necessitate an explanation in terms of separate, modular computations of phonology, meter, syntax, semantics, and pragmatics. Thus, syntactic form is not determined by meaning in a functionalist-determinist way; rather, both components are autonomous, and the application of their processes is strictly ordered (syntactic processes precede semantic ones).[2] A similar domain-specificity characterizes metrical and phonological form: we present a number of regularities in metrical form, and then show that they can be explained more coherently in terms of metrical-specific rules rather than by referring to phonological formalisms. To conclude, we argue that although our evidence warns against blindly accepting bidirectionality, it does nevertheless feed back into linguistics in that it provides an argument for a modular approach to literary linguistics and cognition.

2. Metrical form / phonological form

This section shows that metrical form has regularities that cannot be explained in terms of phonological form or any other kind of linguistic or literary form. We provide an account that explains this fact and details how metrical form relates to linguistic form. In order to do so, we start out by reviewing a number of traditional claims concerning metrical forms and their relation with other components of the grammar.

Many researchers in traditional literary criticism and generative metrics have maintained that metrical form is determined by language, either as a particular setting of linguistic constraints, or as a stylized version of the grammar.[3] The

2. This is the working hypothesis of generative linguistics, established as early as Chomsky (1975 [1955]: Chapter 2) and argued for in many places since.

3. This is found in the tradition of generative metrics: Nespor and Vogel (1986), Hanson and Kiparsky (1996), Golston and Riad (2000), Hayes (2009) a.o. A discussion of the different approaches to literary language can be found in Fabb (2009b), whereas Fabb (2009a) puts forth a different theoretical framework.

central claim in this approach is that most meters are built on a phonological representations proper. Here we show that this is not the case. In fact, the number of phonological syllables in a verseline need not coincide with the number of metrical positions, as the following three lines by Dante show (1a–c):[4]

		1 2 3 4 5 6 7 8 9 10 Δ	
(1)	a.	*Nel méz.zo del cam.mín di nós.tra ví.ta*[5]	(11 syll.)
		'Halfway through our life's strive'	
		1 2 3 4 5 6 7 8 9 10 Δ	
	b.	*La nò.bi.le vir.tù B<u>e.a</u>.trì.c<u>e in</u>.tèn.de*[6]	(13 syll.)
		'the noble virtue Beatrice understands'	
		1 2 3 4 5 6 7 8 9 10 Δ	
	c.	*fe.dìr tor.ne.a.mèn.t<u>i e</u> còr.rer giòs.tra;*[7]	(12 syll.)
		'Tournaments stricken, and the joustings run;'	

(1a–c) are perfectly metrical instantiations of Dante's *endecasillabo* from the *Divina Commedia*. All these lines differ in the number of phonological syllables (as indicated in the examples), yet they still constitute legitimate input for this metrical form, which must count up to exactly ten metrical positions.[8]

Moreover, the same meter requires the diphthong /ea/ to be parsed differently in (1b–c), although it appears in the same environment (C/ea/-'C(C)V-CV): it occupies one single metrical position, namely (7) in (1b), and two positions (4 and 5) in (1c). Traditional metrical theory regarded both cases as the application of conventional optional rules, such as syneresis, in (1a), and dieresis, in (2b). In the former case, two word-internally adjacent syllables may be assigned to 1 single metrical position; in the latter, 1 syllable is counted as two distinct positions. Another common metrical convention in Italian, synalepha, has that position 9 in (1b) counts as one single metrical position, although conjoining two distinct syllables (note also that none of these rules need be implemented in performance). We believe that conventions need to be explained, too, and this

4. Henceforth, word stress is marked in the examples of this section; syllable sequences counting as one metrical position by the effect of synalepha are underlined; extrametrical syllables are marked with Δ for visual convenience.

5. Dante *Inferno*, I, 1. Henceforth, all lines by Dante are quoted from Bosco and Reggio (1994).

6. Dante, *Purgatorio*, XVIII: 73.

7. Dante, *Inferno* , XXII: 6.

8. Note that Italian, as other traditions, names this meter from an often-occurring 11th extrametrical unstressed syllable, though the last obligatory stress is that on the 10th.

one precisely shows that meter does not take as input phonology proper but an underspecified representation of its structure.

Metrical form may display a measurable degree of complexity, which may involve linguistic complexity. However, there is no causal relationship between the two: linguistic complexity does not create metrical complexity, and vice versa, as we show in (2).

(2) *nel tèmpo che colùi che 'l mòndo schiàra*[9]
[at–the time [that who [that the world enlightenIND–3SG]]]
'at the time that the one who the world enlightens'

This line has a highly complex syntactic structure, displaying two recursively embedded sentences. The same structure, however, can be mapped onto an unambiguously iambic metrical pattern, thus showing that the computation of syntactic complexity in a line does not create complexity in the computation of meter, which needs to be explained separately.

The specificity of metrical representation is also evident when we consider the phenomenon known as *caesura*, i.e. a metrical break splitting lines into two parts (*emistiches*). The phenomenon is attested in many different kinds of metrical verse, and it has been often claimed, both in traditional and in generative metrics, to be phonologically driven, i.e. isomorphic to phrase boundaries (e.g. Nespor and Vogel 1986: 279; Esposito 2003: 88–90). The following example is a long line by Eugenio Montale (1981: 926), written in a compound meter (phonological phrasing is marked by "$[\ldots]_\phi$"):

(3) *[ad un amore]*$_\phi$ *[di donne / barbute,]*$_\phi$ *[a un vano farnetico]*$_\phi$
to a love of women / bearded, to a vain gibber
'to some bearded women's love, to some gibberish talk'

Caesura in (3) is indicated by the slanted line in the example. We can see that its placement disregards phonological phrase boundaries, which in this case are split in two in order to conform to syllable counting (the branching node in ϕ, corresponding to the adjectival complement *barbute*, is cut from its head).[10] This corresponds to the fact that, though more strict metrical traditions may consistently

9. Dante, *Inferno*, XXVI: 26.

10. For other similar examples, see Montale (1981: 925–927). Note that, in standard prosodic phonology, the domain of phonological phrase (ϕ) is mapped from syntactic structure, though it is not isomorphic to it. Thus, our argument may well be extended to the relation meter–syntactic structure.

place caesura at word or phrase boundaries,[11] this is not itself a phonologically driven phenomenon. Instead, it belongs to the specifics of the metrical computation, which necessarily defines line boundaries.

The representation of metrical verse is also traditionally claimed to be based on linguistic rhythm. In (4a–b) we show two common stress patterns in Italian *endecasillabo*:

```
              *           *     *     *
          *   *  *  *   *  *  *  *   *  *Δ
(4)  a.   Nel mézzo del cammín di nóstra víta
          *   *         *     *       *
          *   *  *  *  *  *  *    *   *  *  Δ
     b.   ésta sélva selvággia e áspra e fórte[12]
          'this forest, savage, rough, and stern'
```

The two examples display quite a varied distribution of lexical stresses, which may fall on even as well as on odd positions. Traditional metrical studies simply account for this as an uncontrolled variation in rhythm, allowing lines to be built on feet of different length. Generative metrics makes instead the assumption that *endecasillabo* is built on an underlying template displaying a 5-iambic-feet structure. This is then matched onto language strings that may depart from the template itself; both metrical rules, such as trochaic inversion, occurring twice in the first 4 positions in (4b), and prosodic rules are then hypothesized to re-establish the iambic alternation, thus allowing the line to be perceived as metrical.[13]

In these stress patterns, then, we can see a significant deviation from the assumed iambic template, while the lines are still perfectly metrical. Here we follow the non-templatic approach to meter proposed by Fabb and Halle (2008): they argue that meter is a set of rules and conditions controlling the length of the line by computing over an underspecified representation of phonological constituents (syllables); the well-formed computation produces a bracketed grid, and checks it against the specific condition imposed by the meter.[14] We provide an example

11. As it is the case for French alexandrin; see Fabb and Halle (2008: 144–147).

12. Dante, *Inferno*, I, 4.

13. See for instance Nespor and Vogel (1986). In particular, phonological metrics assumes: (i) that meter is organized in architectonic fashion, thus partly resembling the prosodic hierarchy; (ii) that, in Italian, the foot and the phonological phrase are the relevant constituents for matching linguistic stress onto the metrical template.

14. The system works similarly to Finite State Automata (cf. Partee et al. 1990). For an application of such a device to the computation of prosodic structure, see Idsardi (2009).

of the set of rules for the specific pattern of Italian *endecasillabo* in (5a), and show how this scans across the line in (5b):

(5) a. *Endecasillabo*

i. Gridline (GL) 0: starting just at the R edge, insert iterative R parenthesis, form binary groups, heads R.

ii. GL1: starting just at the L edge, insert iterative L parenthesis, form binary groups, heads L.
The last (right-most) group must be incomplete – unary.

iii. GL2: starting just at the L edge, insert iterative L parenthesis, form binary groups, heads L.
The last (right-most) group must be incomplete – unary.

iv. GL3: starting just at the L edge, insert iterative L parenthesis, form binary groups, heads L.

b.

```
  ésta  sélva  selvággia e  áspra e forte                GL
) *   * ) *    Δ)  *   * )   *     * )   *     * )  Δ      ←0
     (Δ   *      ( *         *                (*           →1
        ( *                  *                (*           →2
        ( *                                   (*           →3
          *                                                 4
```

The rules in (5) work as an input-output system, generating the actual grid from a projection of the verseline onto successive GLs. They involve a general projection rule that promotes heads of the groups to the upper GLs, and the setting of a few parameters, such as the direction in which parentheses are inserted, the number of elements grouped, and the position of the head in each group. Thus, the form is directly generated from a representation of phonology that, for instance, may leave unprojected some phonological elements and project others.

A further empirical advantage of modeling meter as a form-specific computation is the possibility of scanning loose forms (Fabb and Halle 2008: 69–96). In some varieties (typically those with longer lines), meter does not control all the syllables in line. This happens by virtue of a special non-iterative rule that first marks a restricted category of stresses (*stress maxima*) in the line.[15] This influences the way grouping is achieved, in that, here, non-adjacent asterisks force the iterative rules to skip a unit, and restart the grouping from the next available sequence (the phenomenon is marked by ↓):

15. For this meter, *maxima* are defined as lexical stress preceded and followed by an unstressed syllable. For visual convenience, they are marked with square brackets.

```
            1    2    3  4  5        6     7 8 9  10 11 12
(6)   a.    E    tu   seguìssi       le    fràgili_architettùre[16]
            'And if you would follow the frail architectures'
                          ↓       ↓
      b.    E tu seguìssi le fràgili_architettùre        GL
            (*  *](*  *]*(*  *]*(*  *(*  *](            ←0
            (*    (*      *     (*    *(                ←1
            (*           (*           *(                ←2
            (*                        *(                ←3
                                      *                   4
```

Other interesting examples come from the attempts at reproducing classical meters in modern languages. In (7a–b) we scan up to GL0 a line from Goethe's *Römische Elegien*, where iterative parentheses build L-headed ternary groups:

```
(7)   Oftmals hab'ich auch schon in ihren Armen gedichtet    GL
      ([*   * [*   *    *(   [*   *[* * [*   *   *([*   *   0→
```

When non-iterative brackets break the sequence, groups generated by iterative rules may be of different length. Thus, though in these cases the linguistic rhythm of the line may seem closer to its metrical form, a simple consideration of the computation shows that there is still a clear distinction between rhythm and meter.[17]

Loose meters of various forms abound in many literary traditions: adopting specific rules to count syllables in the way shown above allows us to account for poetic forms as well as traditions that were deemed to lack meters.

In (6) to (7) we have described metrical forms by referring to a representation that is distinct from that of the phonology of the line. In our approach, this constitutes further evidence that meter is not a particular organization of linguistic rhythm, but an input-output system that computes over certain units – in the case of poetry, these consist of a projection of syllables – and groups them building the verseline.

All the forms we have been discussing so far display various mismatches between the form of meter and that of the language in its various components. We have shown how a coherent theory of metrical form must account for this by putting forth a separate explanation from that which linguistic theory seeks for

16. Montale (1981), *Notizie dall'Amiata*, II: 1. Note that the line has 12 metrical positions, and it is traditionally named *tredecasillabo* (i.e. 13-syllable line). Versace (2009) and, more extensively (2010), claim that such lines are in a loose variant of *endecasillabo*.

17. See Fabb and Halle (2008: 94). A similar scansion for an Italian loose dactylic hexameter can be found in Versace (2010).

the components of the grammar proper. This receives a simpler theoretical justification if we assume that metrical form is parsed separately from phonological, syntactic-logical form, as well as from any other kind of linguistic form; in other words, if the computations of metrical form are assumed to be form-specific, which is only possible if the architecture of the mind is assumed to be modular.[18] In what follows, we show how similar benefits may arise from applying the same hypothesis to different kinds of literary forms.

3. Deviant forms in English poetry

Deviation has long been understood as one of the fundamental ways in which literary language differs from ordinary language (see Attridge 1988), but there have been few attempts to explain how the deviant structures are interpreted formally. In this section we sketch the questions that need to be answered in engaging with these issues, and suggest that the processes involved in the interpretation of some examples of deviant sentences also implicate a modular approach to literary language. Consider the following poem by E. E. Cummings, 'me up at does':

(8) Me up at does

out of the floor
quietly stare

a poisoned mouse

still who alive

is asking What
have i done that

You wouldn't have

Probably the most striking feature of the poem is its chaotic syntax; this is a hallmark of cummings' poetry, and this particular text is an extreme example of the kind of deviant literary language that we should seek to analyze when considering what 'literary language' might be. Yet despite this apparent chaos, the sentence is almost entirely unambiguous: that is, it is given the same interpretation as the well-formed alternative "a poisoned mouse, who, still alive, is asking what I have done that you wouldn't have, does quietly stare up at me out of the floor"; the only

18. This idea is strongly supported by experimental facts, according to which speech rhythm is processed in different brain areas than those related to other linguistic components (e.g. Geiser et al. 2008). Furthermore, a modular architecture of the mind is often taken as a starting point in designing experiments on these issues. See, for instance, Endress et al. (2009).

possible ambiguity is with the adverb 'still', which has little consequence for the interpretation of the whole sentence. Indeed it seems that the poem is basically a reordered version of this well-formed sentence, involving the movement of at least four parts of the text into non-canonical positions.

We can ask, then, how one reaches an unambiguous interpretation of the meaning of this sentence in spite of the radical reordering of its parts. One answer would be that the deviant sentence is interpreted like any other, with the form of the reordered sentence mapping onto the same meaning as the normal one. In this regard it would be the same as for normal sentences with stylistic variants, such as 'That car, I like', which is understood as an example of 'focus movement'; in this case, the form maps onto the meaning by way of some rules from the grammar. Specifically, we might argue that these more highly marked constructions in cummings' poem, which are not found in ordinary language, are part of some 'poetic grammar' (Austin 1984; Fitzgerald 2007), along with the rules for poetic inversion and other such unusual reorderings in English language poetry. The poetic grammar is understood as a version of the ordinary language grammar that encompasses the extra rules required to derive the forms that occur in poetry, and these approaches are typically characterized by a non-modular approach to language.[19] Assuming bidirectionality without question, we would argue that cummings' language provides evidence for poetic grammar as part of the speakers' linguistic competence.

However, the notion of a 'poetic grammar' is highly problematic, as there are no criteria for which rules or constructions could be part of this grammar and which could not; rather, one would simply stipulate a rule where a deviant text receives an interpretation, proposing structures that do not generalize beyond a single text. We have an example of this problem in the first line of the poem, "me up at does": specifically, movement of a pronoun ('me') out of a preposition phrase ('up at') is unattested in English and unlike other focus movements. In addition, it is an instance of movement out of a constituent that has already been focus-moved to the front of the sentence, which is also unattested in English for any kinds of constructions (since it is a 'derived island'; see Rochemont and Culicover 1990). If the grammar never works like this, it seems unwise to propose that it happens to do so just in this instance. Therefore an explanation in terms of rules of 'poetic grammar' seems doomed to failure if one proposes to define the grammar with any degree of rigor.

19. For example, Fitzgerald's (2007) proposal, in the framework or Optimality Theory, involves the interaction of metrical and syntactic constraints; such a conception of syntactic operations is in principle impossible in a standard generative approach to syntax. Thoms (2010: Chapter 3) provides an extended critique of this approach.

These theoretical problems may be put aside, however, if we were to deny that the structures we see are in fact the output of a set of formal rules and instead assume that these structures are derived on the basis of functional relations. Working within the framework of Cognitive Grammar, Chen (2003) proposes such an account for inversions in poetic language; he argues that such inversions are examples of a more generally attested and functionally derived Ground-before-Figure model of focus comprehension. However, it is very difficult to see how this analysis could be extended to the kind of examples seen in the cummings poem, since it displays so many different structural rearrangements within a single sentence, some of which are embedded in others; in such examples, it is effectively impossible to determine what is the ground and what is the figure, or what is focused and what is not,[20] and in these cases motivating the structural relations between the different parts of the sentences on functional grounds becomes problematic. Although the inversions in a lot of more traditional poetry may be said to have a specific effect on discourse structure or prominence, it is in the nature of avant-garde texts such as cummings' poem to resist and reject such deterministic views of the use of language in poetry, and it is for these reasons that these deviant texts are of interest for formal analyses of literature.

If one still hopes to motivate the re-orderings in the texts by some other formal system, one might propose that the literary forms, the meter and rhyme scheme, impose this structure on the sentence; this would be another example of a non-modular approach to the problem. However, such a solution is untenable. It is highly unlikely that rhyme could motivate syntactic operations; to accept this we would have to explain why syntactic relations between rhymed lines are absent in the vast majority of texts. It is plausible that the pieces of the text have been re-arranged in the given way to produce the rhyme scheme, but this does not entail any relation between rhyme and interpretation; one could re-arrange a perfectly grammatical sentence into a text that rhymes but has no coherent interpretation, as the following re-arranged version demonstrates:

(9) still is who does

out of up stare
You wouldn't floor

quietly mouse

You wouldn't alive

20. Relevant to this discussion is the work of Constable and Aoyama (1999), which shows that there are no clear correlations between distortions in word order (i.e. optional movements, inversions) and focus, as determined by the distribution of related prosodic properties.

done asking What
the Me up that

have poisoned have

If rhyme were able to provide information that allowed the recovery of the original form of the sentence, we would expect this text to be interpretable in the same way as the original version of the poem.

We can conclude, then, that the sentence does not in fact have a single syntactic structure that maps onto the meaning, and that the non-modular approaches to the problem overviewed above are untenable.[21] How, then, does one reach an unambiguous interpretation, in the absence of this? We propose that the interpretation is created by the interaction of separate linguistic processes: well-formed sub-parts of the text are parsed with discrete syntactic structures and meanings, and pragmatic inferencing on the basis of these pieces of language produces the correct interpretation. In this explanation, syntax, semantics, and pragmatics are each assumed to have formal autonomy from one another; specifically, the output of the syntactic component is the input for the semantic component, and the logical forms that constitute the output of this component form the input for pragmatic inferencing. While all these components interact in the creation of meaning, they do so in an ordered fashion; thus, meaningful relations established by pragmatics or semantics do not create syntactic relations, since they do not constitute inputs for this component. This is a modular approach to cognition, and it handles the problems posed by deviant sentences without any additional stipulation.

To clarify how this would work, consider how cummings' text would be interpreted. There are clearly many pieces of the text that are syntactically well-formed in themselves, such as the separate phrases *a poisoned mouse*, *out of the floor*, and the whole relative clause headed by *who*. In addition to this, syntactic relations between phrases are indicated by morphological inflection: most significantly, the third person agreement on the finite auxiliary *does* indicates that the noun phrase *a poisoned mouse* is the subject of the predicate. We can say that these fragments of syntax map onto partial semantic structures; the subject-predicate relation between the verb phrase and the subject *a poisoned mouse* provides a larger structural template for the form of the sentence, and pragmatic inferencing on the basis of this evidence picks out the matching combination of this template and the other complete phrases, which is the sentential meaning delivered by the text.

21. For extensive discussion of empirical and theoretical problems for such approaches, see Thoms (2008, 2010).

This analysis entails that a text does not need to have a fully specified syntactic structure to map onto a meaning, obviating the need to create new rules for the interpretation of apparently formal relations in the text.[22] This is in itself a theoretical advantage, as it allows us to explain interpretive phenomena without appealing to stipulations that have no generality beyond the specific text: that is, we do not need to posit a 'complement of a preposition focus movement rule' to deal with structures like *me up at does*, the empirical motivation for which would almost certainly be restricted to this sole text. But it also has empirical advantages: for example, not only does it allow us to understand how some deviant sentences do receive an unambiguous interpretation, it also allows us to explain why some texts do not receive an unambiguous interpretation. This involves assessing the kind of residual structure parsed in a given text: if there is enough to provide a substantial blueprint, interpretation may be possible. This kind of predictive theory can be schematized and falsified in accordance with further empirical research, and this is surely an important requirement for any theoretical work on literary language.

4. Conclusions

In this chapter we have seen two different challenges to bidirectional analyses of formal properties of literary texts. First, we saw that metrical form has a number of properties that cannot be collapsed into the domain of phonology, and this provided evidence for assuming that meter is the output of a specialized computation, in which the form it produces is parsed separately from any other literary or linguistic form. Second, we saw that the formal characteristics of deviant syntax in poetry cannot be derived either from a simple extension of the syntactic system or from a system in which the syntax is functionally derived from other factors; rather, such explanations are difficult to maintain, and instead the data seems to require an explanation in which interpretation is derived from the interaction of separate computations. In both cases, straightforward bidirectionality is problematized: poetic meter does not provide direct evidence for the formal characteristics of the phonology, and deviant poetic syntax does not provide us with direct evidence for how the syntax works.

---

22. A different approach is adopted in Thoms (2010), where it is argued that the parts of the sentence are united in a single LF representation, but they are not combined in the standard manner in the narrow syntactic derivation. Despite this difference, the logic of the argument is the same as what is presented here: explanation of the facts relies crucially upon separating and ordering different stages of the linguistic computation.

Nevertheless, this does not mean that the analysis above cannot contribute to bidirectional research at all. Rather, we argue that it provides clear evidence for a theory of forms (linguistic or literary) as the products of separated domains; that is, modular. We have seen that assuming modular computation allows us to provide a more precise account of complexity in literary forms and their interactions. In our proposal, literary work appears to be the combination of different outputs, which then necessarily imply the simultaneous computations of different input systems. Metrical form, phonology, syntax, semantics, and pragmatics interact and produce bundles of features that we tag as literary forms.

The claims we have been gathering are part of an ongoing research project on the forms of literature; our theoretical assumptions as well as the evidence presented here suggest that literary processes offer a fruitful ground for explaining how the mind apprehends forms more generally. That is, the forms that human beings perceive or produce, of which literary forms constitute a subset, can be best understood as having very different features: their explanation thus supports a modular account of the mind. This in turn may provide a more promising avenue for pursuing explanatory questions about literature in future research.

References

Attridge, Derek. 1988. *Peculiar Language: Literature as Difference from the Renaissance to James Joyce*. Ithaca, NY: Cornell University Press.

Austin, Timothy. 1984. *Language Crafted: A Linguistic Theory of Poetic Syntax*. Bloomington, IN: Indiana University Press.

Bosco, Umberto and Giovanni Reggio (eds.). 1994. *Dante Alighieri. La Divina Commedia*. Firenze: Le Monnier.

Carruthers, Peter. 2006. *The Architecture of the Mind*. Oxford: Oxford University Press.

Chen, Rong. 2003. *English Inversion. A Ground-before-Figure Construction*. Berlin and New York: Mouton de Gruyter.

Chomsky, Noam. 1975 [1955]. *The Logical Structure of Linguistic Theory*. New York, NY: Plenum Press.

Constable, John and Hideaki Aoyama. 1999. Word length frequency and distribution in English Part II. An empirical and mathematical examination of the character and consequences of isometric lineation. *Literary and Linguistic Computing* 14 (4): 507–535.

Cummings, E. E. 1994. *Complete Poems, 1904–1962*. London: W.W. Norton & Co.

Endress, Ansgar D., Marina Nespor and Jacques Mehler. 2009. Perceptual and memory constraints on language acquisition. *Trends in Cognitive Sciences* 13 (8): 348–353.

Esposito, Edoardo. 2003. *Il verso: forme e teoria*. Roma: Carocci.

Fabb, Nigel. 2009. Why is verse poetry? *PN Review* 189: 52–57.

Fabb, Nigel and Morris Halle. 2008. *Meter in Poetry: A New Theory*. Cambridge: Cambridge University Press.

Fitzgerald, Colleen. 2007. An optimality treatment of stylistic inversions in English poetry. *Language Sciences* 29: 203–217.
Fodor, Jerry A. 1983. *The Modularity of Mind*. Cambridge, MA: MIT Press.
Fodor, Jerry A. 2000. *The Mind Doesn't Work That Way*. Cambridge, MA: MIT Press.
Geiser, Eveline, Tino Zaele, Lutz Janke and Martin Meyer. 2008. The neural correlate of speech rhythm as evidenced by metrical speech processing. *Journal of Cognitive Neuroscience* 20 (3): 541–552.
Golston, Chris and Tomas Riad. 2000. The phonology of Classical Greek meter. *Linguistics* 38 (1): 99–167.
Hanson, Kristin and Paul Kiparsky. 1996. A parametric theory of poetic meter. *Language* 72: 287–335.
Hayes, Bruce. 2009. Faithfulness and componentiality in metrics. In *The Nature of the Word. Essays in Honor of Paul Kiparsky*, Kristin Hanson and Sharon Inkelas (eds.), 113–149. Cambridge, MA: MIT Press.
Idsardi, William J. 2009. Calculating metrical structure. In *Contemporary Views on Architecture and Representations in Phonology*, Eric Raimy and Charles E. Cairns (eds.), 191–212. Cambridge MA; London : MIT Press.
Montale, Eugenio. 1981. *Tutte le poesie. L'opera in versi*. Torino: Einaudi.
Nespor, Marina and Irene Vogel. 1986. *Prosodic Phonology*. Dordrecht: Foris.
Partee, Barbara H., Alice ter Meulen and Robert Wall. 1990. *Mathematical Methods in Linguistics*. Dordrecht: Kluwer.
Rochemont, Michael S. and Peter Culicover. 1990. *English Focus Constructions and the Theory of Grammar.* Cambridge: Cambridge University Press.
Sperber, Dan. 2002. In defense of massive modularity. In *Language, Brain, and Cognitive Development*, Emmanuel Dupoux (ed.), 47–57. Cambridge, MA: MIT Press.
Thoms, Gary. 2008. When sentences are not sentences: evidence against poetic grammar. *Online Proceedings of the Annual Conference of the Poetics and Linguistics Association (PALA)*. <http://www.pala.ac.uk/resources/proceedings/2008/thoms2008.pdf> (last accessed March 13th, 2011).
Thoms, Gary. 2010. *Poetic Language: A Minimalist Theory*. PhD dissertation, Strathclyde University, Glasgow.
Versace, Stefano. 2010. *Due teorie generative della forma metrica e il verso di Eugenio Montale*. PhD dissertation, Università degli Studi di Milano.
Versace, Stefano. 2011. On the meter-phonology interface. Evidence from Italian. In *Philological Research*, Gilda M. Socarrás and Almitra Medina (eds.), 25–42. Athens: ATINER.

Cognitive science meets language pedagogy

Alexander Ziem
Heinrich-Heine-Universität, Düsseldorf

Bringing together insights from cognitive schema theories and studies on vocabulary acquisition in a foreign/second language (L2), this chapter singles out the importance of semantic schemas for teaching and learning word meanings. Its main theoretical hypothesis is that cognitive schemas underlying semantic representations of lexical meanings are effective didactic tools for teaching vocabulary. Relevant language-pedagogical concepts are discussed, such as "explicit teaching" and "rich instructions". Having introduced basic elements of schemas, it is shown how schema-based illustrations could help to explain sense relations between words as well as the emergence of new meanings in complex constructions.

Keywords: lexical meaning, semantic schemas, sense relations, teaching, vocabulary acquisition

1. The impact of cognitive schema-theories on vocabulary teaching

In the field of foreign/second language (L2) acquisition, the central role of vocabulary teaching has been ignored for a long time. It was only in the early 1990s that some studies explicitly faced this issue (e.g. Arnaud and Béjoint 1992, Nation 1990). During the past decade, however, it has become commonplace that learning vocabulary has crucially contributed to attaining a high level of linguistic proficiency. Knowing vocabulary is not only essential for grammar acquisition; it also increases the comprehensibility of the language input. This, in turn, improves the structural understanding of a foreign language (Ellis 1997).

Vocabulary learning and teaching is a highly demanding task, not only in terms of cognitive efforts but also in terms of necessary preconditions. There is converging evidence that word meaning is complex in nature. Several language pedagogical and psycholinguistic studies in the field of vocabulary acquisition provide overwhelming evidence for the hypothesis that a lot of different factors such as word structure, collocation, frequency, word class, salience, but also extralinguistic parameters such as social settings and contexts have a considerable impact on word learning (Ellis 1997, Ellis and Laporte 1997, Ellis and Ferreira-Junior 2009, Nagy 1997, Schmitt 2000: 23). Similarly, it has become a truism in cognitive linguistic research that lexical meanings can neither be adequately learned by means of "raw" definitions nor by merely studying contexts in which a lexical item usually occurs. Rather, in addition to contexts, mapping processes as well as the embodied nature of linguistic meaning must be considered in order to explain the emergence of lexical meanings (Fauconnier 1997: 34–71, Geeraerts 2006: 227–326, Johnson 2007: 111–206).

Bringing together insights from cognitive schema theories and studies on vocabulary acquisition in L2, this article singles out the importance of semantic schemas as didactic tools for teaching and learning word meanings. In doing so, it correlates psychological and cognitive-linguistic insights in the nature of meaning construction, on the one hand, and language-pedagogical issues in the domain of vocabulary teaching and learning, on the other hand. In terms of the mutual applicability of cognitive semantics and vocabulary teaching and learning, it is argued that language pedagogy could considerably benefit from theoretical concepts such as "schema", "frame", and "blending"; in turn, however, the didactic application of schemas and frames does not have discernable repercussions on cognitive schema theories, particularly because these theories have not been designed to provide didactic tools in the realm of language teaching and acquisition.

From a cognitive and pedagogical point of view, it is first argued that vocabulary teaching is most efficient if it accounts for several culture-specific and cognitive factors (Section 2). An "encyclopedic" account of lexical meaning suggests a schema-driven approach to teaching vocabulary (Section 3). Indeed, the discussion of the pedagogical concepts of "explicit teaching" and "rich instruction" reveals a strong need for schemas in the domain of vocabulary teaching (Section 4). Against this background, basic elements of semantic schemata are introduced and put into practice (Section 5). Further detail is given on how schemas could help to illustrate word meanings, including sense relations, contextual modulation, and complex constructions such as metaphor and compound words (Section 6).

2. The complexity of word meaning

2.1 What does it mean to know a word?

In studies on vocabulary acquisition one often finds word meaning, word use, lexical concept, and lexical reference sharply distinguished from one another. In his influential book *Learning Vocabulary in Another Language* (2001: 27), Nation, for example, differentiates between different knowledge types that a person must master in order to know a word; among them are knowledge aspects concerning the spoken form, the written form, word parts, associations, collocations, constraints on use (including grammatical function, pattern in which the word occurs, and pattern in which the word must be used), concepts, and referents. In an earlier book, Nation (1990: 31) presents a somewhat different list. Here, "use", "concepts", and "referents" are not characterized as distinguished knowledge types. Instead, "register" and "grammatical behavior" are mentioned.

With the term "concept" Nation obviously addresses the information content made available when a word is used. However, it remains unclear to what extent "referents", that is, entities being referred to, differ from "concepts". Do "referents" exist outside of the language user's mind? Or are they rather conceptual entities? Considering the nature of abstract nouns, an anti-conceptual view clearly becomes problematic, since understanding the meaning of abstract nouns always involves establishing cognitive reference to an evoked concept (cf. Ziem 2008: 288–298). Hence, referring and conceptualizing are two sides of the same coin. The entity being referred to is a mental concept in that its properties are objects of mental manipulation (in the case of abstract nouns) or depend on psychophysical preconditions of human perception.

Separating usage from meaning is equally problematic. On the one hand, psycholinguistic investigations on word processing have shown that context information have tremendous effects on the representation of lexical meanings (for an overview see Ziem 2010). Cognitive psychologists like Barsalou (1999, 2003) and Zwaan (2004) therefore argue that an explanatory account of lexical representation has to account for the context-sensitive and encyclopedic nature of word meaning. In addition, recent cognitive-linguistic studies provide evidence for the hypothesis that words have no intrinsic meaning (e.g. Fillmore 1985, Lakoff 1987, Fauconnier and Turner 2002, Evans 2009). Word meanings are rather functions of both the context in which a word is embedded and the cognitive processes that help to match contextual information with conventionalized lexical meanings.

For illustration, consider, the word *hypotenuse* mentioned in the pioneering work of the cognitive linguists Charles Fillmore (1985: 228) and Ronald Langacker (1988: 59). To access its lexical meaning, a language user has to activate background knowledge about geometrics, e.g. about the properties of a triangle (including knowledge about one-dimensionality opposed to multi-dimensionality and about the nature of a line between two points opposed to an infinite line), as well as knowledge about properties of a right-angled triangle in particular. Fillmore argues that language users have to mentally access such knowledge "chunks" via frames, in order to capture the lexical meaning of a word in context. Frames are schematic knowledge structures that motivate the use of a word, and frames are also cognitive structures evoked by a word; as such, they facilitate meaning comprehension.

Note, however, that cognitive processes, such as selecting, focusing and abstracting, are also involved in the construction process of lexical meaning. In the case of *hypotenuse*, for example, we only focus on the side opposite to the triangle's right angle while abstracting from other pieces of information such as the length of the side, the triangle's overall size or its color. Blending theory, first developed by Fauconnier and Turner (2002), as well as Lakoff's (1987) theory of idealized cognitive models (ICM) put emphasis on such cognitive processes underlying meaning constructions even in the case of seemingly simple and well known expressions. Lakoff gives the often cited example of the word *mother*, which may, depending on the context, give rise to many different, albeit correlated meanings. Depending on contextual information, the expression *mother* could refer to (i) a woman who gives birth, (ii) a genetic mother (who is not necessarily the birth mother), (iii) a woman who is the wife of the father, (iv) a woman who nurtures and raises the child, or (v) the closest female ancestor. Lakoff argues persuasively that all these lexical meanings exist as categories although (i) is taken to be the central case from which the others are metonymically derived. As such, they emerge as idealized cognitive models of the entities being referred to by *mother*.

Coming back to Nation's (2001) list of knowledge types, it should be obvious by now that the list does not provide a full account of knowledge types involved in word comprehension, nor does it help to explain how word knowledge is acquired. In this respect Schmitt and McCarthy (1997: 4) critically remark:

> The list gives us an inventory of ideal native-like knowledge, but it does not tell us *how* this knowledge is acquired. […] When considering this list, it is also important to remember that the categories of word knowledge are separated for convenience's sake; in reality, the different categories are interrelated, and it is probably best to think of vocabulary knowledge as an integrated whole from which we can artificially separate various kinds of word knowledge for research or discussion purposes.

For vocabulary teaching and learning it is thus much more important and relevant to gain insights into the cognitive mechanisms and cognitive structures that motivate meaning construction.

In this article, I adopt a constructionist, cognitive-linguistic approach to word meaning. Such a framework has proved to be helpful for foreign language teaching in numerous ways (cf. Boers and Lindstromberg 2006). Basic cognitive-linguistic premises have helped to give a better and clearer account of learning processes not only in the first language (Tomasello 2003) but also in L2 (Ellis 2003, O'Grady 2003). The premises encompass the view of language as an integral part of human cognition (Langacker 1987), the adoption of a usage-based model, the assumption that language can be exhaustively described as a system of form-meaning parings (or symbolic units in Langacker's terminology) and, with respect to semantics, the view that meaning is encyclopedic in nature and the object of multiple conceptualization processes including metaphoric and metonymic shifts.

2.2 What does it mean to teach a word?

As illustrated above, knowing the meaning of a word involves activating a vast amount of (possibly culture-specific) background information against which an actual lexical meaning is profiled. To become cognitively relevant, this background knowledge is "pre-packaged" in mental schemata that allow language users to access relevant semantic information fast and efficiently. On top of background knowledge, multiple metonymical shifts as well as semantic selection and focusing processes lead to new word meanings. Such shifts of lexical meanings are the rule rather than the exception.

If semantic variation is persuasive, an insight which is also increasingly accepted in language pedagogy (Nagy 1997), it is misleading and, from a didactic vantage point, devastating to draw a line between the meanings of a word, the usages of a word and the concepts that the words make available. Instead, meaning, usage, and concept are inextricably intertwined in such a way that only the usage of a word in a concrete context gives access to its lexical meaning. In psycholinguistic terms this means, as Verspoor (2008: 265) puts it, that

> an activation of a word may activate any type of association, be it social, cultural, psychological, linguistic (e.g. collocational), or pragmatic. It would also mean that the system is in constant flux because a currently activated notion may influence existing concepts.

For the sake of improving vocabulary teaching, there are a number of important conclusions to be drawn from the dynamics and complexity of word meaning envisaged above. Given that lexical meanings cannot be nicely decomposed into context-free clusters of semantic features without abstracting from possibly essential meaning aspects, it is necessary to account for

- *culture-specific background information*: Many (if not all) word meanings are related to background information that are essential for vocabulary learning and teaching. Without having a concept of adoption, for example, the meaning of *surrogate mother* remains intransparent for language learners.
- *metaphoric and metonymic meaning potentials of words*: In non-artificial ordinary language use, many word meanings undergo metonymic and/or metaphoric shifts; for language learners it is crucial to get acquainted with the systematic polysemy of words and the mechanisms prompting meaning extensions.
- *sense relations*: Word meanings are not encapsulated but semantically related to many other lexical meanings. Most important are hyponomic/hyperonomic, antonymic, synonymic, and meronymic sense relations, since they help to semantically specify the word in question.
- *degrees of semantic entrenchment*: Language use leads to semantic change and different degrees of semantic entrenchment. Instead of appealing to lexical core meanings, the concept of a deeply entrenched default meaning needs to be introduced and put into practice.
- *the impact of contextual information*: The actual meaning of a word is only in part determined by entrenched meaning aspects; it is also a function of context information that gives rise to meaning modulation.

I consider these aspects essential for vocabulary teaching. Ignoring any of them gives rise to many well-known problems of vocabulary acquisition. If, for example, the metaphoric or metonymic meaning potential of a word was ignored, language learners would probably tend to underestimate the scope of the entrenched non-metaphoric/non-metonymic lexical meaning. The meaning of *mother*, for instance, would then be limited to the idealized cognitive model of "birth mother". If, in turn, metaphoric and metonymic meaning extensions of *mother* were taken into account, notwithstanding the fact that they differ in terms of entrenchment, the category "mother" would be overextended. To use and comprehend the word adequately, one has to keep in mind that *mother*, meaning "woman who is the wife of the father", is less entrenched and only applies to cases in which the term *step-mother* is roughly synonymous to the term *mother*. Overall, vocabulary teaching should ideally encompass not only word definitions but also rich information about possible uses and typical contexts in which the word occurs.

But how can language users be encouraged to learn such "rich" word meanings? And which tasks do they have to master in order to be successful in learning rich meanings? Vocabulary acquisition is an incremental process involving at least three different, albeit intertwined cognitive tasks (see Aitchison 2003: 86–117). A closer look at these tasks may help to assess which aspects should be taken into account in vocabulary teaching. First, language learners have to master what Aitchison calls the "labeling task". That is, a highly underspecified concept is established and "labeled" with a certain phonological form. A language learner has successfully mastered the labeling task if the same phonological form triggers the same concept whenever the form is encountered. In the light of the "packaging task" then, the triggered concept becomes more specific. More pieces of information are conceptually integrated. The concept becomes more distinct. Equally important is, finally, the "network-building task". After having established a rich "information package", it is essential to relate it to other concepts that are already well-entrenched in the language learner's mind. The network-building task involves, for example, the mastery of sense relations (cf. Section 6.1). It is often necessary to learn new words in order to build up an extensive semantic network. In the case of *mother*, setting up meronymic relations involves knowledge of body-part terms such as *head*, *leg*, *arm*, *hand* etc. The more meronyms a language user knows, the richer the concept of mother becomes.

It is worthwhile noticing that the semantic network is principally open in scope. It follows that vocabulary learning is a process of approximation; strictly speaking, it never comes to an end. For vocabulary teaching it is thus important to select the most salient concepts to which the word meaning in question is interrelated. In Section 5.2 I argue that these concepts are represented as "default values" instantiated in a semantic schema. Furthermore, it seems to be essential that teachers make use of didactic tools in order to support language learners to master the labeling, packaging, and network-building task. Semantic schemas could serve well as powerful didactic tools.

3. The need for semantic schemata as didactic tools in vocabulary teaching

3.1 Experiential semantics and meaning-meaning motivations

In the field of language pedagogy, recent cognitive approaches stress that conceptual motivation may help to enhance language learning processes. It is one of the key insights of cognitive-linguistic research that language is, at least to some extent, a non-arbitrary system of signs. There are two good reasons why the concept of motivation has become the focus of attention in cognitive studies on vocabulary teaching (Boers and Lindstromberg 2006, Boers and Lindstromberg 2008: 17–37). First, any kind of linguistic non-arbitrariness may help language learners to systematically study linguistic structures. Additionally, semantic motivation helps language learners to remember lexical meanings faster and more effectively.

Radden and Panther (2004) differentiate between three types of motivation, meaning-meaning connections, meaning-form connections, and form-form connections. So far, most cognitive-linguistic investigations have focused on meaning–meaning motivation. While one research strand addresses polysemy, particularly semantic extensions of prototypical word meanings to peripheral word meanings by means of metaphor, metonymy, and image-schematic transformations, a second group of studies deals with conceptual metaphor (in the tradition of Lakoff and Johnson 1980) and concentrates on the non-arbitrariness of figurative idioms (for an overview see Boers and Lindstromberg 2006: 309–313). However, another ubiquitous subtype of meaning meaning connection has not been investigated yet; nonetheless it seems to be of major importance for vocabulary teaching: meaning is generally motivated by encyclopedic knowledge about the world, that is, by (extralinguistic) experiences of a language user. The fact that contemporary cognitive approaches to vocabulary teaching do not elaborate this subtype in more detail is all the more surprising since experience-based motivation of meaning applies to linguistic phenomena of all kinds. This finding goes back to Lakoff's so-called experientialist view of semantics, according to which there is no "objective" level of semantic description; linguistic meaning is rather grounded in human embodied experience (Lakoff 1988). Recent experimental studies about the mental representation of words provide additional evidence for the experientialist view, particularly for the relevance of sensorimotoric and affective experiences on lexical meaning (Vigliocco et al. 2009: 224–230).

If word meaning thus also "reflects" the experiential and thus cultural background of language users, vocabulary teaching needs to take account of recurrent cultural experience. They, too, motivate word comprehension. In this vein Schmitt remarks: "[t]o use the potentially vast amount of encyclopedic knowledge in real time requires that it be organized and controlled in some way. This is especially true in that the mind draws up more meaning information than what is finally used" (Schmitt 2000: 28).

It is at this point where cognitive schemas inevitably come into play. To cope with the vast amount of data input, the human conceptual system relies on cognitive schemas that reduce the flow of input information to basic and relevant information (Barsalou 1992, Fillmore 1977, Minsky 1975, Rumelhart 1980; for an overview see Ziem 2008: 247–282). Schemas provide structured information about (imagined) real-world entities including their relations to other entities.

In the cognitive sciences, it is a truism that schemas are important building blocks of a cognitive system. Following the pioneering work of the psychologist

Bartlett (1932), many scholars have provided empirical evidence for the essential role of schemas in human cognition (for an overview see Barsalou 1992). Barsalou (1992: 21) even argues that schemas are the general format of knowledge representation. Undoubtedly, schemas are the format in which word meanings are stored in our mental lexicon. They underlie and motivate word meanings. Given this basic role of cognitive schemas, it seems to be plausible to employ schemas as didactic tools for vocabulary teaching as well. Although schemas are not the subject of didactic theories yet, they seem to be of major pedagogical interest. By means of schemas it is possible to benefit from experientially grounded meaning-meaning motivations. The schema theory introduced in Section 5 allows integrating relevant background knowledge into the modeling of word meaning.

3.2 Implicit and explicit teaching

The relevance of implicit and explicit teaching in the domain of vocabulary acquisition is a long-standing issue in language pedagogy. If language learners, as described above, principally undergo cognitive processes of labeling, packaging, and network-building when learning new words, how can they be pedagogically supported? Which didactic tools are appropriate for teaching rich word meanings? In contemporary language pedagogy, there is a strong tendency to prioritize a mixture of explicit and implicit study of vocabulary. Implicit learning involves learning – or better: guessing – meaning from context. Although context data are very helpful and even necessary to gain access to a full-fledged lexical meaning, guessing word meaning from context remains extremely ineffective. Krantz (1991: 130) reports that only 5% of all unknown words encountered in a text are learned implicitly. Moreover, contextually inferred word meanings are highly speculative, and since there is no authority that approves or corrects inferred meanings, idiosyncratic meaning aspects are often remembered and stored in long-term memory. Most approaches to vocabulary teaching therefore tend to focus on activities for the explicit study of vocabulary. These activities include, but are not limited to, explicit explanations of word meaning, exercises addressing lexical meaning, the use of dictionaries, and attempts to make language learners think about lexical meaning (Nation 2001: 34).

Teaching activities, however, are only effective if language learners have enough vocabulary to start making use of words (and their meanings) that they encounter for the first time. Schmitt (2000: 145) argues that a well-structured vocabulary program should always start with explicit teaching of new words. In line with Schmitt's view, Nagy (2005) additionally emphasizes that it is not enough to provide mere word definitions, since they abstract from context data and from the semantic flexibility and potential vagueness of lexical meaning; rich background knowledge must be taken into account, too:

> More attention must be given to activating relevant background knowledge, explaining relationships among related concepts, providing graphic representations of these representations, and distinguishing between examples and non-examples of the new concept. (Nagy 2005: 4)

"Relevant background knowledge" should also include knowledge about the metaphoric and metonymic usage potential of words as well as knowledge about degrees of semantic entrenchment. For Nagy it is beyond doubt that acknowledging schemas underlying word meaning helps to maximize the effectiveness of teaching. But when it comes to the question *how* semantically rich concepts should be taught, Nagy's suggestions remain vague. To help students become better word learners, he proposes (i) modeling what to do when one encounters an unfamiliar word, (ii) teaching how to make use of word parts and context clues, and (iii) providing strategies for dealing with unfamiliar words (Nagy 2005: 6). But how should word parts and context clues be included in the teaching process of word meanings? And what kind of strategies may help to infer meanings of unfamiliar words? The lack of answers indicates a strong need for schemas as didactic tools for vocabulary teaching.

Addressing the same issues, Nation (2001) mentions three intentional word study strategies: (i) studying word parts, that is, roots and affixes, (ii) using dictionaries, (iii) using word cards. While studying word parts should help to associate roots and affixes with known roots and affixes, dictionaries may provide new bits of semantic information and help to reinforce those meaning aspects that the language learner is already familiar with. In Nation's opinion, using word cards is the most efficient method to learn vocabulary intentionally.

Note, however, that relying on these three didactic tools prioritized by Nation implies some major problems. First, neither of them takes account of the encyclopedic nature of lexical meaning. Dictionaries only provide information about highly entrenched meanings; yet, they do not guarantee exhaustivity, and they may also cover rather marginal meaning aspects that could be deliberately neglected. Second, they tell us nothing about either the metaphorical or metonymical potential of lexical items or about supposedly peripheral meaning aspects that may well be relevant in specific contexts in which a lexical item is encountered. Third, most crucially, none of these tools provide a strategy to infer relevant semantic aspects from context data or co-textual information.

Overall, the effective use of the tools suggested by Nagy (2005) and Nation (2001) presupposes that language learners are already, at least rudimentarily, familiar with the word meanings at issue. Looking up word meanings in dictionaries may even prevent language learners from integrating contextually relevant meaning aspects into the overall lexical meaning. To support the integration process, a didactic tool is needed that takes account of contextually triggered meaning variation. Before schemas are introduced, the next section argues that rich semantic instruction is essential for any language learner who starts learning a new language.

4. Arguments for deploying "rich instruction" in vocabulary teaching

The term "rich instruction" goes back to Beck, McKeown, and Omanson (1987) who use the notion to designate a teacher-centered method to support the process of vocabulary acquisition. In this sense, the term "rich instruction" is roughly synonymous with the notion "intensive instructions" introduced by Nagy (2005: 3). Rich instruction can be provided by teachers, but it may also be elaborated by language learners on their own, for example with the help of gap filling exercises or in group discussions. Beyond this, the term "rich instruction" also has a significant methodological impact. As a theoretical term it draws attention to the fact that a language learner must have already learned a considerable amount of words *before* he or she is capable of acquiring new words implicitly. That is, at the initial state of L2 acquisition, language learners are dependent on rich instruction provided by the teacher (or someone else).

Undoubtedly, it is a long-term goal to master a large number of words, but there are good reasons why teachers should spend most of the class time on providing detailed semantic information of selected words. Nation (1990: 14ff.) points out that the 2,000 most frequent words of a language cover approximately as much as 80% of any text. Although an average native speaker masters about 15,000–20,000 low frequency words, these words cover only 5% of a text. On this basis, Nation holds that intentional, explicit learning should comprise at least a quarter of the total teaching time. The statistical data reported by Nation indicate that it is inevitable to extensively deploy rich instructions in foreign language learning.

Meara (1995) argues even more radically when he says it is worth the effort to explicitly teach and learn all of the 2,000 most frequent words of a language. Due to the enormous pedagogical importance of these words, he suggests to teach all of them right at the beginning of a language course. In his view, knowing these words is a sound basis for L2 acquisition in general. Meara's suggestion is undeniably far-reaching, not only in terms of didactic implications but also with respect to motivational aspects of language learning. At first sight, it seems to be doubtful whether language learners can be motivated to solely concentrate on vocabulary learning for a long period of time. If, however, as Schmitt (2000: 143) hypothesizes, language learners acquired fifty words per week, 2000 words could be learned within forty weeks of school. Indeed, this seems to be an attainable goal, keeping in mind that the didactic tools mentioned by Nagy and Nation as well as the schema theory introduced here may help to teach word meaning in an interesting and varied fashion.

Another argument for deploying rich instruction in vocabulary teaching concerns the inseparability of vocabulary and grammar. Neither of them is distinct from the other because knowledge about word classes as well as valency information form integral parts of word knowledge. Vocabulary knowledge enhances grammar acquisition, in that knowing words in a text allows capturing the complex meaning of an entire text passage; as a result, grammatical patterns become more transparent, too. In short, knowing the most frequent words increases the comprehensibility of a language considerably, including its grammatical patterning (Ellis 1997).

In sum, the discussion of issues in vocabulary teaching has revealed three important points. First, given the evidence of cognitive-linguistic findings in the field of lexical semantics, it is important to acknowledge the inherent complexity of word meaning. The meaning of a word cannot be properly separated from the background knowledge and multifarious experience of language users. Second, it is essential for language beginners to learn the most frequent 2000 words of a foreign language right at the start of a language course. The more thoroughly and extensively word meanings are acquired, the more effectively the further language acquisition process will become. Third, there is neither a didactic tool that helps teachers to model word meaning in context nor a tool that allows language users to infer word meaning from context data. In the next sections, a schema-theoretical approach to word meaning is introduced that offers a new perspective on this issue.

5. Modeling word meanings with the help of schemas

During the last three decades, various cognitive schema theories have been developed by many scholars including Barsalou (1992), Fillmore (e.g. 1977, 1985), Gärdenfors (2000), Lakoff (1987), Langacker (1987), Rumelhart (1980), and Schank and Abelson (1977). Although they tackle similar questions, particularly issues concerning the nature of the human capacity to mentally store, retrieve, and represent knowledge structures of different kinds, and although they offer seemingly comparable answers, their approaches differ substantially in terms of terminology. Terms like "schema", "scene", "script", "scenario", "frame", "cognitive model", "domain", and "conceptual space", to mention but a few, are used likewise to describe (some aspects of the) human cognitive representation structures. Taylor (2003: 90f.) is perfectly right when he states:

> The terminology in this area is confusing, partly because different terms may be used by different authors to refer to what seems to be the same construct, or the same term may be used to refer to very different constructs. Furthermore, it is not at all clear that it is possible to make clean conceptual distinctions in this area.

In the following sections I prefer the theory-neutral term "schema". These sections are a crash-course in schema theory (for a detailed discussion see Ziem 2008: 283–366). Unlike the alternative expressions mentioned above, "schema" is a cover term in that the entities denoted by "script", "domain", "frame" etc. are schematic in nature. Linguistically speaking, the term "schema" is thus a hyperonym. The denoted concept is more abstract, but it covers all prototypical characteristics of the other concepts. For the sake of brevity, I first present a summary of the most important characteristics of cognitive schemas. By doing so, I concentrate on those characteristics that prove to be highly relevant for didactic applications.

5.1 Reference: Conceptualization as multifarious categorizations

People learn vocabulary for practical reasons. Vocabulary learning helps to improve text comprehension skills. The more words language learners know, the more accessible texts become. A large vocabulary also enhances both the users' oral and written expressive power. To be successful in vocabulary acquisition, language learners need to learn to which conceptual entity a word in use refers. Usually, this is a demanding cognitive task since it requires the activation of background knowledge. To refer to an entity can thus be equated with the activation of a complex cognitive schema. Fillmore (1982: 111) defines a semantic schema, what he calls a "frame", as a system of interrelated concepts:

> By the word 'frame' I have in mind any system of concepts related in such a way that to understand any of them you have to understand the whole structure in which it fits; when one of the things in such a structure is introduced into a text, or into a conversation, all of the others are automatically made available.

A word is used in order to establish a cognitive reference to a conceptual entity. To be successful, the entity must be qualified in numerous respects. The word *car*, for example, gives access to a concept that is qualified in terms of size, shape, color, functions, parts, and so on. Hence, linguistic reference always entails implicit predications. It was one of the key insights of Searle (1969) that both reference and predication are basic functions of a language. They allow language users to attain rich meanings through a poor stimulus, namely the mere word form. Following Croft (1991) reference leads "the hearer to identify as what is being talked about", while predications concern "what the speaker intends to *say about* what he is talking about (the referent)" (Croft 1991: 52, emphasis in original). Hence, vocabulary teaching is efficient if the schema evoked by a lexical item is maximally specified, that is, if many predicates are assigned to the conceptual entity being referred to.

In sum, the basic and (at least within the cognitive-linguistic paradigm) uncontroversial assumption is that language users have to cognitively master a reference and predication task in order to capture a contextually specified lexical meaning. In all schema-theoretical approaches the process of schema "activation" (Minsky 1975, Barsalou 1992) or, in Fillmore's terms, "evocation" (Fillmore 1985: 232), is equivalent to the establishment of cognitive reference. If, for example, the word form *car* evokes a car schema in the language user's mind, the linguistic form prompts him or her to conceptualize a particular entity. Hence, meaning is conceptualization. To linguistically refer to an entity means to activate, or evoke, an adequate schema.

5.2 Elements of semantic schemas: Slots, values, and default values

Conceptualizations rely on multifarious categorizations. Encountering the word *car*, for example, immediately prompts us to activate a schema that is specified in terms of color, size, functions, parts, owner, model etc. These rather abstract categories are specified by other, less abstract categories: The car may be a red BMW 316, having four doors, a spoiler, and much more extra equipment, and its owner may be Peter Peterson.

Henceforth, elements like "parts", "color", "size", "functions" etc. are called "slots". Slots are categories that indicate in which respects a linguistically evoked concept like "car" could be specified. This terminology goes back to Minsky (1975), but the same term is also used by some cognitive linguists (see, for example, Coulson 2001: 35). The cognitive psychologist Barsalou (1992) prefers the term "attribute", which is basically synonymous with "slot", notwithstanding the fact that Barsalou emphatically stresses the recursive character of schema-attribute structures. That is, on a higher level of abstraction, attributes are also schematic entities, just like the schema in which they are integrated. Following Barsalou, this holds recursively. The schema evoked by the term *color*, for example, provides slots like "green" and "red".

For didactic purposes it is worthwhile noting that Minsky (1975) as well as Fillmore (1977: 64) propose to equate slots with questions addressing the entity a word refers to: Which parts does the car consist of? What color is the car? What is its size, and what are its functions? It is principally impossible to limit the number of slots. It depends on the background knowledge of the language user which slots will be activated. Car mechanics could possibly pose an infinite number of questions, whereas laymen would probably concentrate on the most salient slots of the evoked car-schema. Vocabulary teaching should also concentrate on the salient slots by providing rich background knowledge that supports the learning process.

Once the slots are specified, a schema becomes more distinct. Thereby it is distinguished from similar concepts, here, for example, from concepts such as "tractor" or "bus". Hyperonymic expressions, like "vehicle" or "means of transport", evoke a schema comprising fewer specifications, whereas hyponomic expressions, like "van" or "Mercedes", evoke a schema comprising more and more specific specifications. It is well known from prototype theory, however, that basic level categories like "cars" are the most distinct categories in a language. Subordinated categories are richer, but their slots and values are less salient. Consequently, due to their high cognitive salience, basic level categories should, if possible, be taught and learned first.

Linguistically, any specification takes the form of a predication.

(1) The car…
 a. is green/red/blue etc. (slot: color)
 b. has four wheels. (slot: parts)
 c. is 4.40 meters in length. (slot: size)
 d. functions as a vehicle. (slot: function)

(2) a. The car which is Peter's. (slot: owner)
 b. The car is Peter's.

(3) a. The [red/green/old etc.] car… (slot: color)
 b. The car is red/green/old etc.

When we think of a car, we build up a cognitive model by inferring encyclopedic information that fits that model. We activate a schema elaborated by means of predicates such as (1a–d). Textually transparent predicates, like (1a–d), are called "values" or "fillers". They are expressed in the – oral or written – text, either explicitly, like in (1a–d), by means of predicates assigned to a reference object, or implicitly, like in (2a) and (3a) by means of constructions that could be transformed into predicates (cf. (2b) and (3b)).

When an entity is introduced into a text world, it is never fully specified. Rather, many slots are left unspecified, since a text usually does not offer many predicates assigned to the entity in question. Nonetheless, in (4) the activated car schema appears to be semantically "rich", although only the slot "owner" is specified. All other pieces of information are inferred.

(4) Peter's car is stolen.

For comprehension, the activated car-schema prompts the reader or hearer to assign so-called default values to those slots that are provided by the schema. Default values are inferred knowledge elements (predicates). They are automatically available as soon as a schema is activated. Whereas default values are inferred on the basis of entrenched conceptual knowledge, values are information chunks provided by the text. Usually, texts offer only a few values. Default values are thus essential for word comprehension.

Schemas (and their elements slots, values, default values) can be fully described in terms of propositions (Lönneker 2003: 63–66). As mentioned above, propositions include both reference and predication (Searle 1969: 39–46). A schema is the entity that a proposition refers to, whereas the predication specifies a slot of the schema. A predication is therefore an instantiation or categorization: A value or default value is instantiated in a slot provided by the evoked schema. Figure 1 interrelates Searle's terms with schema-theoretical terms (see Searle 1969: 39–46).

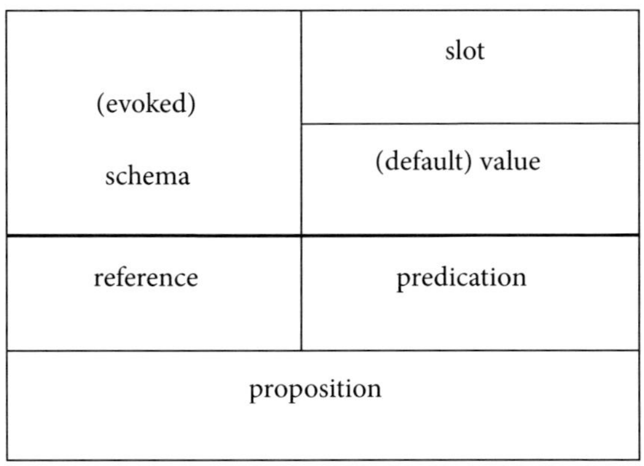

Figure 1. Elements of semantic schemas: comparing terminological differences

5.3 Type and token entrenchment

Schemas are "principle[s] of categorization" (Taylor 2002: 43). They either enforce or weaken categorization relations. Enforced value-slot-instantiations give rise to so-called "cognitive routines" (Langacker 1987: 163) or "cognitive entrenchment". Entrenchment is a function of frequency. That is, the more often an activated schema prompts us to specify a particular slot with a particular (type of) value, the more cognitively accessible and thus salient this slot and value (type) become.

In the literature two types of cognitive routines are distinguished: type entrenchment and token entrenchment.

> Each of these gives rise to the entrenchment of different kinds of linguistic units. While token frequency gives rise to the entrenchment of instances, type frequency gives rise to the entrenchment of more abstract schemas.
>
> (Evans and Green 2006: 118)

In the case of schemas, type entrenchment occurs when a particular schema gains strength due to a high frequency of values that elaborate on it. Hence, type entrenchment equates with schema entrenchment. Any time a value occurs, the whole schema is evoked in the language user's mind. Accordingly, a high number of values instantiated in the same slot gives rise to schema entrenchment.

Token entrenchment, on the other hand, appears to be equally important. Token entrenchment indicates which kinds of values are likely to become default values. A predicate (value) is entrenched if it is recurrently instantiated in the same slot. Depending on the context, the word *handle*, for example, may evoke a *door-*, *bucket-*, *knife-*, or *cup*-schema, since the slot "part of" and its prototypical values are most salient. This is due to a high token or type frequency. Figure 2 illustrates the case of the *handle*-schema linked to the cup-schema. Elements in rectangles are slots, circled elements are default values (that may, recursively, also evoke a schema). Bold lines indicate a high degree of entrenchment.

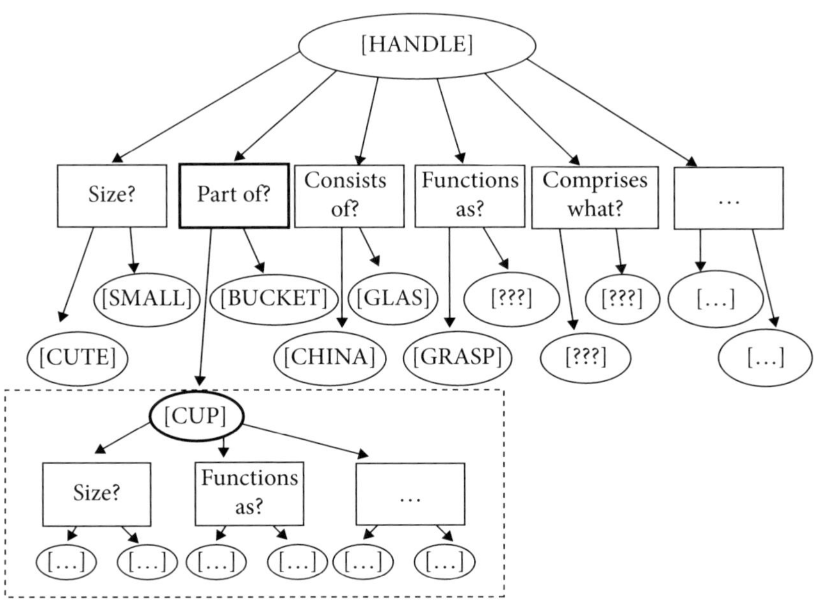

Figure 2. Entrenched elements of the *handle*-schema

Beyond token and schema entrenchment, I suggest another schema-specific type of entrenchment: slot entrenchment. Slot entrenchment concerns an intermediate level of abstraction. As we have seen, slots are categories subsuming a set of instances. Different values may thus specify the same slot. For illustration, consider once again (1a). It is arbitrary what color a car has; it is not arbitrary, however, that a car *is* colored. Many different predicates may elaborate this slot, and if this is the case, the slot "color" gets entrenched more deeply. Analogously, the category "handle" is not only a part of the *cup*-schema but also of many other schemas, including the *bucket*-, *jug*-, *door*-, *broom*-, and *mug*-schema. If the term *handle* is recurrently used to denote a part of either these entities, the slot "part" likewise becomes entrenched. Here, we can see that slot entrenchment coincides with schema entrenchment (but not with value entrenchment). Value entrenchment, on the other hand, coincides with both schema and slot entrenchment.

To teach vocabulary efficiently it is important to keep in mind that elements of lexical schemas differ with respect to their degree of entrenchment. When teaching vocabulary, it is reasonable to focus first on salient slots and values. Less entrenched slots and values, however, should not be disregarded entirely, since salience is also a function of context. Any slot and/or value may become semantically relevant, if contextual information triggers a particular slot and/or value.

5.4 Schema blending

To create new and more complex schemas, two (or more) schemas can be integrated into one another. A language provides many possibilities to merge schemas, including word formation, syntactic integration (e.g. by means of connectors such as *if, when, but, although*), metaphor, and metonymy. In cognitive linguistics, the issue of schema integration is closely related to the work of Gilles Fauconnier and Mark Turner (1998, 2002). Under the heading of "blending" or "conceptual integration", Fauconnier and Turner explore the cognitive principles underlying blending processes. One, or maybe *the* central hypothesis of their pioneering work is that schema blending cannot be adequately explained with the help of unidirectional mappings from elements of a source domain to a target domain. The new schema ("blend") rather contains elements, that is, slots and default values, provided by any of the input schemas. Additionally, Fauconnier and Turner stress that a blend also covers emergent information that is not inherited from any of the input schemas.

It is beyond the scope of the present article to provide detailed information about Turner's and Fauconnier's blending theory.[1] In the context of vocabulary

1. For this purpose see, for example, the special issue on conceptual blending in *Cognitive Linguistics* 11 (3/4), edited by Seana Coulson and Ted Oakley (2000).

teaching, however, it is worthwhile mentioning that schema blending could be neatly integrated into the schema-theoretical framework presented above (for details see Ziem 2008: 31–35, 421–438), since conceptual schematicity plays an important role in conceptual blending. Although Fauconnier and Turner prefer the term "(mental) space", they consider mental spaces to be schematic in nature (Fauconnier and Turner 1998, Turner 2008). Mental spaces share with schemas all the basic properties elucidated above. For illustration, confer the schema representation in Section 6.2. It elucidates how the emergence of complex word meanings can be illustrated for didactic purposes. This holds for both the metaphorical and/or metonymical potential of lexical items and the potential of lexical items to participate in word formation processes.

Hence, as I shall argue here, Fauconnier's and Turner's way of visualizing conceptual blending may be fruitfully deployed as a didactic method or tool to illustrate (the emergence of) complex word meanings in context. Such illustrations help to bridge the gap between theory-induced linguistic analyses and their didactic applicability. Note, however, that the relationship necessarily remains uni-directional: There is no substantial transfer from language pedagogy and didactics back to blending theory, since the latter is not designed to serve pedagogical purposes.

6. Teaching vocabulary by means of schemas: Words and words in context

In this section, I flesh out how semantic schemas can help to illustrate word meanings. As didactic tools schemas help language learners to remember and retrieve lexical meanings more effectively. From a didactic point of view, deploying semantic schemas offers various advantages. First, as a reaction to McCarthy's (1984) critique that context information is often marginalized though essential, semantic schemas allows teaching "rich" word meanings. If a text provides implicit or explicit predications specifying the entity to which the word in question refers, these predications function as values. As such they can be represented in schemas (see Figure 2). Values semantically specify the concept. Second, to infer word meaning, language learners may employ schemata on their own, once elements of schemas (slot, value, values, and default values) and the principle of entrenchment are introduced. Schemas thus support autodidactic vocabulary learning. Third, schemas acknowledge the fact that language learners bring in their own semantic inferences when inferring word meanings. Since the slot structure of a schema is not arbitrary, slots help language learners to "control" the (number of) potential inferences. The car-schema, for example, does not provide a slot for "duration"; only schemas derived from the event-schema offer such a slot. Finally, it is convenient to use schemas, since the schema activation relies on both declarative and procedural knowledge. Haastrup (1994: 131) criticizes that the impact of procedural knowledge on vocabulary learning is often underestimated or even denied. A schema-theoretical approach, however, is centered on the idea that slots (manifesting themselves as questions in schemas) are devices to infer lexically relevant semantic information on the basis of background knowledge. Hence, slots help language learners to solve the task of meaning determination.

In other words, it is argued that schemas could play an important role in teaching lexical meanings. In this way, schemas are heuristic didactic tools adapted from cognitive approaches to (lexical) meaning. Again, however, the implementation of schemas in the domain of vocabulary teaching and learning does not have a substantial impact on theories of cognitive schemas in a language user's mind; rather, this influence, too, remains uni-directional.

6.1 Lexical schemas and sense relations

For language learners it is important to acquire knowledge about sense relations between words. The more sense relations a language learner knows, the more distinct the involved concepts get in his or her mental lexicon. In practice, however, teachers often introduce new words by means of antonymic and/or synonymous expressions, insinuating that the terms are complete antonymies or complete synonymies respectively. But this is hardly – if ever – the case. Schmitt (2000: 29) remarks:

> Teachers commonly define new words by giving synonyms and antonyms, for example, *eavesdrop = listen* or *shallow = not deep*. However, they must be aware that very few words are completely synonymous or exact oppositions, and so such definitions will only be inexact representations of the word's true meaning.

In a schema-driven approach, it is possible to differentiate between subtle semantic differences. Although the schemas evoked by synonymous words provide the same slot structure, they often differ with respect to both default values and values instantiated in the slots. In the case of *shallow* and *not deep*, schemas facilitate visualizations of these differences. They elucidate that *shallow* and *not deep* are clear instances of partial synonymy. In Figure 3 the *shallow*-schema is compared with the *not-deep*-schema.

Note that the figure does not display two full-fledged schemas; it covers only those aspects that are essential to grasp important semantic similarities and differences. The bold lines indicate default values. While both the slot structure of the schemas and some default values are identical, the meanings of *shallow* and *not*

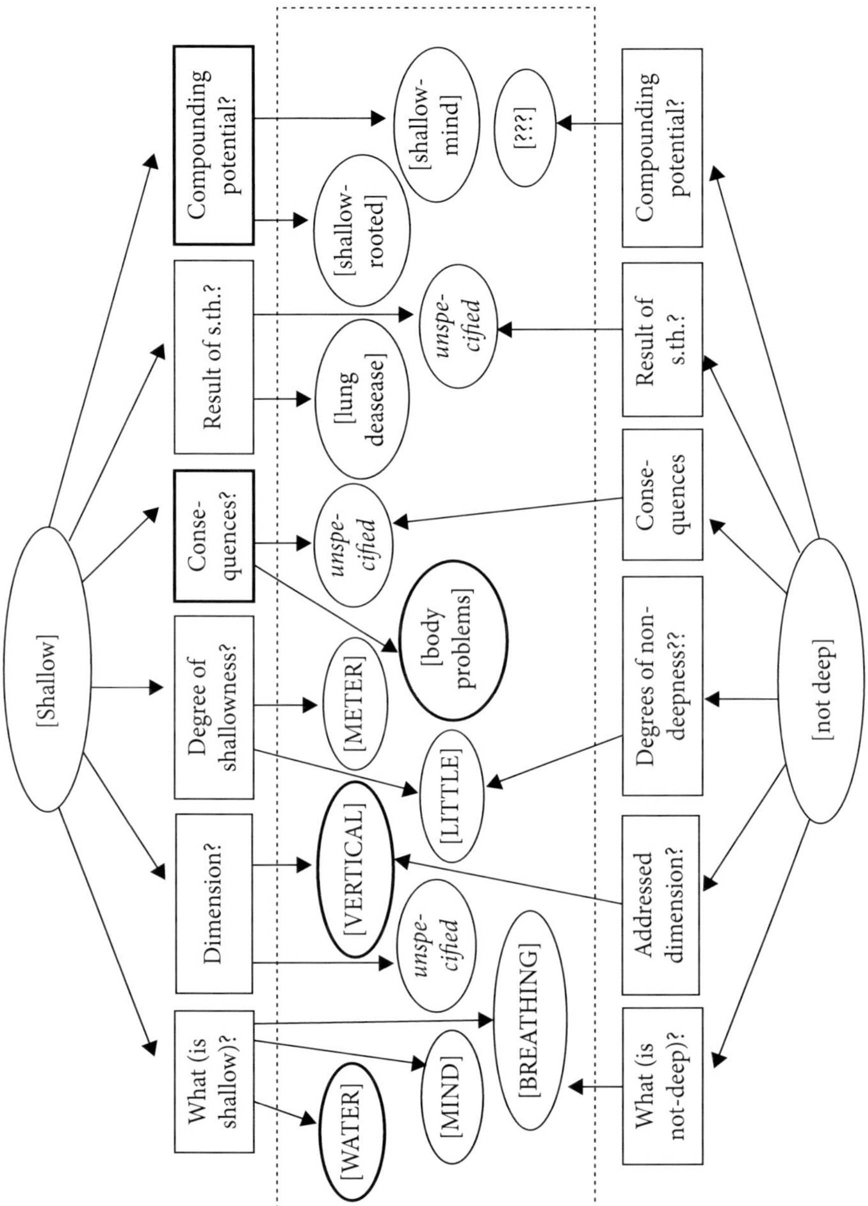

Figure 3. The schemas *shallow* and *not deep* in comparison

deep also differ in some respects. Most notably, at least from a didactic point of view, the words vary in their range of application. While *shallow* may be used to modify abstract concepts, such as "mind", *not deep* cannot be used in that way, and while, in turn, *not deep* may be used to modify nouns such as *cleansing*, *shallow* cannot, compare Examples (5) and (6).

(5) a. ?a (not) deep mind
b. a shallow mind

(6) a. ?shallow cleansing
b. (no) deep cleansing

(7) a. shallow-brained
b. ?not-deep-brained

In class, these simple examples may suffice; they demonstrate important semantic differences between *shallow* and *not deep*. On top of that, it is worthwhile noting that *not deep*, in contrast to *shallow*, cannot be used analogously to form compound-like constructions; for illustration, see Example (7).

By means of schemas, it is also possible to consider meaning aspects of *shallow* exhibited by given context data. In the following extract of an article from the *New York Times*, for example, two values additionally specify the *shallow*-schema. In Figure 3, the values are highlighted in grey.

(8) Rapid shallow breathing
A normal breathing rate for an adult at rest is 8 to 16 breaths per minute. For an infant, a normal rate is up to 44 breaths per minute. Tachypnea is the medical term that your doctor would use to describe your breathing if it is too fast, particularly if you have rapid, **shallow breathing due to an associated lung disease** or other medical cause. […] (*New York Times*, November 10, 2009; http://health.nytimes.com/health/guides/symptoms/rapid-shallow-breathing/overview.html)

In (8), deviating from prototypical usages, the entity to which the property "shallow" is assigned is not a concrete object but a kind of bodily action or process, namely breathing. Moreover, it is made explicit what gives rise to the "shallowness (of breathing)": it is the result of a lung disease.

Beyond (partial) synonymies, schemas could generally help to illustrate sense relations, including antonymies, meronymies, hyperonymies, and hyponymies. In the case of the antonomy *cold/warm*, for example, the evoked schemas provide not only the same slot structure but also the same set of default values – with one exception: In one slot (here: concerning temperature degree) different default values are instantiated, belonging to opposite sides of a scale.

In the case of meronymic relations, say, for example, between *hand* and *finger*, two schemas are correlated in such a way that one schema (here activated by *finger*) functions as a default value instantiated in a slot (here: "part of") that is provided by the other, subordinate schema (evoked by *hand*).

Consider, finally, the schemas evoked by hyperonymically correlated words. In these schemas, different default values are instantiated. Additionally, both schemas differ in the number of slots they provide. The *car*-schema, for instance, comprises more slots than the schema evoked by the hyperonym *vehicle*. More precisely, the *vehicle*-schema includes neither the slot "type of engine" nor "steering wheel", or "type of gasoline", to mention but a few. Moreover, those slots shared by both schemas are specified on different levels of abstraction. In the *car*-schema, for instance, the slots "conductor", "user", or "range of application" contain default values such as "driver", "people", and "on public streets". These elements are also potential values for the *vehicle*-schema; yet, they only remain potential, since buses, motorbikes, planes etc. are also vehicles, and the corresponding schemas vary substantially in terms of slot specifications.

6.2 Complex schemas: Metaphors, metonymies, compounds and anaphora

Although the last section has focused on sense relations between words, it also implicitly aimed at illustrating how word meaning could be represented by means of schemas. Basically, schemas allow modeling meaning on all levels of abstraction, including meanings of complex idiomatic units (Ziem and Staffeldt in press).

But that is only half of the story. Lexical items usually do not come alone; they are rather integrated in a text. That way, they establish relations to other words and, as a result, lexical meanings are modified. This holds for metaphorically and metonymically used words but also for compound constructions and meaning modifications due to (associative-)anaphoric relations (see Ziem 2009). In what follows, I present a schema-theoretical illustration of the metaphorical use of *shallow* in the compound-like construction *shallow-witted*. The figurative usage of *shallow* exhibited here exemplifies the metaphoric potential of the expression. If language learners become acquainted with the construction mechanisms involved in the construction *shallow-witted*, they can infer meanings of many other constructions, including *shallow knowledge*, *shallow learning*, *shallow mind*.

Figure 4 is a prototypical illustration of a complex meaning construction. The complex meaning of *shallow-witted* emerges from the conceptual integration of the *shallow*-schema into the *witted*-schema. Although *witted* is not a free morpheme, it serves as the head in that it evokes the *wit*-schema, which is specified subsequently. For didactic purposes, it is indispensable to describe the emerging complex

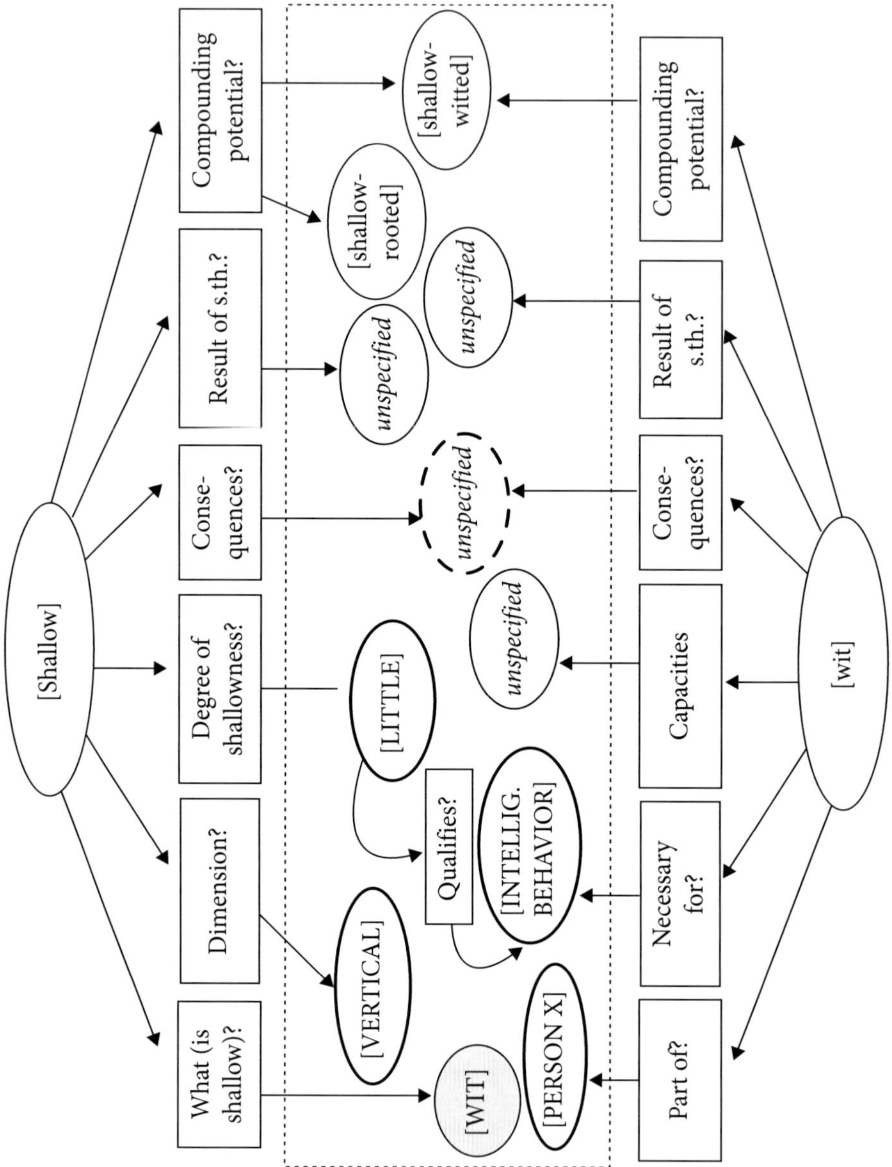

Figure 4. Building a complex schema: the case of *shallow-witted*

meaning simply but adequately. One has to keep track of two points. The first point concerns the integration process. It has to be fleshed out that "wit" is conceptualized as a quantifiable object having a vertical dimension. In the complex schema the height of this "object" provides information about the capacity of intelligent behavior. "Shallow" specifies this vertical dimension. As an instance of the conceptual metaphor MORE IS UP respectively LESS IS DOWN it qualifies intelligent human behavior as "little". The most important metaphorical mapping thus concerns the qualification of intelligent behavior by means of the quantification "little".

The second point is equally important. With blending the *shallow-* and *wit-*schema, knowledge emerges that neither of the schemas provides: The meaning of *shallow-witted* comprises an evaluative component. Calling someone "shallow-witted" implies that the person addressed lacks certain intellectual capacities. Usually, this evaluation has certain consequences, for example in such a way that one should not trust or believe somebody because he or she is shallow-witted. The broken line in Figure 4 indicates the emergent nature of this default value. Without contextual information, however, the emergent default value remains unspecified.

For the purpose of vocabulary teaching, it is not necessary to introduce full-fledged semantic schemas. In the case of complex schemas, it is most important to focus on the process of conceptual integration and on emergent knowledge aspects. Moreover, it might be helpful to single out the relevance of the most important default values, such as "vertical" in the shallow-schema and "person" in the wit-schema.

7. Conclusions

In this article, I have introduced semantic schemas as a didactic tool for teaching vocabulary. Addressing the issue from a cognitive perspective, I argued that schemas could help language learners to acquire semantically rich word meanings. Once language learners are acquainted with the properties of semantic schemas (slots, values, default values), schemas might also help to better infer word meanings in context. Beyond the scope of lexical meaning, a schema-theoretical approach might also illustrate the emergence of sentence and even text meaning (Löbner 2009, Fillmore 1984, Holly 2001). Using cognitive schemas as heuristic devices for teaching and learning vocabulary may thus be considered as a prime example for the uni-directional applicability of cognitive-linguistic concepts in the domain of language pedagogy.

Although schemas may well be integrated into communicative and other teaching activities (Nation 2001: 114–142, Nation and Newton 1997: 241–248), such as listening and reading tasks, its scope of application is not unlimited. Using schemas is more time-consuming than using other techniques. Hence, for low-frequency words other didactic tools do a better job. Since schema representations encourage deep semantic processing and thus help language learners to establish rich word meanings, it is reasonable to deploy schemas for selected high-frequency words. Once language learners have acquired a set of, say, approximately 1,000 high-frequency words, other words can be learned efficiently with the help of context clues.

References

Aitchison, Jean. 2003. *Words in the Mind: An Introduction to the Mental Lexicon* (3rd ed.). Oxford: Blackwell.

Arnaud, Pierre and Henri Béjoint. 1992. *Vocabulary and Applied Linguistics*. London: Longman.

Barsalou, Lawrence W. 1992. Frames, concepts, and conceptual fields. In *Frames, Fields and Contrasts. New Essays in Semantics and Lexical Organisation*, Adrienne Lehrer and Eva Feder Kittay (eds.), 21–74. Hillsdale, NJ: Erlbaum.

Barsalou, Lawrence W. 1999. Perceptual symbol systems. *Behavioral and Brain Sciences* 22: 577–609.

Barsalou, Lawrence W. 2003. Situated simulation in the human conceptual system. *Language and Cognitive Processes* 5 (6): 513–562.

Bartlett, Frederic. 1932. *Remembering. A Study in Experimental and Social Psychology*. Cambridge: Cambridge University Press.

Beck, Isabel L., Margaret G. McKeown and Richard C. Omanson. 1987. The effects and uses of diverse vocabulary instructional techniques. In *The Nature of Vocabulary Acquisition*, Margaret G. McKeown and Mary E. Curtis (eds.), 147–163. Hillsdale, NJ: Erlbaum.

Boers, Frank and Seth Lindstromberg. 2006. Cognitive linguistic applications in second or foreign language instruction: rationale, proposals and evaluation. In *Cognitive Linguistics: Current Applications and Future Perspectives*, Gitte Kristiansen, Michel Archard, René Dirven and Franciso J. Ruiz de Mendoza Ibáñez (eds.), 305–355, Berlin: Mouton de Gruyter.

Boers, Frank and Seth Lindstromberg (eds.). 2008. *Cognitive Linguistics Approaches to Teaching Vocabulary and Phraseology*. Berlin: Mouton de Gruyter.

Coulson, Seana. 2001. *Semantic Leaps: Frame-Shifting and Conceptual Blending in Meaning Construction*. Cambridge: Cambridge University Press.

Coulson, Seana and Ted Oakley (eds.). 2000. Special Issue: Conceptual Blending. *Cognitive Linguistics* 11 (3–4).

Croft, William. 1991. *Syntactic Categories and Grammatical Relations*. Chicago: The University of Chicago Press.

Ellis, Nick C. 1997. Vocabulary acquisition: word structure, collocation, grammar, and meaning. In *Vocabulary: Description, Acquisition and Pedagogy*, Michael McCarthy and Norbert Schmidt (eds.), 122–139. Cambridge: Cambridge University Press.

Ellis, Nick C. 2003. Constructions, chunking, and connectionism: The emergence of second language structure. In *The Handbook of Second Language Acquistion*, Catherine J. Doughty and Michael H. Long (eds.), 63–103. Malden, MA: Blackwell.

Ellis, Nick C. and Fernando Ferreira-Junior. 2009. Construction learning as a function of frequency, frequency distribution, and function. *Modern Language Journal* (93): 370–385.

Ellis, Nick C. and Nadine Laporte. 1997. Contexts of acquisition: effects of formal instruction and naturalistic exposure on second language acquisition. In *Tutorials in Bilingualism: Psycholinguistic Perspectives*, Anette M. B. de Groot and Judith F. Kroll (eds.), 53–83. Hillsdale, NJ: Erlbaum.

Evans, Vyvan. 2009. *How Words Mean: Lexical Concepts, Cognitive Models and Meaning Construction*. Oxford: Oxford University Press.

Evans, Vyvan and Melanie Green. 2006. *Cognitive Linguistics. An Introduction*. Edinburgh: Edinburgh University Press.

Fauconnier, Gilles. 1985. *Mental Spaces. Aspects of Meaning Construction in Natural Languages*. Cambridge, MA: MIT Press.

Fauconnier, Gilles. 1997. *Mappings in Thought and Language*. Cambridge: Cambridge University Press.

Fauconnier, Gilles and Mark Turner. 1998. Conceptual integration networks. In *Cognitive Science* 22 (2): 133–187.

Fauconnier, Gilles and Mark Turner. 2002. *The Way We Think. Conceptual Blending and the Mind's Hidden Complexities*. New York: Basic Books.

Fillmore, Charles J. 1977. Scenes-and-frames semantics. In *Linguistic Structures Processing Vol. 5*, Antonio Zampolli (ed.), 55–81. Amsterdam: North Holland.

Fillmore, Charles J. 1982. Frame semantics. In *Linguistics in the Morning Calm*, The Linguistic Society of Korea (ed.), 111–137. Seoul: Hanshin Publishing Company.

Fillmore, Charles J. 1984. Lexical semantics and text semantics. In *New Directions in Linguistics and Semiotics*, James E. Copeland (ed.), 123–147. Houston: Rice University Studies.

Fillmore, Charles J. 1985. Frames and the semantics of understanding. In *Quaderni di Semantica* 6 (2): 222–254.

Gärdenfors, Peter. 2000. *Conceptual Spaces. The Geometry of Thought*. Cambridge, MA: MIT Press.

Geeraerts, Dirk. 2006. *Words and Other Wonders. Papers on Lexical and Semantic Topics*. Berlin: Mouton de Gruyter.

Haastrup, Kirsten. 1994. On word processing and vocabulary learning. In *Kognitive Linguistik und Fremdsprachenerwerb*, Wolfgang Börner and Klaus Vogel (eds.), 129–147. Tübingen: Narr.

Holly, Werner. 2001. Frame als Werkzeug historisch-semantischer Textanalyse. Eine Debattenrede des Chemnitzer Paulskirchen-Abgeordneten Eisenstuck. In *Politische Kommunikation im historischen Wandel*, Hajo Dieckmannshenke and Iris Meißner (eds.), 125–146. Tübingen: Stauffenberg.

Johnson, Mark. 2007. *The Meaning of the Body. Aesthetics of Human Understanding*. Chicago: The University of Chicago Press.

Krantz, Gosta. 1991. *Learning Vocabulary in a Foreign Language: A Study of Reading Strategies*. Göteborg: Acta Universitatis Gothoburgensis.

Lakoff, George. 1987. *Woman, Fire and Dangerous Things. What Categories Reveal about the Mind*. Chicago: The University of Chicago Press.

Lakoff, George. 1988. Cognitive semantics. In *Meaning and Mental Representations,* Umberto Eco, Marco Santambrogio and Patrizia Violi (eds.), 119–154. Indiana, IN: Indiana University Press.

Lakoff, George and Mark Johnson. 1980. *Metaphors We Live By*. Chicago: The University of Chicago Press.

Langacker, Ronald W. 1987: *Foundations of Cognitive Grammar.* Vol. 1: *Theoretical Prerequisites.* Stanford, CA: Stanford University Press.

Langacker, Ronald W. 1988. A view of linguistic semantics. In *Topics in Cognitive Linguistics*, Brygida Rudzka-Ostyn (ed.), 49–90. Amsterdam: Benjamins.

Löbner, Sebastian. 2009. Evidence for frames from human language. Talk presented at the *Second Conference on Concept Types and Frames in Language, Cognition, and Science*, Düsseldorf/Germany, August 24–26, 2009.

Lönneker, Birte. 2003. *Konzeptframes und Relationen. Extraktion, Annotation und Analyse französischer Corpora aus dem World Wide Web*. Berlin: Aka.

McCarthy, Michael. 1984. A new look at vcabulary in EFL. *Applied Linguistics* 5 (1): 12–22.

Meara, Paul. 1995. The importance of an early emphasis on L2 vocabulary. *The Language Teacher* 19 (2): 8–10.

Minsky, Marvin. 1975. A framework for representing knowledge. In *The Psychology of Computer Vision*, Patrick H. Winston (ed.), 1–25. New York: McGraw-Hill.

Nagy, William. 1997. On the role of context in first- and second language vocabulary learning. In *Vocabulary: Description, Acquistion and Pedagogy*, Norbert Schmitt and Michael McCarthy (eds.), 64–83. Cambridge: Cambridge University Press.

Nagy, William. 2005. Promoting students' vocabulary development: An overview. Unpublished manuscript, Seattle Pacific University.

Nation, Paul. 1990. *Teaching and Learning Vocabulary*. New York: Newbury House.

Nation, Paul. 2001. *Learning Vocabulary in Another Language*. Cambridge: Cambridge University Press.

Nation, Paul and Jonathan Newton. 1997. Teaching vocabulary. In *Second Language Vocabulary Acquisition: A Rationale for Pedagogy*, James Coady and Thomas Huckin (eds.), 238–254. Cambridge: Cambridge University Press.

O'Grady, William. 2003. The radical middle: nativism without Universal Grammar. In *The Handbook of Second Language Acquistion*. Catherine J. Doughty and Michael H. Long (eds.), 43–62. Malden, MA: Blackwell.

Radden, Günter and Klaus-Uwe Panther. 2004. Introduction: reflections on motivation. In *Studies in Linguistic Motivation*, Günter Radden and Klaus-Uwe Panther (eds.), 1–46. Berlin: Mouton de Gruyter.

Rumelhart, David E. 1980. Schemata: The building blocks of cognition. In *Theoretical Issues in Reading Comprehension. Perspectives from Cognitive Psychology, Linguistics, Artificial Intelligence, and Education*, Rand Spiro et al. (eds.), 33–58. Hillsdale, NJ: Erlbaum.

Schank, Roger C. and Robert P. Abelson. 1977. *Scripts, Plans, Goals and Understanding. An Inquiry into Human Knowledge Systems*. Hillsdale, NJ: Erlbaum.

Schmitt, Norbert. 2000. *Vocabulary in Language Teaching*. Cambridge: Cambridge University Press.

Schmitt, Norbert and Michael McCarthy. 1997. Introduction. In *Vocabulary. Description, Acquisition and Pedagogy*, Norbert Schmitt and Michael McCarthy (eds.), 1–5. Cambridge: Cambridge University Press.

Searle, John. 1969. *Speech Acts. An Essay in the Philosophy of Language*. Cambridge: Cambridge University Press.

Taylor, John. 1993. Some pedagogical implications of cognitive linguistics. In *Topics in Cognitive Linguistics*, Brygida Rudzka-Ostyn (ed.), 201–223. Amsterdam: Benjamins.

Taylor, John. 2002. *Cognitive Grammar.* Oxford: Oxford University Press.

Taylor, John. 2003. *Linguistic Categorization* (3rd ed.). Oxford: Oxford University Press.

Tomasello, Michael. 2003. *Constructing a Language. A Usage-Based Theory of Language Acquisition.* Cambridge, MA: Harvard University Press.

Turner, Mark. 2008. Frame blending. In *Frames, Corpora, and Knowledge Representation*, Rema Rossini Favretti (ed.), 13–32. Bologna: Bononia University Press.

Verspoor, Marjolijn H. 2008. What bilingual word associations can tell us. In *Cognitive Linguistics Approaches to Teaching Vocabulary and Phraseology.* Frank Boers and Seth Lindstromberg (eds.), 261–289. Berlin: Mouton de Gruyter.

Vigliocco, Gabriella, Lotte Meteyard, Mark Andrews and Kousta Stavroula. 2009. Toward a theory of semantic representation. *Language and Cognition* 1–2: 219–247.

Ziem, Alexander. 2008. *Frames und sprachliches Wissen. Kognitive Aspekte der semantischen Kompetenz.* Berlin: Mouton de Gruyter.

Ziem, Alexander. 2009. Frames im Einsatz: Aspekte anaphorischer, tropischer und multimodaler Bedeutungskonstruktion im politischen Kontext. In *Wissen durch Sprache: Theorie, Praxis und Erkenntnisinteresse des Forschungsnetzwerkes „Sprache und Wissen"*, Ekkehard Felder and Marcus Müller (eds.), 209–246. Berlin: Mouton de Gruyter.

Ziem, Alexander. 2010. Welche Rolle spielt der Kontext beim Sprachverstehen? Zum Stand der psycholinguistischen und kognitionswissenschaftlichen Forschung. In *Kontexte und Texte: Soziokulturelle Konstellationen literalen Handelns*, Peter Klotz, Paul R. Portmann and Georg Weidacher (eds.), 59–84. Tübingen: Niemeyer.

Ziem, Alexander and Sven Staffeldt. In press. Compositionality and embodied meanings of phraseologisms: the case of somatisms. In *Converging Evidence – Discussing and Extending the Methodological Tool-Kit of the Linguist* [working title], Doris Schönefeld (ed.). London: Equinox.

Zwaan, Rolf. 2004. The immersed experiencer: Toward an embodied theory of language comprehension. In *The Psychology of Learning and Motivation Vol. 44*, Brian H. Ross (ed.), 35–62. New York: Academic Press.

The conceptualization of personality
Converging and diverging evidence

Juliana Goschler
Universität Hamburg

In this chapter, I present observations of recurrent motifs and metaphors in literature that point to a culturally conventionalized idea of the heart as the seat of emotions and personality. A corpus study shows that this is not only a phenomenon of literary language, but also a systematically and frequently used metaphor in everyday language. However, the results of a questionnaire study suggest that this linguistic mapping is not a reflection of a "folk model" of human personality. I conclude that the relation between metaphors and folk models is more complex than sometimes assumed, and that approaches and methods from different disciplines have to be combined in order to understand connections between art, language, culture, and folk beliefs.

Keywords: corpus study, culture-specific concepts of the heart, everyday language, folk models, questionnaire study

1. Introduction

In the 1998 Hollywood movie *Meet Joe Black*, a young woman named Susan meets a man with whom she falls in love at first sight. Shortly after their first encounter, the man is killed in a car accident without her knowing. Later, Susan meets Death, impersonated in the body of that very young man, at her father's house. Her father has made a deal with Death: he will not die at this time, and in return he will give Death "a tour through life". Thus, Death in the body of the young man is introduced to all his friends and relatives as "Joe Black". During his time on earth, Death falls in love with Susan, and through the course of the film, they get closer and closer and end up making love. However, Death realizes that Susan is not in love with him, but with the dead man whose body he uses. In the end, he sends the young man back to the living, so that he and Susan can be together. The film ends with the happy couple watching fireworks, and Susan indicating that she knows that the young man and Joe Black are two different men.

This plot raises questions about life and death and answers them in an artistic context, but many puzzling implications of the narrative remain unaddressed: What exactly is the nature of a love like Susan's, that does not focus on the personality and the immaterial features of a man, but on his body? Since she is not presented as a superficial person, is the audience assumed to think that the core of a person is localized in the body? If so, where exactly? And if not, is it localized somewhere else?

In German, at least linguistically, the personality of a person is often metaphorically described as being localized in the heart: this is reflected in recurrent heart motifs and metaphors in literature and everyday language. But where do people really think the core of a person is located? Are the heart metaphors and motifs reflections of a deeper conceptual mapping of personality traits onto the human body or parts of it? And how can we determine the exact nature of the connection between linguistic and literary expressions and folk models and beliefs?

I argue that it is only a combination of approaches from the humanities (such as studies of motifs in literature and other genres) and empirical methods based in the social sciences that can uncover the exact relations and bi-directional influences between art and literature, language, and people's beliefs.

I begin by presenting evidence that there is, in fact, a strong tradition of heart motifs and metaphors in German culture. Most of these motifs suggest that the idea of the core personality traits of a person as being located in the heart is deeply embedded in German literary and artistic traditions. I then show, on the basis of a quantitative corpus study, that the cultural influence of these metaphors extends to everyday language. The final part of my investigation consists of a questionnaire study that directly focuses on people's beliefs about personality traits and their location in the human body. Taken together, the results from these three parts of my chapter yield a mixed picture: while there is a connection between literature and art on the one hand and everyday language on the other, neither domain is directly related to conceptualization. Therefore, only a combination of different empirical methods enables us to make progress in uncovering the relation between language and thought.

2. Heart motifs and heart metaphors in German music and literature: General observations

The heart motif in German has a long tradition and can be found in the oldest preserved works of literature. In several epics, the heart of a protagonist plays a central role. Sometimes the heart is literally taken from someone, e.g. as proof of death, sometimes as the most precious part, which is given to a lover. For example, in Konrad von Würzburg's *Das Herzmaere* (thirteenth century), a dying knight orders his heart to be sent to his mistress. In Ulrich von Etzenbach's *Alexander* (1278/83), the heart of one of the protagonists literally breaks from sorrow. Sometimes, the heart is described as an object with magical power – it can drive away pain and heal deadly diseases, as in Heinrich von Neustadt's *Apollonius von Tyriant* (1300) or E.T.A. Hoffmann's *Ignaz Denner* (1816/17), or it can cause people to fall in love, as in Andreas Gryphius's *Cardenio und Celinde* from 1657. The motif of the "heart of stone" is the central topic of the fairytale *Das kalte Herz* (1827) by Wilhelm Hauff, where the main protagonist exchanges his heart for a stone and in consequence becomes successful, but loses the ability to be happy. These examples, as different as they may be, nevertheless show very clearly that the idea of the heart as the seat of emotions – especially love, empathy, and courage –, as the central body organ, which somehow represents the person itself and therefore can have magical power, goes back at least to the Middle Ages. The heart motif is not restricted to the epic genre. In songs that have been preserved from the Middle Ages, the heart motif is already used to describe feelings of love, as in Albrecht von Johansdorf's *Kreuzlieder* (*Min herze und min lip diu wellent scheiden*), and it is still very much alive in current songs through all styles of German music. Thus, it is safe to assume that the heart motif is deeply embedded in German culture and has been present in German literature and music for at least 750 years.

3. General language patterns: Heart metaphors in German

The general idea that the heart is the seat of emotions and personality is also reflected in the use of metaphors in everyday language. In fact, the literary motif may well be based on poetic extensions of frequent conventionalized patterns in everyday language. This connection between literary motifs and everyday metaphors is described as a reflection of a very basic feature of our conceptual system in Conceptual Metaphor Theory (Lakoff and Johnson 1980, Lakoff and Turner 1989, Lakoff 1987, Lakoff 1993). This theory is based on the idea that metaphors in language are just one of several manifestations of general mappings of one conceptual domain (the source domain) onto another (the target domain). These mappings, according to the theory, enable us to deal with concepts we cannot experience directly. In assuming a conceptual basis for metaphors, Conceptual Metaphor Theory is able to explain why metaphors are pervasive in everyday language as well as in expert discourse, art, poetry, and literature. This theoretical approach assumes that metaphors occurring in these different contexts are all essentially identical in nature, that "[t]o understand poetic metaphor, one must understand conventional metaphor [...] [P]oetic metaphor, far from being ornamentation, deals with central aspects of our conceptual systems". (Lakoff and Turner 1989: 214f.)

3.1 Corpus study: Heart metaphors in everyday language

In order to systematically investigate the role of heart metaphors in everyday German, I conducted a corpus study. The need to employ corpus linguistic methods for an empirically properly grounded metaphor analysis has been forcefully argued by Deignan (1999, 2005) and Stefanowitsch (2004, 2005, 2006), who show that these methods yield more valid and reliable results than introspection or an unsystematic collection of examples. On the basis of the PUBLIC corpus (available from the *Institut für Deutsche Sprache* (IDS) in Mannheim, Germany), a continually expanding monitor corpus of texts from German, Austrian, and Swiss newspapers, a concordance for the word *Herz* ('heart') was created. The word occurred 55,170 times in the corpus. A sample of 500 randomly selected occurrences was manually coded for metaphorical versus literal uses.

Table 1. Literal and metaphorical use of *Herz* in the corpus sample

	absolute frequency in the corpus sample	relative frequency in the corpus sample
literal use (meaning the body organ)	106	21.2%
literal use (icon/deck of cards)	19	3.8%
proper names	36	7.2%
metaphorical uses	339	67.8%

In the sample, *Herz* is used in a strictly literal way – referring to the body organ – in 21.2% of the cases (see Table 1). In 3.8% of the cases it is used to refer to the iconic shape or the hearts suit in a deck of playing cards. In another 7.2% *Herz* is a proper name or part of a name. The rest – 67.8% – are metaphors or other idiomatic uses of the word *Herz*.

In 8.8% of all uses of *Herz*, the word is part of the idiomatic construction *the heart of* NP, as in *das Herz der Stadt* ('the heart of the city'), *das Herz der Region* ('the heart of the region'), and other similar instances of the construction. In this metaphorical use of *Herz*, the heart stands for the center, the core, or the vital part of something. These expressions use the source domain HEART to describe abstract units that emerge when people live and work together in social or geographical networks (like institutions, industries, cities, or regions) in terms of the human body. Since this metaphorical mapping is frequent in everyday language, according to the theoretical assumptions of Conceptual Metaphor Theory, a systematic conceptual mapping can be assumed as the basis of these expressions.

The other metaphorical uses of *heart* (59% of all uses) were expressions denoting things related to feelings or characteristics. These metaphors have different source domains like (VALUABLE) OBJECT, PERSON, LOCATION, CONTAINER, or SUBSTANCE. Most of these expressions are highly conventionalized and some of them are more or less opaque, so that it is difficult to name a source domain at all, for example in the expression *Hand aufs Herz!* ('cross my heart; be honest'), a request for absolute honesty (likely derived from the ritual of putting the hand to the heart area of the chest while swearing an oath).

Table 2. Metaphors of different source domains in the corpus sample

source domain	**absolute frequency of occurrence in the sample**	**relative frequency of occurrences in the sample**
object	37	7.4%
person	46	9.2%
container	36	7.2%
location	53	10.6%
substance	85	17%
unclear/opaque	38	7.6%

As becomes clear from Table 2, the heart is often described as an OBJECT (see also Niemeier 1997, 2000) that can belong to someone, given away, be lost, captured, or broken:

(1) a. unser Freund verdient ein weibliches **Herz**, das ihm ganz angehöre (GOE/AGM.00000)
'our friend deserves a female heart which fully belongs to him'

b. ich schenk mein **Herz** (M89/905.16897)
'I give my **heart**'

c. verliert sein **Herz** an die Details (R97/SEP.71010)
'loses his **heart** to details'

d. erobert sich damit nebenbei das **Herz** einer Frau (E97/JUL.16519)
'he captures the **heart** of a woman'

e. Du brichst mein **Herz** entzwei (O99/APR.45190)
'You break my **heart** in two'

If the heart is described as an object, "hardness" of the heart is something bad, maybe because nothing has an effect on something hard, therefore, a *heart of stone* is undesirable:

(2) a. Wer will denn schon eine Regierung mit einem **Herz** aus Stein? (E98/MAI.11990)
'Who would want a government with a **heart** of stone?'

b. daß die Leute ein **Herz** haben wie Kieselstein (P98/APR.18122)
'that people have a **heart** like (a) pebble'

In personifications (THE HEART IS A PERSON), hearts are metaphorically described as having feelings or being able to think or act like people:

(3) a. das mitfühlende **Herz** einer Frau (R99/DEZ.104346)
'the empathic **heart** of a woman'

b. das **Herz** hat mir gesagt, ich soll fahren (V99/JAN.00952)
'the **heart** told me to go'

Another frequently used source domain is CONTAINER, which feelings can be metaphorically put into or let out of:

(4) a. Sanftmut ins **Herz** geträufelt (M95/506.03478)
'Meekness trickled into the **heart**'

b. Vertraut ihm jederzeit, und schüttet euer **Herz** bei ihm aus! (A99/JUL.508686)
'Trust him always and pour out your **heart**!'

Within this CONTAINER metaphor, a heart can be "open", meaning that the person is able to experience positive feelings towards others or because of pleasant events:

(5) Bruno Dinner (...) ging dabei so recht das **Herz** auf (A98/JUL.529897)
'Bruno Dinner's **heart** opened'

Accordingly, a "big heart" is good, since there is space for many positive feelings:

(6) Monika Webers **Herz** ist gross, viele haben darin Platz (E98/FEB.04025)
'Monika Weber's **heart** is big, many people fit in'

Metaphorically, feelings can also be experienced "around" the heart, thus implying that the heart is a region of the human body where emotions are perceived, or that it is the LOCATION of feelings itself:

(7) nicht gerade fröhlich ums **Herz** (A00/MÄR.22783)
'not happy around the **heart**'

In some cases, OBJECT or CONTAINER metaphors are mixed with a TEMPERATURE metaphor, where WARMTH stands for positive and empathic feelings, and COLDNESS for the lack of them:

(8) a. Das **Herz** kurz erwärmt an einem alten Hadern voll Sehnsucht (O00/JAN.08861)
'The **heart** quickly warmed by an old story full of longing'

b. ihr **Herz** erfüllte sich mit jener Freude und Wärme (I99/DEZ.59077)
'her **heart** filled with joy and warmth'

The most frequent source domain is SUBSTANCE, where the heart is something one can have more or less of. In most of these metaphors, which are almost exclusively used in the construction [*mit/ohne* (adv) *Herz*], *Herz* means something like feeling, enthusiasm, or courage:

(9) a. Die junge Mannschaft spielt mit viel **Herz** (A97/NOV.40551)
'The young team plays with a lot of **heart**'

b. Wir haben ohne **Herz** gespielt (R99/APR.28829)
'We played without **heart**'

c. Mit **Herz**, ohne Angst (O99/JUN.91899)
'**with** heart, without fear'

d. Ein Gorilla mit **Herz** und Humor (I99/JUL.28419)
'a gorilla with **heart** and humour'

Most of these metaphors are highly conventionalized. Some of them have been in use for several centuries and have been listed in dictionaries since the eighteenth century (Stockinger 1996), and some of them obviously correspond to the literary motifs I briefly described above.

Thus, the great majority of all occurrences of the word *Herz* is used as a part of an idiomatic or metaphorical expression describing feelings, personalities, or personal characteristics of people, whose core characteristics are thus localized in their bodies or, more precisely, in their hearts. This quantitative analysis confirms the impression that the heart as the seat of emotions and personality is not just a literary motif, but also a highly frequent and conventionalized, yet still productive pattern in everyday language. The other pattern, where things are conceptualized as living bodies with the heart as the center or most important part, can be interpreted similarly as instances of a conceptual mapping of the domain BODY on more abstract things. Here, too, a kind of "core feature" is associated with the heart.

3.2 Corpus study: Brain metaphors in everyday language

Yet all this might be just one of several ways of speaking about people, their feelings and their personalities. Since the majority of members of the German speech community is aware of the fact that the brain is the organ where all cognitive and emotive processes take place, one might assume that the word *brain* would be equally employed in talking about people's characters and personalities. To test this, I created a concordance for the word *Gehirn* 'brain' and again coded a sample of 500 occurrences (out of a total of only 8,901 occurrences in the corpus) as metaphorical or literal, with a subsequent coding for source and target domains for all metaphorical uses. The first striking difference to the use of *Herz* is that the cases where the word *Gehirn* does not refer to the actual organ were very rare, accounting for only 7.4% of the sample. Most of these cases were instances of the pattern [*das Gehirn* $NP_{Genitive}$] ('the brain of NP'), where *brain* stands metaphorically for some central instance of an organization, a machine, or a factory, as in *das Gehirn der Fußballmanschaft* ('the brain of the football team') or *das Gehirn der Armee* ('the brain of the army'). The brain is used here to describe the thinking, planning, or intelligent part of something – this metaphorical pattern is parallel to those metaphorical expressions containing *Herz* that I described above. However, the other pattern, where feelings and personality of people are described as being localized in their bodies is not present in expressions containing *Gehirn* (for a more thorough analysis of metaphors with brain as the target domain see Goschler 2005, 2008) – thus, if feelings and personality traits are described metaphorically, it is almost exclusively the heart that is used as their metaphoric location.

3.3 Conceptual Metaphor Theory and the Embodiment Hypothesis

Conceptual Metaphor Theory as introduced by Lakoff and Johnson (1980) explains the ubiquity of metaphors in language with the existence of a conceptual metaphor underlying linguistic metaphors. Applied to the general observations and the results of the corpus study, this would suggest that the linguistic metaphors of the heart as the location of personality traits and feelings reveal a conceptual process whereby people actually *think* of the heart as the center of a person.

Lakoff and Johnson (1980, 1999) argue that many conceptual metaphors are grounded in bodily experiences such as the front-back and up-down orientation, temperature, hunger, etc. Based on this theory, some researchers have argued that the folk model of the heart as the seat of emotions and personality is grounded in bodily experiences, for example the experience of blood rush, pulse rate, rising of body temperature in connection with emotions. These experiences motivate a conceptual mapping that is reflected in linguistic metaphors (Kövecses 1986, 1990, 2000, Niemeier 1997, 2000, 2008). Therefore, the combination of Conceptual Metaphor Theory and the Embodiment Hypothesis offers an explanation for the frequent use of heart metaphors in German – the systematic and frequent use of these metaphors is due to an underlying folk model that is embedded in bodily experiences and cultural traditions.

4. Concepts of heart, brain, and personality

Conceptual Metaphor Theory and the Embodiment Hypothesis offer an elegant explanation for the linguistic and literary patterns I described so far. However, this theory makes far-reaching assumptions about the nature of the human conceptual system solely on the basis of observations of linguistic behavior. But do people really think the way they speak? Is the use of heart metaphors in literary and everyday language enough evidence for the claim that people conceptualize the heart as the seat of feelings and personality?

In a discussion of folk models of the heart and the brain, one should not forget that the nature of these abstract concepts has been a topic of serious academic discussion for more than 2000 years and is thus at least as old as our subconsciously used linguistic metaphors. In Greek philosophy, this was considered a central question, for example in the works of Aristotle and Plato, who both try to localize personality in the body, although they do not agree on whether the heart or the brain is the right place to look for it. And even those philosophers who thought that the heart was the center of all mental, emotional, and spiritual processes and thus the most important part of a human being, did not necessarily think about the heart in the terms suggested by the metaphors discusses above: Augustinus, for example, did see the heart as the seat of personality, but he is explicit about the fact that by *heart* he does not, in this context, refer to an anatomical entity, to the "little fleshy part behind our ribs" ("illa particula carnis nostrae, quae sub costis latet", quoted in Fuchs 1995: 39). In scientific contexts, the heart has been described as nothing more than a muscle for over a hundred years. From the eighteenth century onwards, even ordinary people knew that the heart does not literally contain feelings; this was already recognized as a metaphor (Stockinger 1996). On the other hand, this scientific and popular knowledge about the human body and its organs did not completely eliminate the idea of locating emotions and personality traits in a person's heart. Besides the Greek tradition, the particular symbolism of Christianity has had a far-reaching influence. Heart symbols reached the height of their popularity in the popular religious *Erbauungsliteratur* (edifying literature) of the seventeenth and eighteenth centuries (Fuchs 1995). In the literature of Germany in the eighteenth and nineteenth centuries, the heart as the seat of soul and feelings remained a central image (Stockinger 1996), and some uses of the motif with ethno-medical roots continued in the nineteenth century.

However, beginning in the nineteenth and continuing throughout the twentieth century, the systematic scientific investigation of the brain and the rise of neurology as a scientific discipline have uncovered facts about the brain, thought, and cognition that had formerly been objects of mere philosophical speculation. Today, it is common knowledge that all cognitive and emotional processes are neural processes. Therefore, one could expect that even though the idea of the heart as the seat of personality is still visible in language, people actually think differently than their usage of metaphors would suggest.

In order to assess German speakers' beliefs about the location of personality traits and emotions more directly, I conducted a questionnaire study in which subjects had to judge a number of hypothetical situations that involved hidden decisions about whether it is a person's heart or the brain that holds their personality.

4.1 Subjects

Subjects were 39 undergraduate students at Humboldt-University Berlin and the University of Hamburg. They were between 19 and 40 years old. To avoid an educationally and/or culturally biased sample, four groups were assembled from students who major in law, economy, biology, and the humanities.

4.2 Procedure and material

Subjects had to answer ten questions of similar form, eight distractor items and two test items, exemplified in (10) and (11).

(10) ITEM A ("HEART"):
Thomas leidet an einer schweren Herzkrankheit, die sein Leben bedroht. Ihm wird das Herz von Karl transplantiert, der bei einem Unfall verunglückt ist und dessen Gehirn keine Aktivität mehr zeigt. Die Transplantation ist erfolgreich. Welche Person überlebt?
nur Thomas / hauptsächlich Thomas / Thomas und Karl gleichermaßen / hauptsächlich Karl / nur Karl

('Thomas is suffering from a serious heart disease threatening his life. He receives Karl's heart as a transplant; Karl had an accident after which his brain no longer showed any activity. The transplantation is successful. Which person survives?
only Thomas / mainly Thomas / Thomas and Karl equally / mainly Karl / only Karl')

(11) ITEM B ("BRAIN"):
Peter liegt im Koma. Sein Gehirn zeigt keine Aktivität mehr. Ihm wird das Gehirn von Wolfgang transplantiert, dessen Herz aufgehört hat zu schlagen. Die Transplantation ist erfolgreich. Welche Person überlebt?
nur Peter / hauptsächlich Peter / Peter und Wolfgang gleichermaßen / hauptsächlich Wolfgang / nur Wolfgang

('Peter is in a coma. His brain no longer shows any activity. He receives Wolfgang's brain as a transplant; Wolfgang's heart had stopped beating. The transplantation is successful. Which person survives?
only Peter / mainly Peter / Peter and Wolfgang equally / mainly Wolfgang / only Wolfgang')

Subjects had to choose their answers from the set of five possible answers provided for each item. They were encouraged to answer even if they felt unsure about the "correct" answer. The experimenter did not answer any additional questions regarding the procedure, the content of the questions or the expected answer.

4.3 Results

The answers to the first question show unambiguously that almost all participants think that it is the recipient of the heart who survives (Table 3). Clearly, people do not think of the heart as the body organ that determines the personality and identity of a person. The answers given are consistent with the fact that in transplantation medicine, the recipient of a new heart is indeed still seen as the person he or she was before – he/she has the same name and legal status as the potential recipient before. The donor, on the other hand, is considered dead. Despite literary motifs and systematic metaphorical patterns, then, people do not locate the core of personality inside the heart, instead, they choose their answers on the basis of established scientific facts.

Table 3. Answers given to item A (Surviving a heart transplantation)

	only donor	**mainly donor**	**both**	**mainly recipient**	**only recipient**
Law students	0	0	0	0	10
Humanities students	0	0	0	1	10
Biology students	0	0	0	0	8
Economy students	0	0	0	1	9
Total	0	0	0	2	37

More surprising were the answers to the second question, which differ considerably. Here, almost the half of the participants again decided that it is the recipient of the brain who survives, the rest is more or less equally distributed over the other possible answers (see Table 4).

Thus, the brain is not seen as the seat of personality by the majority of participants either – because if it was, one would expect that the answer to question B would be that the donor of a brain transplantation is the survivor, because his personality is transferred together with the brain. Yet, only six of the participants in my study thought so, whereas the others chose answers incompatible with scientific mainstream assumptions.

Table 4. Answers given to item B (Surviving a brain transplantation)

	only donor	mainly donor	both	mainly recipient	recipient
Law students	4	1	0	0	5
Humanities students	0	3	0	3	5
Biology students	1	1	2	1	3
Economy students	1	1	3	0	5
Total	6	6	5	4	18

4.4 Discussion

The results show that people do not actually think of a body organ as the heart or the brain as the seat of personality, but rather seem to localize the core of a person in the body in a more general way. The great majority of subjects agrees that the recipient, not the donor of a transplanted heart is the person who survives the transplantation. They do not think that the core of a person is literally located in the heart (and therefore transplanted with the actual organ). Thus, literary motifs and everyday metaphorical idioms do not directly reflect folk beliefs. What is surprising, though, is that roughly half of the subjects do not believe that personality is located in the brain, either, since they think that if one's brain was transplanted into another body, the person the body belonged to would be the survivor of this transaction, whereas the owner of the brain is considered dead.

There is no significant correlation between the major subject of the students – meaning that their different background, education, and therefore maybe worldview, has no discernible impact on their beliefs about the nature of human personality.

The results are not easy to interpret, and in order to obtain a clearer picture about people's actual beliefs, more studies with more diverse questions are needed. What the results show unambiguously, though, is that assuming a direct correlation between figurative language use and folk beliefs is not appropriate.

5. Bi-directional perspectives: Converging and diverging evidence

We have seen that motifs and metaphors in art and literature indeed correspond with everyday language patterns, thus confirming Lakoff and Turner's idea that literary metaphors are extensions of mappings also apparent in everyday language (Lakoff and Turner 1989). However, these mappings cannot be equated with beliefs that are used to argue about certain topics. Instead, scientific knowledge, philosophical considerations, and religious beliefs can reflect language patterns or run counter to them. In the case of the conceptual connection of heart, feelings, and personality, if one takes these different sources of evidence into account, a mixed picture emerges. It is clearly not appropriate to draw conclusions about folk beliefs from purely linguistic data or from an eclectic collection of examples for this or some other belief. In order to make valid statements about the nature of peoples beliefs, a direct investigation of these beliefs is necessary.

What does this mean for the combination of humanities and cognitive sciences? On the one hand, my study sketches out the potential of an inter- or even transdisciplinary approach to phenomena that have a cultural and a cognitive dimension. I hope to have shown that the inclusion of empirical methods from the social sciences not only adds interesting observations to topics that have been the focus of the humanities, but also actually allows us to confirm or reject long agreed-on assumptions made solely on the basis of literary or linguistic data. They offer a chance of coming to more valid and reliable statements about human concepts and cognition. They also offer us a chance of discovering facts that other methods cannot reveal at all. The results of the questionnaire study show something that is not at all visible in language data, be it literary or everyday language, and that would not have been discoverable by qualitative or quantitative approach to language use.

However, this positive outlook on the combination of methods from different disciplines should not obscure remaining fundamental problems for a 'bi-directional' perspective on language. For the study presented here, rather superficial observations on literary motifs and metaphors were not much more than a motivation for a quantitative corpus study and another quantitative questionnaire study. The question posed at the beginning of this article is after all a question about human conceptualization and as such it was obvious from the outset that it would have to be answered by an investigation of conceptualization. The discussion of literary motifs and metaphors at best adds some interesting observations of what could be side-effects of the phenomenon discussed here. Therefore, the combination of disciplines and methods is indeed a useful extension and offers a broader picture, but the combination is not bi-directional – it is an expansion into one direction. I think that this kind of expansion is useful and necessary. But it is doubtful that it leads to a transformation of one discipline, let alone a merging of two currently distinct disciplines. There may well be considerable overlap in the concrete phenomena investigated here: linguists as well as literary scholars are concerned with language. But the questions they ask are fundamentally different. Linguists are interested in the systematicity and the structure of language in general, not in single instances of language use. Literary studies focus on certain texts or authors and want to describe characteristics of these *instances* of language. Some of the ways in which they do so can certainly be of interest to linguists, and linguistic methods and models can enhance investigations of literature. But in the first case, a literary text is just another data point to the linguist, no better or worse than any other instance of language use.

Conversely, while linguistic models and methods may be useful in the context of literary studies, they will never be more than convenient tools. This fundamental gap becomes even more apparent when we consider the cognitive sciences in general. An overlap of phenomenon areas should never lead us to believe that the questions we ask are the same. Nevertheless, a look into the methods and theories of other disciplines can expand knowledge in one's own field, as I hope to have shown in my study.

6. Conclusions

The basic idea that motivates the plot of the movie *Meet Joe Black* is that a person consists of material – the body – and immaterial qualities – feelings, consciousness, the 'soul' – that together form the individual. The whole motif of someone using another body is based on the conceptual division between the body and the personality – a philosophical tradition also known as dualism. However, in reality, people do have only one body, they cannot leave it and still exist, or even 'use' another body. Thus, the existence and the nature of this division is not something that humans can experience directly. Therefore, conceptual mappings of more concrete domains on these highly abstract questions could be something that enables humans to develop a concept of human personality. If we look at literature and art as well as everyday language, it becomes clear that the notion of the personality being located in the heart is ubiquitous and deeply embedded in German culture. But this fact does not directly reflect people's beliefs. Instead, people do conceptually locate 'personality' in the body of a person, but not in a specific body part or body organ. The results of my study show that the underlying connection, if it exists at all, is much more complex than often assumed.

Moving beyond this peculiar fact about folk models and the conceptualization of personality, my study shows the dangers of jumping to conclusions about cognition and conceptual structures on the basis of observations about literature and language. Qualitative analysis of recurring motifs and frequent metaphors in literature and folk art can point to interesting language patterns and maybe even to underlying conceptualizations and folk beliefs – but this is only a starting point if we are actually interested in those underlying, more general concepts and beliefs. Even the combination of qualitative and quantitative analyses of literary and linguistic data cannot provide a reliable source of knowledge about non-linguistic concepts and beliefs: a more direct investigation of these matters must accompany analyses of language: psychological and sociological methods have to be employed. My investigation shows how qualitative and quantitative analyses of linguistic data can yield a picture of seemingly converging evidence for a general underlying conceptualization, but a more direct investigation of folk beliefs in this case considerably diverges from that picture and shows that bridging the gap between linguistic data and non-linguistic assumptions is absolutely necessary. Thus, I think that this case study offers one possible combination of different methods to overcome the limits of literary studies and linguistics as part of the cognitive sciences.

References

Deignan, Alice. 1999. Corpus-based research into metaphor. In *Researching and Applying Metaphor*, Lynne J. Cameron and Graham Low (eds.), 177–199. Cambridge: Cambridge University Press.

Deignan, Alice. 2005. *Metaphor and Corpus Linguistics*. Amsterdam: Benjamins.

Fuchs, Thomas. 1995. 'Gewogen und zu leicht befunden'. Herz und Gewissen. In *Herz. Das menschliche Herz – Der herzliche Mensch: Begleitbuch zur Ausstellung "Herz" vom 5.10.1995–31.3.1996 im Deutschen Hygiene-Museum Dresden*, Susanne Hahn (ed.), 32–48. Dresden & Basel: Verlag der Kunst.

Goschler, Juliana. 2005. Gehirnmetaphern: Verschiedene Formen der Metaphorisierung? In *Linguistik im Schloss. Linguistischer Workshop Wartin 2005*, Norbert Fries and Svitlana Kiyko (eds.), 25–34. Czernowitz: Bukrek.

Goschler, Juliana. 2008. *Metaphern für das Gehirn. Eine kognitiv-linguistische Untersuchung.* Berlin: Frank & Timme.

Kövecses, Zoltán. 1986. *Metaphors of Anger, Pride, and Love*. Amsterdam: Benjamins.

Kövecses, Zoltán. 1990. *Emotion Concepts*. New York: Springer.

Kövecses, Zoltán. 2000. *Metaphor and Emotion*. Cambridge: Cambridge University Press.

Lakoff, George. 1987. *Women, Fire, and Dangerous Things. What Categories Reveal About the Mind*. Chicago: The University of Chicago Press.

Lakoff, George. 1993. The contemporary theory of metaphor. In *Metaphor and Thought* (2nd ed.), Andrew Ortony (ed.), 202–251. Cambridge: Cambridge University Press.

Lakoff, George and Mark Johnson. 1980. *Metaphors We Live By*. Chicago: The University of Chicago Press.

Lakoff, George and Mark Johnson. 1999. *Philosophy in the Flesh. The Embodied Mind and its Challenge to Western Thought*. New York: Basic Books.

Lakoff, George and Mark Turner. 1989. *More than Cool Reason. A Field Guide to Poetic Metaphor*. Chicago: The University of Chicago Press.

Niemeier, Susanne. 1997. To have one's heart in the right place – metaphorical and metonymical evidence for the folk-model of the heart as the site of emotions in English. In *Human Contact through Language and Linguistics*, Birgit Smieja and Maike Tasch (eds.), 87–106. Frankfurt/Main: Lang.

Niemeier, Susanne. 2000. Straight from the heart: Metonymic and metaphorical explorations. In *Metaphor and Metonymy at the Crossroads*, Antonio Barcelona (ed.), 195–211. Berlin: Mouton de Gruyter.

Niemeier, Susanne. 2008. To be in control: a cool head and a warm heart. The head-heart dichotomy in English. In *Culture, Body, and Language: Conceptualizations of Internal Body Organs across Cultures and Languages*, Farzad Sharifian, René Dirven, Ning Yu and Susanne Niemeier (eds.), 349–372. Berlin: Mouton de Gruyter.

Stefanowitsch, Anatol. 2004. Happiness in English and German: A metaphorical-pattern analysis. In *Language, Culture, and Mind*, Michel Achard and Suzanne Kemmer (eds.), 137–149. Stanford: CSLI-Publications.

Stefanowitsch, Anatol. 2005. The function of metaphor. Developing a corpus-based perspective. *International Journal of Corpus Linguistics* 10: 161–198.

Stefanowitsch, Anatol. 2006. Words and their metaphors: A corpus-based approach. In *Corpus-based Approaches to Metaphor and Metonymy*, Anatol Stefanowitsch and Stefan Th. Gries (eds.), 61–105. Berlin: Mouton de Gruyter.

Stockinger, Ludwig. 1996. 'Herz' in Sprache und Literatur der Goethezeit. Goethe – Novalis – Hauff. In *Das Herz im Kulturvergleich*, Georg Berkemer and Guido Rappe (eds.), 173–209. Berlin: Akademie Verlag.

Cognitive linguistics as a cognitive science*

Anatol Stefanowitsch
Universität Hamburg

Cognitive linguistics is on its way to becoming a cognitive science, but a number of problems remain. The relationship between cognitive linguistics and the core cognitive sciences (psychology and neurology) must be clarified: cognitive linguists can selectively import models and methods from these disciplines as a foundation for their linguistic theories, they can export their own models to these disciplines for empirical testing and integration, or they can transform linguistics into a core cognitive science in its own right. The latter requires a number of changes to the models and practices of cognitive linguistics: it must refocus on its linguistic heritage, adopt a more scientific outlook, gain a higher degree of methodological awareness and restrict its models to linguistic constructs and hypotheses that can be operationalized and falsified.

Keywords: empirical testing, falsifiability, import of models and methods, methodological awareness, operationability

1. Introduction

The social sciences and the humanities are currently experiencing a "cognitive turn" at least as sweeping and profound as the linguistic turn of the last century. Many disciplines are struggling to redefine themselves in light of this development, giving us a host of new research programs referring to themselves as "cognitive anthropology" (D'Andrade 1995), "cognitive sociology" (Zerubavel 1997), "cognitive economics" (Bourgine and Nadal 2004), "cognitive political science"

* This paper grew out of a keynote talk delivered at the conference *Bi-Directional Perspectives in the Cognitive Sciences* conference in Marburg on February 26–27, 2009. I would like to thank the participants of that conference, particularly Juliana Goschler, for their questions and comments, which have helped me to refine my views on the topic.

(Easton 1985), "cognitive religious studies" (Boyer 1994), and, of course, "cognitive poetics" (Stockwell 2002), "cognitive grammar" (Langacker 1987) and "cognitive linguistics" (Lakoff 1987).

In this short, programmatic chapter, I share some of my thoughts on how successful cognitive linguistics has been so far in redefining linguistics as a cognitive science and where it could do better.

2. What is a cognitive science?

In order to discuss in what way and to what extent cognitive linguistics, or any other discipline, can be regarded as a cognitive science, we need to specify what we mean by "cognitive science". I assume the following broad, and hopefully uncontroversial, definition:

> Cognitive science is the interdisciplinary study of mind and intelligence, embracing philosophy, psychology, artificial intelligence, neuroscience, linguistics, and anthropology. (Thagard 2008)

I further assume that by "mind and intelligence" we mean processes like perception, imagination, categorization, reasoning, and memory.

It follows that any discipline that considers itself to be a cognitive science must contribute to the study of the human mind. For two of the disciplines in the definition above, that contribution is self-evident: the human mind and its material basis constitute the principal objects of research of psychology and neurology respectively. I refer to these as "core cognitive sciences".

For other disciplines, that contribution is less direct, but still clearly discernible. Artificial intelligence, for example, is concerned with the study of systems that *simulate* some of the outwardly visible behavior of the human mind, and to the extent that this simulation is accurate, the mechanisms behind it may be treated as models of the mind. Whether these models are realistic in the sense that they capture not just the input and the output, but also the internal processes of the human mind, however, is a question that can ultimately be answered only by empirical research in the core cognitive sciences.

Still other disciplines have a much more indirect relationship to the study of mind and intelligence; their principal objects of research are related to cognition only in that they are, in some sense, products of human minds. The disciplines mentioned in the introduction all fall into this category. Unlike the core cognitive sciences, these disciplines have a double task to fulfill if they want to be part of the cognitive enterprise: first, they must put forward scientific models concerning their object of study (language, literature, culture, society, economy, etc.), and second, they must specify some relationship between that object of study on the one hand and the study of human mind and intelligence on the other.

3. How to become a cognitive science

There are three main ways in which a discipline can specify a relationship between the human mind and an object of study in the social sciences and humanities, and all of these can be found in the recent history of linguistics.

First, a discipline can draw inspiration from the core cognitive sciences by borrowing from them, more or less selectively, models, perspectives and terminology. I refer to this research strategy as the Import Strategy. There is nothing wrong with this strategy, as long as there is a genuine attempt to import more than just the terminology of the cognitive sciences. But since none of the results of this import of ideas ever flows back to the core disciplines, the potential of this strategy for real progress in our understanding of the human mind is limited.

Second, a discipline can use its domain-specific expertise to build the best (most descriptively adequate, most exhaustive) model of its object of study. This model can then be confronted with models of the human mind from the core cognitive sciences in order to investigate if and how these models may be integrated, i.e. what a model of the particular object of study in question may tell us about the human mind and human intelligence. I refer to this as the Export and Integration Strategy. The integration can be attempted by practitioners of psychology and neurology, by adequately trained practitioners of the discipline in question, or collaboratively, by researchers from both fields. There are practical obstacles to this research strategy – diverging terminology, discipline-internal institutionalization, etc. – but I believe that it is, ultimately, more fruitful than a mere one-way import of ideas from psychology and neurology.

Third, a discipline can take its domain-specific expertise and fuse it into the core cognitive sciences, i.e., it can become an integral part of psychology (or, less likely, neurology). In this scenario, a researcher or group of researchers would start out from a domain-specific model of language, economy, literature, etc. and subject it to testing within the theoretical and methodological frameworks of the core cognitive sciences. The results of these tests would then be successively integrated into these models, making them less and less domain specific until they are essentially cognitive science models of a particular sphere of human behavior. This is unlikely to happen to a discipline as a whole, as not all researchers in the social sciences and humanities take a cognitive perspective in the first place, but it can happen to subdisciplines and to research programs by individual research groups or individual researchers.

4. Cognitive linguistics as a cognitive science

Before Leonard Bloomfield's brief behaviorist revolution in linguistics, connections between language and mental processes were naturally assumed by authors such as Karl Brugmann, Ferdinand de Saussure, Franz Boas, Edward Sapir, Otto Jespersen, Herman Paul, and many others. In this sense, linguistics was one of the earliest disciplines deserving the name "cognitive". However, the behaviorist turn interrupted this tradition and after the idea of linguistics as a cognitive science was reintroduced by Chomsky (1959), it developed in an intriguing, but perhaps ultimately harmful direction.

Chomskyan linguistics adds an interesting twist to the relation between language and cognition. It posits a specialized mental module that contains our linguistic competence, an autonomous language faculty that does not share any representational principles with other aspects of cognition and that interacts with them only in the process of linguistic performance, a process that is explicitly placed outside of the domain of linguistic theory. To be quite clear about it: this is an extremely elegant postulate. It allows Chomskyan linguists to do cognitive science simply by doing linguistic analysis (which in their framework means building models to account for their own acceptability judgments on constructed sentences). A Chomskyan linguist does not have to pay attention to the methods of the core cognitive sciences, nor to anything that they have uncovered about any aspect of human cognition. For a linguistic theory with cognitive aspirations, this is a genuinely ingenious theory-building mechanism: it ensures that any discovery about the structure of language automatically becomes a fact about the human mind.

Despite its elegance, there is a fundamental problem with this postulate: it is not falsifiable. Since linguistic competence in this view is unrelated to anything else, observations about the way that it *interacts* with other cognitive processes cannot tell us anything about its nature or even its existence. That does not mean that the postulate is actually false – language does seem unique among our cognitive skills, and so we cannot dismiss the idea of autonomy out of hand –, but it means that we have no way of knowing whether it is false. This makes it fruitless as a departure point for the scientific study of linguistic cognition. Incidentally, it is also fruitless as a departure point for the scientific study of language itself – there is an open-ended number of theories that are consistent with a convincingly large subset of linguistic data to appear plausible, and because of the autonomy postulate – the idea of an encapsulated linguistic competence that cannot be accessed – there are no competence-external criteria that could distinguish between them.

Cognitive linguistics – the family of theories associated with authors like Ronald Langacker, George Lakoff, Leonard Talmy, and Gilles Fauconnier – avoids such a postulate. This makes its relationship to the core cognitive sciences potentially more fruitful but also much more complicated. As already mentioned, in cognitive linguistics, we see evidence for all three strategies discussed above.

4.1 Cognitive linguistics and the Import Strategy

A brilliantly executed example of the Import Strategy is found in Langacker's Cognitive Grammar, as developed in the two volumes of his *Foundations of Cognitive Grammar* (1987, 1991). Langacker himself refers to his work, quite aptly, in my view, as "an exercise in speculative psychology" and he summarizes its relation to the study of the human mind as follows:

> I have adhered rather strictly, in developing my proposals, to the dictates of both psychological plausibility and linguistic necessity; I have relied almost exclusively on *seemingly indisputable cognitive abilities* (e.g. the ability to compare two events and register a discrepancy between them), and I invoke them for linguistic constructs that must somehow be accommodated in any comprehensive and explicit analysis. (1987: 6, my emphasis)

In other words, Langacker builds his linguistic theory entirely on supposedly general cognitive processes, but he is highly selective in choosing the processes he considers relevant: cognitive routinization ("entrenchment"), the ability to recognize similarities and differences, schematization, the ability to track changes over time, and figure-ground perception.

Langacker is not just selective, though. In addition, he characterizes these processes in very general terms not tied to any specific model from the core cognitive sciences. He adopts a kind of informed common-sense perspective on the human mind – informed in that it takes into account some important developments in 20th century psychology (for example, gestalt principles), and common sense in that it defines its concepts in terms that don't require any knowledge in psychology or neurology. Langacker is not alone in this strategy: George Lakoff's *polysemy networks* (Lakoff 1987), Mark Johnson's *image schemas* (Johnson 1987), Leonard Talmy's *force dynamics* (Talmy 1988), and Gilles Fauconnier's *blending* (Fauconnier and Turner 2002) are all examples of such common-sense cognitive processes.

There are good reasons for a selective common-sense approach to the cognitive sciences. Recall that we are talking here about a one-way import of ideas from the core cognitive sciences to theoretical linguistics, with no serious expectation yet that their application to language will yield results that might flow back to the core cognitive sciences. Given this situation, consider the alternative to such a selective approach, i.e. the full-fledged adoption of a particular model of cognition: every time this model is revised or amended, as it doubtless would be, the linguistic theory based on it would have to be revised and amended as well, regardless of whether the linguistic data call for such revisions. Cognitive linguists would be in a position comparable to that of third-party software developers developing software for the Windows operating system: Every time a new version of Windows is released, they have to change their applications to work with this new system, even if the application was working just fine. On the other hand, these developers have no influence on future versions of the OS, for example by suggesting particular functions that would benefit their application.

In a situation where the relation between the core cognitive sciences and linguistics is a one-way street from the former to the latter, then, the kind of selective common-sense psychology we see in the works of Langacker, Lakoff, Talmy and others is the only viable option. It is not without problems as a basis for interacting with psychologists, neurologists, and computational scientists, who do tend to adhere to specific theories of the human mind and who are often exasperated with cognitive linguists because of their fast and loose approach to cognitive science models and terminology. Still, this strategy does have the potential to yield results that will feed back into the core cognitive sciences if and when it is combined with an empirical approach to linguistic data. Such an empirical approach, however, is lacking almost completely in the work of the authors I have mentioned. I return to this issue after discussing the second of the three strategies introduced above.

4.2 Cognitive linguistics and the Export and Integration Strategy

There are a number of examples for the Export and Integration Strategy, where a discipline applies its domain-specific expertise to the analysis of its object of research and then turns the results over to the core cognitive sciences for an investigation of how these results may be accounted for within a general theory of the human mind. I discuss two of them here: metaphor theory and construction grammar.

What is interesting about both of these areas is that they do not start out by importing ideas from the core cognitive sciences in the way that Langacker's Cognitive Grammar or Fauconnier's theory of blending do. Instead, they start out from linguistic analysis. This is not to say that they do not make claims relevant to the cognitive sciences. Take Lakoff and Johnson's Conceptual Theory of Metaphor – it takes them less than one paragraph to get to their central claim:

> We have found [...] that metaphor is pervasive in everyday life, not just in language, but in thought and action. Our ordinary conceptual system, in terms of which we both think and act, is fundamentally metaphorical in nature.
> (Lakoff and Johnson 1980: 3)

But if we consider the evidence on which they base these far-reaching claims, we find that it is exclusively linguistic, as they themselves readily admit on the very next page (Lakoff and Johnson 1980:4). Essentially, their evidence consists of two facts: first, that metaphor is pervasive in everyday language, and second, that metaphor is best described in terms of general, highly systematic mappings between concrete and abstract semantic domains. Both of these observations are purely linguistic (and neither of them was new at the time when Lakoff and Johnson pointed them out). So, at first glance, Lakoff and Johnson's strategy is very similar to the Chomskyan strategy: they uncover a fact about language and automatically assume that it is also a fact about the mind. What distinguishes their strategy from Chomsky's is that they do not posit an encapsulated language faculty, and that therefore their claims are, in principle, testable. They make predictions about the human mind in general, not just about the part of it that is concerned with language, and thus, behavioral evidence can be brought to bear on the issue.

Based on what we currently know about metaphor in the human mind, I think it is fair to say that the original claims by Lakoff, Johnson and others are too sweeping – matters are considerably more complex. But we can also clearly say that these claims are right in their broad outline: we do have empirical evidence for a correlation between linguistic metaphors and general patterns of thought.

Interestingly, not all of this evidence was adduced by cognitive linguists aware of metaphor theory: it is well-established in Social Psychology, for example, that people associate moral failure with physical pollution: experiments show that people have the urge to wash when they have thought or talked about disagreeable, unpleasant or morally questionable acts (Fairbrother et al. 2003, Zhong and Liljenquist 2006) and, conversely, they have a higher tolerance for such acts if they wash their hands first (Schnall et al. 2008). It is suggestive to explain these results with reference to linguistic metaphors like the following:

(1) the *stain* of guilt/sin/illegitimacy
impure thoughts/soul/character
a *smutty* manner/joke/film/novel
a *filthy* word/mouth/mood/mind
an *unclean* thought/spirit/mind
a *dirty* mind/look/word/secret
to *contaminate* a relationship
to *taint* someone's reputation
to *wash away* someone's sins
to feel *soiled* by an experience
to *cleanse* oneself of negativity
to *pollute* someone's mind/thoughts

What these studies leave open is the causal mechanism underlying such correlations: do we think of disagreeable things as DIRTY because there are linguistic metaphors that make this connection, or are the linguistic metaphors simply an expression of a preexisting mental connection? It is here that the speculative-psychology approach of the leading theoreticians in cognitive linguistics reaches its limits – empirical research is necessary within the field of cognitive linguistics itself.

Before I turn to a discussion of empirical research, let me briefly discuss my second example: construction grammar. Note that, just like metaphor theory, construction grammar was not inspired by models from the core cognitive sciences, but by purely linguistic considerations: authors like Lakoff (1970) and Fillmore (1985) became increasingly aware of the extent to which the grammar of natural languages was suffused by irregularity and idiosyncrasy. Construction grammar was developed as a framework capable of capturing these idiosyncrasies and the generalities that, of course, also exist, by a uniform mechanism: the grammatical construction. Construction grammarians vary in their willingness to make claims about the human mind based on their theoretical considerations, but Lakoff (1987) and Goldberg (1995) clearly assume that their grammatical analyses reflect the way in which grammatical knowledge is mentally represented.

Again, we are a long way from a complete understanding of the nature of grammatical constructions, of how constructions emerge, develop and disappear in language change, of how they are learned, of how they are represented in the mind and of how they are put to use in producing and comprehending language. But, as in the case of metaphor theory, we can safely say that we have empirical evidence of their existence and of their relevance to a cognitively plausible characterization of grammatical knowledge (see the early work by Gibbs 1995; work by Michael Tomasello, Elena Lieven, and their associates, e.g. Tomasello 2003, Dąbrowska and Lieven 2005; recent work by Goldberg 2006; and my own work on negative evidence, Stefanowitsch 2006, 2008).

5. Empirical research in cognitive linguistics

In the final section of this short chapter, I discuss the issue of empirical research in cognitive linguistics. In my view, the adoption of a strictly empirical mind-set is by far the most important obstacle currently standing between cognitive linguistics as a cognitive science and cognitive linguistics as actually practiced.

Currently, cognitive linguistics as a whole cannot be classified as a cognitive science quite simply because it is not a science. Most cognitive linguists are happy to rely exclusively on the Import Strategy: they import, first-hand or, more typically, second-hand via authors like Lakoff and Langacker, selected perspectives and partial models from the core cognitive sciences, analyze linguistic data in terms of these models, and leave it at that. This unfortunate fact is responsible for an endless series of somewhat dull and repetitive talks at national and international cognitive linguistics conferences, talks about selected metaphors for LOVE in Chinese, Russian, or Old Norse, about the polysemy of the English preposition *over*, about the most visually pleasing way to draw a network for the resultative construction, about the prototypical meaning of this or that Korean verb and about the embodied way in which we talk about the human body. More importantly, the self-sufficient reliance of cognitive linguists on a few conceptual imports and pseudo-imports from the core cognitive sciences is responsible for the fact that very few people in the cognitive sciences pay attention to cognitive linguistics.

Someone following the development of the field closely may be less pessimistic with regard to the role of empirical research: if we look at the conference programs of cognitive linguistics conferences over the past fifteen years, there seems to be a growing interest in empirical research (see also Geeraerts 2006): from the mid-1990s onwards, second-generation researchers in cognitive linguistics began to turn to psycholinguistic methods (and some psycholinguists began to take an interest in cognitive linguistics research issues, such as metaphor and polysemy). And from the late 1990s onwards, second- and third-generation cognitive linguists began to turn to corpus-based methods (see Tummers et al. 2005; Stefanowitsch in press). The results generated by these empirical researchers are often spectacular and they are certainly highly relevant to theoretical considerations at the heart of cognitive linguistics.

In the domain of metaphor, for example, there is now evidence for more than just a correlation between metaphorical language and metaphorical thought: it seems that there is a causal direction from metaphorical language to metaphorical thought. Lera Boroditsky has presented experimental results suggesting that speakers of English, who talk about time as moving horizontally from front to back also reason about time in this way, while speakers of Chinese, who talk about time moving vertically from top to bottom reason about time correspondingly (Boroditsky 2001). In a corpus-based paper, I have uncovered language-specific differences between the metaphors associated with near-synonyms such as *joy* and *happiness* (Stefanowitsch 2004), and Ben Bergen and his colleagues have provided psycholinguistic evidence suggesting that such differences might affect non-linguistic cognition (Tseng et al. 2005).

But there are a number of reasons, both substantial and institutional, why the emergence of an empirical tradition in cognitive linguistics is cause for cautious hope at best. One institutional problem is that the non-empirically-minded majority of researchers in the field are quite happy to pay attention to empirical results when these results are spectacular, and when they confirm dearly-held beliefs of the cognitive community. But most empirical research is *not* spectacular and just as often as not it contradicts some dearly-held belief.

With respect to the lack of spectacularity, take work by Gerard Steen and his work group on metaphor identification (e.g. Steen 2007), or my papers on metaphorical pattern analysis (e.g. Stefanowitsch 2004, 2006). These papers are not particularly dazzling – and I mean that as a compliment. We are grappling with a difficult issue: how to tell that we are dealing with a metaphor and how to classify a metaphor once we have identified it. Given the soaring heights of abstraction and generality at which the theoretical discourse on metaphor takes place, this issue is bound to seem dull and uninspiring. But it has to be done, if we are to make any progress on the deeper questions concerning the relationship between metaphorical language and metaphorical thought.

With respect to the contradiction of dearly-held beliefs, consider again the question of whether metaphorical language causes metaphorical thought or vice versa. In spite of the encouraging results in Boroditsky (2001) just mentioned, this question has not been conclusively answered. For example, Chen and colleagues (Chen et al. 2007) were unable to replicate Boroditsky's results regarding the vertical metaphor of TIME. However, this fact has received virtually no attention in the field of metaphor research, while Boroditsky's study is still widely cited (at the time of writing, Google Scholar lists a mere eleven citations for Chen et al.'s study, compared to 209 for Boroditsky's – this difference seems too large to be explained merely by the fact that Boroditsky's study was published six years earlier than Chen et al.'s).

A second reason for tempering our optimism regarding the empirical turn in cognitive linguistics is that the growing number of empirical researchers in cognitive linguistics is not due to a large-scale conversion of researchers already active the field. Despite all massive efforts to provide training opportunities in the form of methodology workshops at national and international conferences and textbooks, collections and monographs on methodology (e.g. González-Márquez 2007), the increase in empirical research is almost exclusively a result of new researchers joining the field. It is my impression that this influx of empirical researchers is currently still offset by an even faster-growing group of non-empirical linguists adopting the cognitive linguistics framework. In any case, the empirical researchers are, and will be for the foreseeable future, a minority. And a minority, that is not as widely respected in the field as one might expect, given that they present the best shot the field has for becoming a cognitive science on equal terms with psychology, neurology and artificial intelligence.

This lack of interest is especially unfortunate in light of the fact that cognitive linguistics needs its young empirical researchers more than these researchers need cognitive linguistics. There are, after all, well-established traditions of psycho-linguistic and corpus-linguistic research that precede the emergence of cognitive linguistics by at least thirty years. And cognitively inspired empirical research is generally well-received in these traditions, so that empirically-minded cognitive linguists typically have at least one foot firmly planted in one of these traditions. If theory development in cognitive linguistics continues to proceed largely independently of the empirical research done at the fringes of the field, it seems inevitable that these young researchers will abandon the field, leaving behind a framework that may be inspired by the cognitive turn, but that remains a discipline of the humanities, caught in a hermeneutic circle of intuitions about intuitions.

6. Concluding thoughts

Summing up the discussion in the preceding sections, we can say that in order to become a cognitive science, cognitive linguistics as a discipline must accomplish at least four tasks.

First, cognitive linguistics must become more linguistic. It must rediscover a focus on detailed linguistic analysis, continuing the tradition of linguistic argumentation but replacing intuitive grammaticality judgments by the most advanced corpus-based methods available.

Second, cognitive linguistics must become more scientific. It must ensure that its fundamental notions can be operationalized and that its claims can be phrased as falsifiable hypotheses. The work by its leading theoreticians may serve as an inspiration, but it can be integrated into scientific models in cognitive linguistics only to the extent that it lends itself to operationalization and falsification (cf. Stefanowitsch 2010, Section 3.1).

Third, and as a consequence of the second task, cognitive linguistics must become more methodologically aware. It must adopt strict observational and experimental protocols and state-of-the-art statistical methods (see e.g. Gries 2006).

Fourth, cognitive linguistics must become less theoretically ambitious. As cognitive linguists become more linguistic in their analyses, more scientific in their outlook, and more methodologically aware, they will have to restrict themselves more strictly than is currently the case to linguistic constructs that can be operationalized and research hypotheses that can be falsified. This will place large parts of current models outside of the scope of a cognitive linguistic research program and it will take years of detailed research to reach scientifically sound models of the same scope (cf., again, Stefanowitsch 2010).

Many individual researchers in the field of cognitive linguistics have already accomplished these feats and are doing cutting-edge research – sometimes dull and sometimes spectacular, sometimes confirming central tenets of the theorists that initiated the discipline, and sometimes disconfirming them. The potential of these loosely connected researchers to transform the field as a whole into a cognitive science is there. Whether it is actualized, remains to be seen.

The above remarks are essentially a recipe for turning cognitive linguistics into a science. What would make this science a cognitive one would be its groundedness in a framework that treats language first and foremost as a product of and a process in the human mind, one that interacts with other processes, such as perception, learning, memory, etc.

Turning cognitive linguistics into a cognitive science can involve both the Import Strategy and the Export/Integration Strategy, as long as the results of both strategies feed into a cycle of empirical testing and refinements of the relevant models on the basis of these results. Thus, the potential for true bi-directionality is built into the process.

In keeping with the topic of the current volume, I would like to conclude my programmatic sketch with a thought on the importance of such a bi-directional approach, returning once more to metaphor theory. Note that metaphor theorists usually assume a directionality from mind to language with respect to the mappings they postulate. For example, Lakoff (1993: 218) argues with respect to the metaphor TIME IS SPACE that

> [i]n our visual systems, we have detectors for motion and detectors for objects/locations. We do not have detectors for time (whatever that could mean). Thus, it makes good biological sense that time should be understood in terms of things and motion.

Let us leave aside the question of whether we really do not have detectors for time in our brains and focus on the causal relationship postulated by Lakoff: our brains have particular circuits and lack others, so the circuits we have are co-opted for tasks for which we do not have circuits. The fact that we *talk* about time in terms of space is an epiphenomenon of this neurological fact.

Now consider again Boroditsky's results about horizontal and vertical metaphors of time discussed above, that show that speakers of languages with horizontal time metaphors reason about time on the basis of a horizontal model of time while speakers of languages with vertical time metaphors reason about time on the basis of a vertical model. If her results can be confirmed, they would seem to support Lakoff's general claim regarding the conceptual basis of metaphor.

However, note that Boroditsky's results must be explained in terms of a causality that clearly goes into the opposite direction from the one Lakoff proposes: the fact that we think about time in a particular way seems to be a consequence of the way we talk about it. While Lakoff posits a causal relationship from mind to language, results like Boroditsky's suggest a causal relationship from language to mind.

This is not to suggest that this contradiction could not be resolved in a sufficiently sophisticated model. There are metaphors that clearly suggest a directionality from thinking to speaking. For example, it is at least questionable that the expressions instantiating the MENTAL IMPURITY IS PHYSICAL POLLUTION metaphor discussed above are widespread and frequent enough in English to cause the behavior observed in the experiments I mentioned. Thus, it is likely that our understanding of immorality in term of pollution, like the TIME IS SPACE mapping, is independent of language and that the linguistic expressions instantiating it are a consequence rather than a cause of such a mapping.

Other metaphors are just as systematic, but too language-specific to be explained in terms of independent mental mappings. For example, in English and German, there is a metaphor HEART IS THE SEAT OF EMOTIONS/PERSONALITY (see, for example, Goschler, this volume; Niemeyer, this volume). In Indonesian, the corresponding metaphors see the liver as the seat of emotions and personality (e.g. Siahaan 2008). Clearly, whether speakers localize emotions in the heart or in the liver (or in neither of the two organs) is a purely linguistic (and cultural) issue and is not derived from independently existing mental mappings, just like the question whether time is seen as moving vertically or horizontally.

A model that can integrate both kinds of metaphors would have to be formulated in terms of general, language-independent conceptual mappings and cognitive constraints on possible mappings (best investigated by psychologists) on the one hand, and variation in the presence or absence as well as the precise nature of their linguistic coding in particular languages (best investigated by linguists). Thus, researchers from two disciplines would have to work together in order to build a complete model of metaphor in language and mind. This would be true bi-directionality, with each discipline importing and exporting its results to and from the other.

References

Boroditsky, Lera. 2001. Does language shape thought? Mandarin and English speakers' conceptions of time. *Cognitive Psychology* 43: 1–22.

Bourgine, Paul and Jean-Pierre Nadal (eds.). 2004. *Cognitive Economics. An Interdisciplinary Approach*. Heidelberg: Springer.

Boyer, Pascal. 1994. *The Naturalness of Religious Ideas: A Cognitive Theory of Religion*. Berkeley, CA: University of California Press.
Chen, Jenn-Yeu. 2007. Do Chinese and English speakers think about time differently? Failure of replicating Boroditsky (2001). *Cognition* 104: 427–436.
Chomsky, Noam A. 1959. A review of B. F. Skinner's verbal behavior. *Language* 35: 26–58.
D'Andrade, Roy G. 1995. *The Development of Cognitive Anthropology*. Cambridge: Cambridge University Press.
Dąbrowska, Ewa and Elena Lieven. 2005. Towards a lexically specific grammar of children's question constructions. *Cognitive Linguistics* 16: 437–474.
Easton, David. 1985. Political science in the United States: Past and present. *International Political Science Review* 6: 133–152.
Fairbrother, Nichole, Sarah J. Newth and Stanley Rachman. 2003. Mental pollution: Feelings of dirtiness without physical contact. *Behaviour Research and Therapy* 43: 121–130.
Fauconnier, Giles and Mark Turner. 2002. *The Way We Think. Conceptual Blending and the Mind's Hidden Complexities*. New York: Basic Books.
Fillmore, Charles J. 1985. Syntactic intrusions and the notion of grammatical construction. *Proceedings of the Annual Meeting of the Berkeley Linguistic Society* 11: 73–86.
Geeraerts, Dirk. 2006. Introduction: A rough guide to cognitive linguistics. In *Cognitive Linguistics. Basic Readings*, Dirk Geeraerts (ed.), 1–28. Berlin: Mouton de Gruyter.
Gibbs, Raymond W. Jr. 1995. Idiomaticity and human cognition. In *Idioms: Structural and Psychological Perspectives*, Martin Everaert, Erik-Jan van der Linden, André Schenk and Robert Schreuder (eds.), 97–116. Mawah, NJ: Lawrence Erlbaum.
Goldberg, Adele E. 1995. *Constructions. A Construction Grammar Approach to Argument Structure*. Chicago: The University of Chicago Press.
Goldberg, Adele E. 2006. *Constructions at Work. The Nature of Generalization in Language*. Oxford: Oxford University Press.
González-Márquez, Mónica, Irene Mittelberg, Seana Coulson and Michael J. Spivey (eds.). 2007. *Methods in Cognitive Linguistics*. Amsterdam: Benjamins.
Gries, Stefan Th. 2006. Some proposals towards more rigorous corpus linguistics. *Zeitschrift für Anglistik und Amerikanistik* 54: 191–202.
Johnson, Mark. 1987. *The Body in the Mind. The Bodily Basis of Meaning, Imagination, and Reason*. Chicago: The University of Chicago Press.
Lakoff, George. 1970. *Irregularity in Syntax*. New York: Holt, Rinehart and Winston.
Lakoff, George. 1987. *Women, Fire and Dangerous Things. What Categories Reveal About the Mind*. Chicago: The University of Chicago Press.
Lakoff, George. 1993. The contemporary theory of metaphor. In *Metaphor and Thought* (2nd ed.), Andrew Ortony (ed.), 202–251. Cambridge: Cambridge University Press.
Lakoff, George and Mark Johnson. 1980. *Metaphors We Live By*. Chicago: The University of Chicago Press.
Langacker, Ronald W. 1987. *Foundations of Cognitive Grammar*. Vol. 1: *Theoretical Prerequisites*. Stanford, CA: Stanford University Press.
Langacker, Ronald W. 1991. *Foundations of Cognitive Grammar*. Vol. 2: *Descriptive Application*. Stanford, CA: Stanford University Press.
Schnall, Simone, Jennifer Benton and Sophie Harvey. 2008. With a clean conscience: Cleanliness reduces the severity of moral judgments. *Psychological Science* 19 (12): 1219–1222.

Siahaan, Poppy. 2008. Did he break your heart or your liver? A contrastive study on metaphorical concepts from the source domain ORGAN in English and in Indonesian. In *Culture, Body, and Language. Conceptualizations of Internal Body Organs across Cultures and Languages*, Farzad Sharifian, René Dirven, Ning Yu and Susanne Niemeier (eds.), 45–74. Berlin: Mouton de Gruyter.

Steen, Gerard. 2007. Finding metaphor in discourse: Pragglejaz and beyond. *Language, Culture and Representation* 5: 9–25.

Stefanowitsch, Anatol. 2004. HAPPINESS in English and German: A Metaphorical-pattern Analysis. In *Language, Culture, and Mind*, Michel Achard and Suzanne Kemmer (eds.), 137–149. Stanford: CSLI.

Stefanowitsch, Anatol. 2006. Negative evidence and the raw frequency fallacy. *Corpus Linguistics and Linguistic Theory* 2: 61–77.

Stefanowitsch, Anatol. 2006. Words and their metaphors: A corpus-based approach. In *Corpus-based Approaches to Metaphor and Metonymy*, Anatol Stefanowitsch and Stefan Th. Gries (eds.), 63–105. Berlin: Mouton de Gruyter.

Stefanowitsch, Anatol. 2008. Negative evidence and preemption: A constructional approach to ungrammaticality. *Cognitive Linguistics* 19: 513–531.

Stefanowitsch, Anatol. 2010. Empirical cognitive semantics: some thoughts. In *Corpus-Driven Cognitive Semantics. Usage-Based Approaches to Conceptual Structure*, Dylan Glynn and Kerstin Fischer (eds.), 355–380. Berlin: Mouton de Gruyter.

Stefanowitsch, Anatol. In press. Cognitive Linguistics meets the corpus. In *Converging and Expanding Trends in Cognitive Linguistics*, Mario Brdar, Milena Žic Fuchs and Stefan Th. Gries (eds.), Amsterdam: Benjamins.

Stockwell, Peter. 2002. *Cognitive Poetics. An Introduction*. London: Routledge.

Talmy, Leonard. 1988. Force dynamics in language and cognition. *Cognitive Science* 12: 49–100.

Thagard, Paul. 2008. Cognitive Science. In *The Stanford Encyclopedia of Philosophy* (Fall 2008 Edition), Edward N. Zalta (ed.), <http://plato.stanford.edu/archives/fall2008/entries/cognitive-science/>.

Tomasello, Michael. 2003. *Constructing a Language. A Usage-based Theory of Language Acquisition*. Harvard: Harvard University Press.

Tseng, Meylysa, Yiran Hu, Wen-Wie Han and Benjamin Bergen. 2005. "Searching for Happiness" or "Full of Joy"? Source domain activation matters. *Proceedings of the Annual Meeting of the Berkeley Linguistics Society* 31: 359–370.

Tummers, José, Kris Heylen and Dirk Geeraerts. 2005. Usage-based approaches in Cognitive Linguistics. A technical state of the art. *Corpus Linguistics and Linguistic Theory* 1: 225–261.

Zerubavel, Eviatar. 1997. *Social Mindscapes: An Invitation to Cognitive Sociology*. Cambridge, MA: Harvard University Press.

Zhong, Chen-Bo and Katie Liljenquist. 2006. Washing away your sins: Threatened morality and physical cleansing. *Science* 313: 1451–1452.

Index